Darwin's Roadmap to the Curriculum

Darwin's Roadmap to the Curriculum

Evolutionary Studies in Higher Education

Edited by
DAVID SLOAN WILSON
GLENN GEHER
HADASSAH MATIVETSKY
ANDREW C. GALLUP

Oxford University Press is a department of the University of Oxford. It furthers
the University's objective of excellence in research, scholarship, and education
by publishing worldwide. Oxford is a registered trade mark of Oxford University
Press in the UK and certain other countries.

Published in the United States of America by Oxford University Press
198 Madison Avenue, New York, NY 10016, United States of America.

© Oxford University Press 2019

All rights reserved. No part of this publication may be reproduced, stored in
a retrieval system, or transmitted, in any form or by any means, without the
prior permission in writing of Oxford University Press, or as expressly permitted
by law, by license, or under terms agreed with the appropriate reproduction
rights organization. Inquiries concerning reproduction outside the scope of the
above should be sent to the Rights Department, Oxford University Press, at the
address above.

You must not circulate this work in any other form
and you must impose this same condition on any acquirer.

Library of Congress Cataloging-in-Publication Data
Names: Wilson, David Sloan, editor. | Geher, Glenn, editor. | Mativetsky, Hadassah, editor. |
Gallup, Andrew (Andrew C.), 1985– editor.
Title: Darwin's Roadmap to the Curriculum: evolutionary studies in higher education/
edited by David Sloan Wilson, Glenn Geher, Hadassah Mativetsky, Andrew C. Gallup.
Description: New York, NY : Oxford University Press, [2019] |
Includes bibliographical references.
Identifiers: LCCN 2018048098 | ISBN 9780190624965 (hardcover : alk. paper)
Subjects: LCSH: Human evolution—Study and teaching (Higher)
Classification: LCC GN281.4.E928 2019 | DDC 599.93/80711—dc23
LC record available at https://lccn.loc.gov/2018048098

9 8 7 6 5 4 3 2 1

Printed by Sheridan Books, Inc., United States of America

CONTENTS

Acknowledgments ix
Contributors xi

SECTION 1 | **EVOLUTIONARY STUDIES IN HIGHER EDUCATION**

CHAPTER 1 Darwin-Inspired Curricula: The EvoS Revolution in Higher Education 3
Glenn Geher, David Sloan Wilson, Andrew C. Gallup, Hadassah Mativetsky, and Richard H. Holler

CHAPTER 2 Evolutionary Studies in Higher Education: Interdisciplinarity and Student Success Embodied 13
Glenn Geher, Rosemarie Sokol-Chang, Jennifer Waldo, David Sloan Wilson, and Hadassah Mativetsky

CHAPTER 3 Building an Evolutionary Studies Program at a Small Liberal Arts College 29
Susan M. Hughes

CHAPTER 4 The Evolutionary Studies Program: Perspectives from the Flies on the Wall 41
Nicole Wedberg, Kian Betancourt, Richard H. Holler, and Vania Rolón

SECTION 2 | **EVOLUTIONARY STUDIES EMBODIED WITHIN DISCIPLINES**

CHAPTER 5   Extraordinary Claims, Extraordinary Proof   61
Thomas G. Nolen

CHAPTER 6   Charles Darwin and Selection in Relation to Sex in the Colors of Monkeys   97
Sandra Winters, Megan Petersdorf, and James P. Higham

CHAPTER 7   The New Human Science: Sound, New Evolutionary Theory Gives Us Ultimate Causal Understanding of Human Origins, Behavior, History, Politics, and Economics   117
Joanne Souza and Paul M. Bingham

CHAPTER 8   Controversies Surrounding Evolutionary Psychology   157
Glenn Geher and Vania Rolón

CHAPTER 9   Evolution, Religion, and Other Meaning Systems   179
David Sloan Wilson

CHAPTER 10   From Physical Culture to the Primal Life: Evolutionary Health Movements in Historical Context   193
Hamilton M. Stapell

CHAPTER 11   From Genetic Evolution to Engineering Optimization   219
Yaser Khalifa

SECTION 3 | **APPLIED EVOLUTIONARY STUDIES**

CHAPTER 12   The Role of Evolutionary Studies in Education for Sustainable Development   249
Dustin Eirdosh and Susan Hanisch

CHAPTER 13   Evolutionary Approaches to Health Issues and Behaviors Across the Life Course   273
Daniel J. Kruger and Jessica S. Kruger

CHAPTER 14   Integrating Evolutionary Thinking into Medical Education and Curricula   303
Barbara Natterson-Horowitz and Daniel T. Blumstein

CHAPTER 15 How Evolutionary Studies Enables People to Think Outside the Box   321
           Gordon G. Gallup Jr., Jennifer A. Stolz, Rebecca L. Burch, and Jennifer A. Bremser

CHAPTER 16 The "EvoS Effect": The Influence of Evolutionary Training on Critical Thinking Skills   363
           Richard A. Kauffman Jr., Ian F. MacDonald, and David Sloan Wilson

CHAPTER 17 Our Evolutionary Underpinnings: The Past, Present, and Future of Evolution Education in the United States   409
           Amanda L. Glaze

CHAPTER 18 Reconciling Evolution with a Christian Identity: A Professional Development Workshop to Reduce Anxiety and Enhance Self-Efficacy for Science Teachers   425
           Patricia H. Hawley, Rachael K. Phillips, and Matthew S. Olson

CHAPTER 19 The Evolutionary Studies Summer Institute at New Paltz: A High-Impact, Condensed, Interdisciplinary Educational Experience for Teachers   447
           Glenn Geher, Aileen Toback, and Nicole Wedberg

CHAPTER 20 Teaching Evolution across the Curriculum: Beyond Campus-Wide Programs   463
           David Sloan Wilson, Glenn Geher, Andrew C. Gallup, and Hadassah Mativetsky

Index   469

ACKNOWLEDGMENTS

The editors are deeply appreciative for the support of our colleagues and students, including all of the contributors to this volume, who have helped with this project in multiple capacities. In particular, we thank Jacqueline Di Santo and Zachary Rausch of the SUNY New Paltz Evolutionary Psychology Lab who were instrumental in helping with the final editing details—we are grateful for all of their efforts. We are also thankful to all of the external reviewers who were solicited for this project, as their feedback was crucial in shaping the final contents of this book. We also greatly appreciate the work of Abby Gross and Katharine Pratt of Oxford University Press who continue to demonstrate why Oxford is such a leader in advancing ideas on a global scale. Finally, we thank Charles Robert Darwin for bravely advancing his insights into the nature of life over a century ago.

We would like to extend special thanks to Michelle Sedberry (Texas's Region 17 Science Specialist) for helping us to understand regional obstacles to the teaching of evolution and the challenges experienced by teachers in West Texas schools. This workshop was partially supported by a grant from the National Science Foundation Grant DEB-1542599 to MSO.

CONTRIBUTORS

Kian Betancourt
Department of Psychology
Evolutionary Studies Program
State University of New York, New Paltz
New Paltz, NY, USA

Paul M. Bingham
Department of Biochemistry and Cell Biology
Stony Brook University
Stony Brook, NY

Daniel T. Blumstein
Department of Ecology and Evolutionary Biology
University of California Los Angeles
Los Angeles, CA, USA

Jennifer A. Bremser
Department of Psychology
SUNY Plattsburgh at Queensbury
Queensbury, NY, USA

Rebecca L. Burch
Department of Human Development
State University of New York at Oswego
Oswego, NY, USA

Dustin Eirdosh
EvoLeipzig Initiative
Global ESD
Leipzig, Germany

Aileen Toback
New York State Master Teacher Emeritus
Lead Science Teacher
Heritage Middle School
Newburgh Enlarged City School District
Newburgh, NY

Andrew C. Gallup
Psychology Program
SUNY Polytechnic Institute
Utica, NY, USA

Gordon G. Gallup Jr.
Department of Psychology
University at Albany
Albany, NY, USA

Glenn Geher
Department of Psychology
Evolutionary Studies Program
State University of New York,
　New Paltz
New Paltz, NY, USA

Amanda L. Glaze
Department of Middle Grades and
　Secondary Education
Georgia Southern University
Statesboro, GA, USA

Susan Hanisch
EvoLeipzig Initiative
Global ESD
Leipzig, Germany

Patricia H. Hawley
Professor
College of Education, Texas Tech
　University
Director, Collaborative for
　Creative Engagement
Lubbock, TX, USA

James P. Higham
Department of Anthropology
New York University
New York, NY, USA

Richard H. Holler
Department of Psychology
Evolutionary Studies Program
State University of New York,
　New Paltz
New Paltz, NY, USA

Susan M. Hughes
Department of Psychology
Albright College
Reading, PA, USA

Richard A. Kauffman Jr.
Department of Psychology
SUNY Oneonta
Oneonta, NY, USA

Yaser Khalifa
R&D Apparatus Electronics
ABB Inc.
Lake Mary, FL, USA

Daniel J. Kruger
Population Studies Center
Institute for Social Research
University of Michigan
Ann Arbor, MI, USA

Jessica S. Kruger
Department of Community
　Health and Health
　Behavior
School of Public Health and
　Health Professions
University at Buffalo
Buffalo, NY, USA

Ian F. MacDonald
Postgraduate Scholar
Biological Sciences
Binghamton University
Binghamton, NY, USA

Hadassah Mativetsky
Department of Biology
Evolutionary Studies Program
Binghamton University
Binghamton, NY, USA

Barbara Natterson-Horowitz
Visiting Professor
Department of Human
  Evolutionary Biology
Harvard University
Cambridge, MA, USA
Professor of Medicine/Cardiology
University of California, Los
  Angeles
Los Angeles, CA, USA

Thomas G. Nolen
Evolutionary Studies Program
Department of Biology
State University of New York,
  New Paltz
New Paltz, NY

Matthew S. Olson
Department of Biological
  Sciences
Texas Tech University
Lubbock, TX, USA

Megan Petersdorf
Department of Anthropology
New York University
New York, NY, USA

Rachael K. Phillips
Developmental and Learning
  Sciences, Educational
  Psychology
College of Education
Texas Tech University

Vania Rolón
Department of Psychology
Evolutionary Studies Program
State University of New York,
  New Paltz
New Paltz, NY, USA

Joanne Souza
Department of Biochemistry and
  Cell Biology
Science Education Program
Stonybrook University
Stonybrook, NY

Hamilton M. Stapell
Department of History
Evolutionary Studies Program
State University of New York
  (SUNY), New Paltz
New Paltz, NY, USA

Jennifer A. Stolz
Department of Psychology
University of Waterloo
Waterloo, ON, Canada

Jennifer Waldo
Department of Biology
Evolutionary Studies Program
State University of New York,
  New Paltz
New Paltz, NY, USA

Nicole Wedberg
Department of Psychology
Evolutionary Studies Program
State University of New York,
  New Paltz
New Paltz, NY, USA

David Sloan Wilson
Department of Biology
Evolutionary Studies Program
State University of New York,
  New Paltz
New Paltz, NY, USA

Sandra Winters
Department of Anthropology
New York University
New York, NY, USA

SECTION 1 | Evolutionary Studies in Higher Education

CHAPTER 1 | **Darwinian-Inspired Curricula**
The EvoS Revolution in Higher Education

GLENN GEHER, DAVID SLOAN WILSON,
ANDREW C. GALLUP, HADASSAH MATIVETSKY,
AND RICHARD H. HOLLER

IN 2003, DAVID SLOAN WILSON offered a class on evolution to undergraduate students at Binghamton. This class differed from any existing courses on evolution at the time. The class, simply called *Evolution for Everyone*, was not designed for students who were studying genetics or paleontology—this was a class that was literally for everyone!

As documented in his book of the same name (Wilson, 2007), David created this course after concluding that college students in general neither appreciate nor understand the basic principles of evolution. He came to this conclusion after surveying a broad range of undergraduate students at his home university of Binghamton. Two basic themes emerged—each of which was disheartening in its own way.

First, with the exception of advanced biology students, knowledge of evolutionary concepts among undergraduates is dismal. This is a problem for many reasons. Partly, this is problematic because Darwin's ideas are not exactly a secret these days. Since Darwin's ideas were famously published in *On the Origin of Species* in 1859, his view has famously changed the world as we know it—including our understanding of what it means to be human. As evolutionary scholars, we see the principles that follow from Darwin's big ideas as powerful for asking questions about all kinds of phenomena. The power of the theoretical perspective provided by Darwin's work (and derivative concepts) cannot be understated.

This said, most college students at a competitive university (Binghamton) in a relatively "progressive" state (New York) were clueless when it came to basic evolutionary principles. This is a problem.

But that's not all. David also asked students to answer questions related to how relevant evolution was to their everyday lives (to their minds). The findings here were also pretty dismal. In particular, he found that, pretty much across the board (biology majors included), students do not see evolutionary principles as relevant to their everyday lives.

David's survey of his students is not alone in finding that we have an evolution problem in higher education. In a recent study on this topic, Bleske-Rechek and Donovan (2015) examined advancement of evolution knowledge in a cohort of about 200 undergraduate students at the University of Wisconsin–Eau Claire. These students were studied across a three-year period—the bulk of their undergraduate experience, and the findings were striking. The only significant predictor of evolution knowledge at the end of the study was a priori evolution knowledge. In other words, how much students knew about evolution at the start of the assessment was the only factor that predicted knowing a lot about evolution at the end. It didn't matter whether a student was a biology major or had majored in any of the other sciences—the curricular experience did not significantly predict increases in the knowledge and application of evolutionary principles across the educational process. Such a misfortune also provides further evidence that our systems of higher education are failures when it comes to evolution education.

In another study related to evolution education in modern American universities, Glass, Wilson, and Geher (2012) surveyed scholars who had published evolution-themed articles in the journal *Behavioral and Brain Sciences* between 2000 and 2004. A primary finding was that about one third of the articles published in that journal during this period were clearly and explicitly connected with evolutionary concepts. So, on one hand, this study provided evidence that evolution has arrived! This said, the bulk of the findings provide much less reason for celebration. In a survey of these authors regarding their own evolution education, a large majority of authors indicated the following:

- Most of the evolution knowledge that they gained was self-taught.
- Evolutionary principles were almost totally absent in their own education.
- An education in evolution at their current university would be extremely difficult for any student to obtain.

Taken altogether, these findings make one thing clear: There is a problem in the training of students to understand and apply evolutionary principles in the higher education of the United States.

The Nuts and Bolts of EvoS

The Evolutionary Studies (EvoS) concept is designed to address these problems—in a way that has the capacity to reach students across all corners of a campus. While EvoS programs vary from campus to campus, they generally include these basic elements:

- A foundational course open to all students (*Evolution for Everyone* or something similar).
- A set of courses across disciplines that highlight applications of evolution across a range of topics.
- And an upper-level seminar course that includes guest speakers from various disciplines, each of whom presents on scholarship related to some aspects of evolution.

Students in the program end up forming a genuine learning community (Wilson, Geher, & Waldo, 2009) and have formed EvoS Clubs at various schools. EvoS students tend to come from a broad array of majors, and they tend to demonstrate not only improved understanding of evolutionary principles but also improved critical thinking skills in general (see Chapter 16). As is described in several of the preliminary chapters in this book, the EvoS model of evolution education has made strides as an interdisciplinary approach to undergraduate evolution education (see Chapter 3).

Evolutionary Studies Embodied within the Disciplines

One of the core ideas underlying the EvoS concept is that evolution has the capacity to shed light on phenomena across all academic areas. Thus, from this perspective, evolution should not be confined exclusively to biology classrooms. The second section of this book provides several focused examples of how evolutionary reasoning sheds light on a broad array of phenomena including:

- All aspects of biology (see Chapter 5).

- Sexually selected traits in non-human primates (see Chapter 6).
- Computer engineering (via genetic algorithms) (see Chapter 11).
- All areas of the behavioral sciences (see Chapter 15).
- Politics, history, anthropology, and economics (see Chapter 7).
- Religion (see Chapter 9).

This section of the book is sure to open the eyes of any reader in terms of how broadly applicable evolutionary ideas are. In psychology (Geher, 2014), a great number of findings have been unearthed since scholars have started using an evolutionary approach in their work. Three such findings are as follows[1]:

1. **Men are more than twice as likely to experience early mortality (death) during young adulthood compared with women (Kruger & Nesse, 2006).** Men are more likely to die than are women at any and all phases of the life cycle. Applying an evolutionary lens, Kruger and Nesse hypothesized that this phenomenon should be exacerbated during young adulthood when males are more likely to be courting mates and, as a result, engaging in male–male (intrasexual) competition. And that's exactly what they found.
2. **Step-parents are, by a large order of magnitude, more likely to engage in filicide (killing of offspring) compared with genetic/biological parents (see Daly & Wilson, 2005).** Filicide is universally seen as horrific. So it would benefit humanity writ large to understand its antecedents. Applying evolutionary-based reasoning, Daly and Wilson reasoned that as step-parents do not share the same genetic investment with offspring as biological parents do, they might be more likely to engage in filicide. And this is, by a large order of magnitude, exactly what they found.
3. **Across all reaches of the world, men show a stronger preference for variety in mates compared with women. Schmitt et al., 2003).** Schmitt et al. hypothesized that across multiple human groups, males would show a stronger preference for variety of mating partners compared with females (as there are fewer evolutionary costs for males in mating with multiple partners compared with the costs for females). Based on one of the world's largest and

---

[1] Some of this text was taken, with permission, from my *Psychology Today* blogpost titled "The Power of Evolutionary Psychology," July 21, 2015. https://www.psychologytoday.com/us/blog/darwins-subterranean-world/201507/the-power-evolutionary-psychology?collection=1077556

most diverse human research samples ever studied, these researchers provided compelling evidence to support their evolution-based hypothesis: Across the world (and across methods of measurement), males demonstrated a stronger preference for variety in sexual partners compared with females.

Without the evolutionary perspective, we simply would not have had the intellectual infrastructure to ask and then answer these important questions about our own humanity. Evolution famously "turns the light on" when it comes to inquiry across a broad range of topics, and this section of the book will highlight precisely that point.

Suffice it to say, evolution is not just for biologists anymore. In fact, as you will see through the chapters of this book, evolutionary principles have demonstrated the capacity to inform our understanding of the world and our place in it.

Evolution and Education

EvoS is an educational initiative at its foundation. The basic premise of EvoS revolves around using what we know about evolution to enhance education across all facets of a standard curriculum. Importantly, given the United States' evolution problem (as previously described), improving evolution education across all levels of education, from grade school on up, is a major goal for those of us who are invested in the objectives of EvoS.

Toward this end, a large section of the book focuses on the evolution/education interface. Herein, several chapters address specific practices for improving evolution education in primary and secondary schools (e.g., see Chapters 17–19). Further, a separate section includes chapters on evolutionary principles applied to important issues, such as sustainability (see Chapter 12), epidemiology (Chapter 13), and medical education and curricula (Chapter 14). These application-based chapters speak to action research and the capacity for taking evolution-based education and making a positive impact on real-world problems.

One chapter that captures the ethos of this section well describes the EvoS Summer Institute at SUNY New Paltz (see Chapter 19). The week-long Institute epitomizes the kind of innovative spirit that EvoS faculty from across various institutions have used to effect positive changes in evolution education.

In essence, the EvoS Institute provides a condensed version of the EvoS curriculum—targeted toward middle school and high school teachers. The faculty of the Institute included several faculty of the SUNY New Paltz EvoS program—each teaching intensive three-hour blocks of their basic content area. For instance, Tom Nolen (biology) provided a basic background in the nuts and bolts of evolutionary principles while Jen Waldo (biology) and Jeff Reinking (biology) gave participants tutorials on experimental genetics. On subsequent days, participants have had sessions led by Ken Nystrom (anthropology), myself (Glenn Geher; psychology), Hamilton Stapell (a historian who is expert on issues of evolution and human health), Andrea Varga (theater arts), and Spencer Mass (biology).

The participants in the EvoS Institute have been primarily teachers who have taught about evolution in their own way for years (such as seasoned high school biology teachers). As elaborated on in that chapter, survey results from participants suggest that they found the Institute extremely enriching—leading most participants to report that they had never thought of evolution in such a broadly applicable manner before taking this curriculum.

The Institute, then, emerges as a way that evolution education within a higher educational context can trickle down to the primary and secondary school levels (a concept also at the core of Chapter 18). Importantly, the education that the participants in that program received was not just a deeper understanding of evolutionary principles but, rather, an arsenal of ways that evolution can be applied across a whole battery of phenomena— including many aspects of everyday life.

The chapters in this section are designed to help advance our understanding of ways to improve evolution education—largely through the highly interdisciplinary and student-focused model of EvoS.

Controversies That Surround Evolutionary Studies

While we, the editors of this volume, clearly think highly of the EvoS model of evolution education, we would be remiss to not mention that there have been ongoing struggles related to rallying support for the EvoS model of higher education. One of the chapters of this book focuses on controversies surrounding evolutionary psychology (see Chapter 8).

Controversies have followed evolution-based scholarship for well over a century now, and past experience suggests that we would be naïve to expect such controversy to vanish any time soon. Controversies, in fact, have come from a variety of sources (Spaulding, Burch, & Lynn, 2014), ultimately impacting the growth of the EvoS model at colleges and universities across the world.

Clearly, no one would expect an EvoS program to emerge at a school such as Liberty University—a school premised on a particular fundamentalist Christian view of the world. This said, many folks are often surprised by the many passionate secular pockets of institutions that are resistant to evolutionary scholarship, yet within the halls of the academy, such resistance and discordance are much more immediate and relevant issues that EvoS faculty and students must confront. For these reasons, several academics with keen interests in evolution have taken strong and proactive steps to advance evolution education locally.

At the University of Alabama, biological anthropologist Chris Lynn and his colleagues worked hard to develop an 18-credit Evolutionary Studies minor. Their EvoS program, housed in Anthropology, surrounds their highly renowned ALLELE (Alabama Lectures in Life's Evolution) speaker series. This series, which has featured such icons as E. O. Wilson and Bill Nye, has been a conspicuous success. Further, the EvoS program at Alabama has served as the nexus of the Southeastern Evolutionary Perspectives Society (SEEPS), providing more synergy for the program there.

This said, attempts to form EvoS programs have not always been met with such support and success. At the University at Albany, then-graduate-student Kristina Spaulding, working in collaboration with Psychology faculty member, Gordon Gallup, valiantly tried to spearhead an EvoS program there. She argued that between faculty in psychology, anthropology, and biology alone, the university surely had the infrastructure to create such a program. Unfortunately, as demarcated in Spaulding et al. (2014), this effort simply met wall after wall after wall of resistance. Part of this resistance was related to the fact that a graduate student, and not a faculty member, was shepherding the project. Further, support from the home department of psychology was not sufficiently strong. At the end of the day, the program at Albany remains in the abstract.

One of the most interesting stories of resistance comes from SUNY Oswego. As demarcated in Spaulding et al. (2014), it was the biology faculty who provided the primary resistance to the EvoS program there. In this particular case, Rebecca Burch, then of the psychology department, developed a proposal for an interdisciplinary EvoS program there. Apparently, all was running along smoothly until faculty of biology department had a chance to voice their opinions. Their basic rationale for rejecting the proposal came down to turf issues. As far as they were concerned, evolution belongs in biology. And that's it.

If you are in the field of academia, these stories likely do not surprise you much. Academia is famously contentious and political—and, regardless of stereotypes of academia being a "liberal" environment, when it comes down to it, change with an academic institution happens painfully slowly—if at all.

Further, several ideas related to EvoS have been framed as essentially politically untenable by many academics. For instance, Geher and Gambacorta (2010) found evidence that many scholars who are outside of the field of evolutionary psychology hold strong resistance to the field of evolutionary psychology, largely due to the fact that evolutionary psychologists have famously documented evidence of evolved behavioral sex differences in humans—a general finding that many feminists and "progressive" academics simply find untenable. This is an issue of the is ought fallacy.

With this in mind, when it comes to controversy, evolutionary studies has its share! From outright rejection by fundamentalist Christians, to skepticism and politics within the academy, to a hostile approach toward the area of evolutionary studies that addresses such phenomena as biological sex differences in behavior, EvoS has no shortage of naysayers!

EvoS in a Crystal Ball

The final chapter of this book is by David Sloan Wilson (the first founder of EvoS) on the future of EvoS (see Chapter 20).

The future of EvoS depends on several factors. In an article on the future of evolutionary psychology in particular, Garcia et al. (2011) make the case that evolutionary psychology essentially will go one of two ways into the future. On one hand, as documented by Garcia et al. (2011), evolutionary psychology is wildly interdisciplinary, with a much higher proportion of scholarship being conducted by scholars from all different kinds of fields contributing to the scholarship in evolutionary psychology compared with in other areas of psychology. As the world becomes more interdisciplinary, the world may become friendlier toward evolution-based approaches.

This said, as is foreshadowed in the prior section, evolutionary psychology is under a tremendous amount of political pressure (Geher, 2006). Forces from across the political spectrum have essentially provided resistance to the work in this field. If the future of academia facilitates such political resistance, it will be hard for evolutionary psychology to maintain its position as an academic discipline.

Two other important factors need to be considered when it comes to the potential futures for EvoS. One of them pertains to the popularity of EvoS coursework among students (see Chapter 2) while another pertains to the popularity of evolution-based scholarship in the media (Fisher, Kruger, & Garcia, 2011). To a large extent, student interest in a field of study cannot be ignored—as student interest is strongly related to course needs, which ultimately (in theory) ties to personnel hires.

Similarly, trends in the media matter, and, as Fisher et al. (2011) point out, in many of the human-related sciences, evolution-based scholarship is disproportionately represented. The work of scholars connected to EvoS is popular, and it is popular among both students as well as the general public.

Conclusion

When it comes to better understanding the world, which is, really, ultimately the goal of any academic curriculum, evolutionary principles are extremely powerful. Evolutionary scholarship has shed extraordinary light on phenomena across all areas of inquiry, and it continues to do so as this chapter is written. Evolutionary scholars are making significant advances in the fields of education (Gray, 2011), medicine (see Chapters 13 and 14), politics (see Chapter 7; Geher et al., 2015), and more.

Darwin's Roadmap to the Curriculum is designed for students of evolutionary studies as well as scholars in this area. Providing an in-depth overview of the field, along with a broad array of applications, this book is designed to provide scholars at all levels with the tools needed to integrate Darwin's big idea into curriculum on any topic.

On behalf of the editorial team, we welcome you to this book and hope that you find the EvoS concept to be as inspiring and thought-provoking as we do.

References

Bleske-Rechek, A., & Donovan, B. A. (2015). Scientifically adrift: Limited change in scientific literacy and no change in knowledge and acceptance of evolution, over three years of college. *Journal of the Evolutionary Studies Consortium, 7*, 21–43.

Daly, M., & Wilson, M. (2005). The Cinderella effect is no fairy tale. *Trends in Cognitive Sciences, 9,* 507–508.

Darwin, C. (1859). *On the origin of species by means of natural selection, or the preservation of favoured races in the struggle for life* (1st ed.). London, England: John Murray.

Fisher, M. L., Kruger, D. J., & Garcia, J. R. (2011). Understanding and enhancing the role of the mass media in evolutionary psychology education. *Evolution: Education and Outreach, 4,* 75–82.

Garcia, J., Geher, G., Crosier, B., Saad, G., Gambacorta, D., Johnsen, L., & Pranckitas, E. (2011). The interdisciplinary context of evolutionary approaches to human behavior: a key to survival in the ivory archipelago. *Futures, 43,* 749–761.

Geher, G. (2006). Evolutionary psychology is not evil . . . and here's why . . . *Psihologijske Teme, 15,* 181–202.

Geher, G. (2014). *Evolutionary psychology 101.* New York, NY: Springer.

Geher, G., & Gambacorta, D. (2010). Evolution is not relevant to sex differences in humans because I want it that way! Evidence for the politicization of human evolutionary psychology. *EvoS Journal, 2*(1), 32–47.

Geher, G., Carmen, R., Guitar, A., Gangemi, B., Sancak Aydin, G., & Shimkus, A. (2015). The evolutionary psychology of small-scale versus large-scale politics: Ancestral conditions did not include large-scale politics. *European Journal of Social Psychology, 46,* 369–376.

Glass, D. J., Wilson, D.S., & Geher, G. (2012). Evolutionary training in relation to human affairs is sorely lacking in higher education. *EvoS Journal, 4*(2), 16–22.

Gray, P. (2011). *Free to learn.* New York, NY: Basic Books.

Kruger, D. J., & Nesse, R. M. (2006). An evolutionary life history understanding of sex differences in human mortality rates. *Human Nature, 74,* 74–97.

Schmitt, D. P., Alcalay, L., Allik, J., Ault, L., Austers, I., Bennett, K. L., . . . Zupanèiè, A. (2003). Universal sex differences in the desire for sexual variety: Tests from 52 nations, 6 continents, and 13 islands. *Journal of Personality and Social Psychology, 85,* 85–104.

Lynn, C. D. (2014). Evolutionary studies' reproductive success and failures: Knowing the institutional ecology. *EvoS Journal, 6*(1), 18–38.

Wilson, D. S. (2007). *Evolution for everyone: How Darwin's theory can change the way we think about our lives.* New York, NY: Delacorte.

Wilson, D. S., Geher, G., & Waldo, J. (2009). EvoS: Completing the evolutionary synthesis in higher education. *EvoS Journal, 1*(1), 3–10.

CHAPTER 2 | # Evolutionary Studies in Higher Education
Interdisciplinarity and Student Success Embodied

GLENN GEHER, ROSEMARIE SOKOL-CHANG,
JENNIFER WALDO, DAVID SLOAN WILSON,
AND HADASSAH MATIVETSKY

SINCE EVOLUTIONARY STUDIES (EVOS) first emerged as a formal academic program at Binghamton University in 2003, the idea of an academic program revolving around Darwin's big ideas across academic disciplines and areas has led to exciting academic outcomes.[1] The idea of EvoS focuses on teaching students about the basic principles of evolution, such as natural and sexual selection, and then guides them toward applications of evolutionary principles into such diverse areas of vertebrate biology, paleontology, and genetics—along with such humanistic and arts-based areas as literary studies, philosophy, engineering and the visual arts. Funded with a large grant from the National Science Foundation (NSF) between 2008 and 2011, EvoS has grown as an international consortium of scholars and students who all share a common vision of allowing Darwin's big ideas to better guide our understanding of the world.

As the primary scholars who worked on an NSF-funded grant to expand EvoS within higher education, we have unique perspectives on these issues—which we summarize here. This chapter, in particular, addresses (a) the highly interdisciplinary nature of EvoS, (b) the unique kinds of

---

[1] This chapter is partly adapted from "Evolutionary Studies in Higher Education: Interdisciplinarity and Student Success," by G. Geher, R. Sokol-Chang, & J. Waldo (2014). *EvoS Journal*, 6(1), 5–11.

student success stories that EvoS programs have cultivated, and (c) the role that EvoS can play in helping bring together traditionally distinct academic areas and perspectives.

EvoS: A Truly Interdisciplinary Approach

Academia is moving toward interdisciplinary forms of education (see Garcia et al., 2011). As a result, we see the growth of such large-scale interdisciplinary academic programs as women's studies, Black studies, American studies, material science, environmental studies, and other interdisciplinary approaches that now represent curricula at many colleges and universities. Started at Binghamton University in 2003, EvoS is something of a newcomer on the scene of interdisciplinary studies. This said, largely based on our prior work related to the NSF grant we received to expand EvoS, there are now around 50 colleges and universities around the world that are explicitly connected to the EvoS Consortium—and some of these institutions, such as Albright College, the University of Alabama, and the University of Missouri, have full-blown curricula in EvoS. Based on the emergence of these EvoS programs at schools across the nation (beyond just Binghamton and New Paltz where these programs first developed), it is fair to say that EvoS is growing. For membership criteria, see Appendix 2.A.

Importantly, when we refer to a school with a *full-blown EvoS program*, such as that found at the University of Alabama, we are referring to a formal academic program (typically a minor or a concentration) offered by the university. The example at Alabama is quite thorough and thoughtfully presented and can be found in Appendix 2.B. Importantly, note that in addition to providing the rigorous and detailed academic program found in Appendix 2.B, Alabama also has a vibrant EvoS Club and concomitant student community that connects with the program. Further, their program hosts a year-round speaker series connected with the EvoS program (called "ALLELE" standing for (**A**labama **L**ectures on **L**ife's **E**volution), which brings in major speakers related to evolution (including Bill Nye). They also (under the leadership of anthropologist, Chris Lynn), starting on February 12, 2016, have launched an annual meeting of evolutionists across disciplines called the SouthEastern Evolutionary Perspectives Society (SEEPS). This multifaceted and multidisciplinary, high-impact approach to evolution education captures what we mean by a "full-blown" EvoS program. Other programs that have "basic" EvoS programs might have, for instance, one course (such as Lebanon Valley College, which has an Evolution for Everyone course) but not a full curriculum in EvoS surrounding it.

EvoS differs from traditional interdisciplinary programs in a significant way. Specifically, most interdisciplinary programs revolve around shared content. For instance, a program in American Studies includes courses from such fields as English, History, and Political Science—all addressing the content of the American experience from the angle of a particular academic approach. EvoS is different. EvoS is an interdisciplinary program based on a shared set of intellectual ideas—the basic ideas of evolutionary theory (such as natural selection) and then connects content from there. As such, students in an EvoS program receive a strong background in evolutionary principles and then they can apply these principles to such diverse content areas as anthropology (biological, cultural, linguistics and archeology), literary studies, social psychology, and more. Once a student has a strong background in EvoS, he or she becomes able to apply and integrate the ideas of evolutionary theory widely across his or her academic experience. For instance, once a psychology student has a strong understanding of evolutionary principles and its applicability, he or she can bring evolutionary-inspired ideas to the understanding of such varied content areas as clinical psychology, social psychology, developmental psychology, neuroscience, and more (see Carmen et al., 2013). For example, a savvy EvoS student in psychology might learn about attachment-related behaviors in a developmental psychology class and come to ask how such behaviors would have made sense as adaptive under ancestral conditions—and he or she could ask how modern conditions (e.g., paid childcare) might mismatch certain features of the ancestral environment that surrounded the development of parent–offspring attachment development. Once an EvoS student has a strong understanding of evolutionary principles, he or she can start to ask questions such as these across all of his or her classes—this is what we mean when we say that EvoS truly opens students' eyes to new ways of seeing things across the academic experience.

Two recent sets of studies speak to the interdisciplinary nature of EvoS and have documented how the evolutionary perspective truly cultivates an approach that is significantly more interdisciplinary in scope relative to other perspectives that are not explicitly rooted in evolutionary ideas. This work, summarized in the following paragraphs, provides evidence that an evolutionarily informed approach to an academic area has strong potential to connect students and scholars with work that spans many different areas of academia.

Garcia et al. (2011) examined the academic departmental homes of first authors of articles in several prominent journals within the behavioral sciences. Some of these journals were explicitly evolution-based in scope (e.g., *Evolutionary Psychology*) while others were more traditional (e.g., *Cognitive Psychology*). Two basic findings emerged. First, the evolution-based journals showed a much higher proportion of authors from areas outside psychology compared with the non-evolution-based journals. So, someone from Anthropology or Biology, for instance, would be much more likely to be a first author of an article in *Evolutionary Psychology* than in *Cognitive Psychology* (or in several other nonevolution-based journals in the behavioral sciences). Second, the total number of disciplines represented by academic homes of authors was much broader for the evolution-based journals than for the others; evolution-based journals included authors from over a dozen disciplines in the sample while the other journals included only typically about two to four disciplines.

In a follow-up study, Geher, Crosier, Dillon, and Chang (2011) examined the literature cited by articles published in evolution-based journals versus a similar set of non-evolution-based journals that were also in the behavioral sciences. This analysis was done by examining the academic affiliations of the first authors of a large sample of articles cited as references in articles that were published in evolution-based versus non-evolution-based journals. The overall findings were very parallel to the findings from Garcia et al. (2011); articles in evolution-based journals were much more likely to draw on literature written by authors from across many different academic areas whereas articles in non-evolution-based journals were not likely to cite literature from disciplines outside the behavioral sciences. So, at least within the behavioral sciences, the evolutionary approach seems to cultivate, or at least go hand in hand with, a truly interdisciplinary approach to academia.

The fact that the field of evolutionary psychology is much more interdisciplinary than are other areas of psychology dovetails with the interdisciplinary nature of EvoS in an important way. Recall that earlier we argue that EvoS is a principle-based (as opposed to content-based) approach to interdisciplinary studies. In other words, we argue that the courses in an EvoS curriculum are connected by a shared set of principles (evolutionary principles) rather than a shared set of content (e.g., as found in a typical program in Asian Studies). We believe that this principle-based approach provides students with a basic set of intellectual tools that can be applied across various academic areas. Based on our analyses of evolution-based journals in psychology, we are seeing the same thing. Evolutionary psychology is not based on a particular content area (e.g., it is not exclusively about cognition, social behavior, development, perception, etc.), but it is, rather, about the application of a basic and powerful set of intellectual ideas (evolutionary principles) across a host of content areas in the field. Thus, the powerful nature of evolutionary psychology as an intellectual area parallels the powerful nature of the EvoS program as an approach to the curriculum.

EvoS and Student Success

EvoS exists to provide new and important opportunities to enhance faculty research and student education. Clearly, there are many ways to gauge student success. The 2012 EvoS Summit (held in October on the campus of the State University of New York at New Paltz) was designed to take stock of the state of affairs regarding EvoS in higher education. Faculty and students across various disciplines came from over 20 different institutions to attend and present at this event. Given the core mission of advancing evolution education, much of the conference focused on how evolution helps advance student learning and intellectual skills.

As we prepared for the EvoS summit, we chose to reflect upon the place that EvoS had in the personal and professional journeys of the students that participated in the NSF-funded EvoS faculty–student research projects over the summers of 2009 and 2010. We asked students to describe where their individual paths led, and how the research conducted for this project influenced their trajectories. In short, this remarkable cohort of students has made the most of the opportunities afforded them—a few of their stories follow.

> **Student Case 1.** Zuchra Zakirova worked with Dr. Jeffrey Reinking (Biology) to investigate the question "When did nuclear receptors evolve the capability to bind heme?" Zuchra graduated from SUNY New Paltz in May of 2010, earning a BS in Cellular and Molecular Biology and a BA in English Literature. Since graduation, she has earned an MSc degree in Genomics and Pathway Biology from the University of Edinburgh (Scotland) and is currently enrolled in a PhD program in Neuroscience from the Open University. When asked to reflect on the place EvoS held in her journey, she replied, "Participating in the EvoS research program helped me grow academically as well as personally, in that it allowed me to dream, make

mistakes, and most importantly learn from them. Undergraduate research is a precious gift, it is able to open up a world of possibilities to a young person, and ignite a passion for science, discovery, and the endless pursuit of answering the question, 'Hmm, interesting... but... HOW does that happen?'"

Student Case 2. Rachael Carmen worked with Dr. Corwin Senko (Psychology) during the summer of 2009 exploring how the ovulatory cycle influences a female's perception of what type of humor style is attractive (self-deprecating vs. other-deprecating). She earned a BA in Psychology with minors in EvoS and History in 2009 and an MS in Psychology in 2013. Rachel has been fantastically busy as a scholar during this time, with 10 publications completed and 4 more in preparation (e.g., Carmen, Guitar, & Dillon, 2012; Carmen et al., 2013; Peterson, Carmen, & Geher, 2013). Her future is likely to involve further academic work, in the form of a PhD program. On the immediate horizon is the opportunity to teach Evolutionary Psychology at New Paltz at the undergraduate level. When asked about the role EvoS played in her education, Rachael replied, "The EvoS program really helped me create a solid foundation to build my writing and teaching on. Taking classes that were outside of my major was a really enjoyable, eye-opening experience for me—and I've recommended it to all the students I've talked to. Every single publication I've gotten was either explicitly related to EvoS or it included some aspect of evolutionary theory within it. It's more than an academic discipline, it's a way to understand the world around us."

Student Case 3. Also in 2009, Jannett Dinsmore worked with Dr. Aaron Haselton (Biology) to look at the effect, if any, of diet on sustained flight in *D. melanogaster*. She completed a BS in Biology in 2009 and an MA in Biology in 2012. She is now a lecturer in the Department of Biology at SUNY New Paltz. Jannett describes her involvement with the EvoS research program as such, "Participating in the EvoS research program provided me with hands-on experience. It gave me the opportunity to take the knowledge that I had gained in the classroom and apply it in a research setting. Instead of just learning about the scientific process, I was able to become an active participant."

Student Case 4. In the Summer of 2010, Aaron Reed followed up on Zuchra's work with Dr. Reinking (Biology) to study the evolution of nuclear receptors. This work was published in the journal *Nuclear Hormone Signaling*. Aaron finished his BS in Molecular and Cellular Biology in 2010, enrolled in a PhD program in Molecular Medicine, at George Washington University, and hopes to work for the Food and Drug Administration upon completion of his degree. Reflecting upon the role of the EvoS research program, Aaron stated, "Academically, this research program exposed me to hands-on, full-time, research at SUNY New Paltz. Many of the molecular biology skills acquired during that period have proven to be useful tools in my independent pursuit of an advanced degree. Personally, it gave me satisfaction when experiments were executed and results were obtained. I also gained an appreciation for the work scientists do."

Student Case 5. Nolan Conaway worked with Professor Alice Andrews (Psychology) and fellow undergrad Leah Manders during the summer of 2010 to conduct an experiment that tested whether scores on a theory of mind assessment would improve in a mating-relevant scenario. Nolan then graduated in the spring of 2011 with a BA in Psychology and minors in EvoS and Philosophy, and in the fall of 2011 began a PhD program at Binghamton University in cognitive psychology. Reflecting on his journey, Nolan states, "The project I worked on gave me a sense of how to actually *do* research. I think that's an important thing for psychology students to learn."

Student Case 6. Working with Dr. Jennifer Waldo (Biology) on the biophysical and biochemical properties of the Dam1 complex, a 10-protein subunit complex that is a component of the kinetochore, Stacey Greagor feels that "The EvoS research program was the foundation of my academic achievements. My research experience helped to build my confidence in the lab and strongly supplemented my studies in the classroom." This work led to two publications—one in *Biochemical and Biophysical Research Communications* and another in *Evolution: Education and Outreach*. Following obtaining her BS in Biology and a minor in EvoS, Stacey was able to quickly find temporary employment doing vaccine research at a large pharmaceutical company. In a little over a year since graduating, she has been able to advance her career to a permanent position with a small, but promising pharmaceutical company.

Student Case 7. In 2010, Dr. Glenn Geher (Psychology) mentored three students in an evaluation of New York State public school health education curricula and efforts to incorporate evolutionary theory in hopes of developing a new sample curriculum. One of the participants, Abigail Kurtz, completed her BA in Psychology in 2011 and then moved to Israel to work with African refugees seeking asylum for a year. She is currently back in the states working at the headquarters for a global humanitarian organization. Her future plans include pursuing a PhD in clinical psychology specializing in refugee trauma/posttraumatic stress disorder. In considering the role of EvoS, Abigail says, "The research program, and EvoS in general, definitely made me a more well-rounded person both academically and personally. I believe I can understand the world in a greater context and truly understand what drives people to act the way they do. Learning and studying evolutionary theories then observing them play out in animal and human behavior is something I will always be fascinated with. The research program exposed me to a lot of ways to address specific factors of human behavior."

Student Case 8. Another student that worked with Dr. Geher (Psychology) in 2010 is Laura Johnsen. Laura graduated from New Paltz in the Spring of 2012 with degrees in Psychology and Theater Arts as well as minors in Anthropology and EvoS. She also completed the requirements for the honors program, including a presentation of her evolutionarily informed costume choices for the musical *Cabaret* at the EvoS seminar series. She is now pursuing a PhD in Anthropology at Binghamton University exploring the interactions between risk-taking and environment. "My participation with the research program was a great experience academically and personally. Academically, it gave me the chance to practice grant-writing skills, learn more about the research process, collaborate with other undergraduate students, and strengthen my presentation skills. It also gave me the chance to learn more about the research interests of students in other departments that I may not have had the opportunity to learn about. Personally, the experience helped me become closer with the two other undergraduate students I worked with, Abbey Kurtz and Stephen Williams. We worked really well together and enjoyed the different challenges that came along with designing a curriculum for high school students."

While all of the students took something slightly different from the program, a recurring theme is the impact of the ability to actively participate in research and the exposure to multiple disciplinary lenses through which to view their education.

Importantly, while we believe that the examples herein are strong and provide a background for seeing student success in the context of EvoS programs, future research could benefit from a more thorough, quantitative, and perhaps longitudinal examination of student success in the area of EvoS.

EvoS as a Tool to Create Grey Academic Boundaries

While the first section of this chapter addresses the unique principle-based approach to curriculum that an EvoS program brings to the table, this section addresses a separate facet of how EvoS illuminates the academic experience. In particular, this section focuses on how a solid EvoS approach can help us better think about potentially divisive intellectual debates such as the "nature/nurture" debate. Beyond the clear outcomes associated with specific student success stories, the EvoS Summit of 2012 allowed us to step back and think about the potential for EvoS to shed important light on the nature of academia itself. Communication between disciplines, or even within disciplines, has traditionally been oversimplified by such binary issues as nature and nurture (see Galton, 1874) or religious creation and evolution (see Scott, 2008). When either side is presented in a black-or-white manner, conversations are halted before beginning. There has been a general sense that one must adhere to a more innate *or* experiential philosophy of the human mind, that human origins can be explained by a greater being *or* scientific explanations.

Scientific disciplines have historically experienced a communicative block over the supposed nature versus nurture debate (see Pinker, 2002). Nature generally represents a genetic and inflexible approach to phenotypes, often characterized by critics as genetic determinism. Nurture generally represents the *tabula rasa* (Locke, 1690/1998) approach—the blank slate—that the mind is created by experience. This approach is often attributed to many, if not all, of the social sciences (see the standard social science model; Tooby & Cosmides, 1992). Such straw creations are often understood simply and only for the goal of tearing down opposing viewpoints. Many emerging fields try to rectify this artificial distinction between nature and nurture, including the fields of epigenetics and evolutionary-developmental biology and psychology (see Buller, 2005, for a commentary on the importance of such an expansion of the evolutionary perspective). EvoS programs also show that when these two philosophies begin to grey, they blend together to address complex issues in complex ways.

EvoS programs have been powerful ways to promote grey boundaries between disciplines. Indeed, the founder of the EvoS movement considers the ways that the different areas of the Ivory Archipelago can be combined by having each contribute to an understanding of life by incorporating evolutionary theory as a basic building block (Wilson, 2007). Within the initial EvoS programs (i.e., Binghamton University and SUNY New Paltz), there has been a consistent blending between the natural sciences, social sciences, and humanities. The EvoS movement intersects with and inspires courses that incorporate a STEAM (science, technology, engineering, arts, and mathematics) approach (see Walker, 2013); seminars speakers that present complex nature *and* nurture understandings (e.g., Massimo Pigliucci, Frances Champagne); and minors and certificate programs that blend disciplines that traditionally operated as separate islands (e.g., biology and literature; psychology and theater).

The true power in blending historically "black-and-white," isolated viewpoints through EvoS programs is overcoming misconceptions about and oversimplifications of evolutionary theory that begin in early education (see Nelson, 2007), promoting integration across disciplines that all share a common goal of understanding how the world and life works, and transforming knowledge into research that can impact the real world.

Conclusion

This chapter speaks to several facets of the EvoS experience that relate to the beneficial effects of an evolutionary approach academic scholarship and student growth. Specifically, we have documented that (a) EvoS fosters an interdisciplinary approach to scholarship that can be empirically documented, (b) EvoS facilitates consistent and strong student outcomes, and (c) EvoS has potential to help academics move beyond traditional debates within academia (e.g., *nature vs. nurture*) to help academic areas work collaboratively in terms of curriculum, student-collaborative research, teaching, and more. The prior section focuses on the many benefits of an EvoS education. Despite these demonstrated benefits, an analysis of how many modern evolutionists perceive the state of evolution education within modern academia, Glass, Wilson, and Geher (2012) provide strong evidence suggesting that much work needs to be done. In a survey of first authors of evolution-themed articles published in the elite journal *Behavioral and Brain Sciences*, the majority of these authors reported (among other things) (a) that their doctoral training institution provided poor opportunities for evolution education, (b) that faculty at their current institution would have a hard time learning about the principles of evolution themselves, and (c) that their current home institution provides little in the way of opportunity for students outside the biological sciences to learn about evolution and its applications. See Glass et al. (2012) for particulars on the frequencies of responses.

Based on a large body of convergent work (e.g., Carmen et al., 2013), we see strong evidence of the utility of evolutionary theory as a foundational set of ideas for any academic experience. By working collaboratively within the bounds of the EvoS Consortium, we hope to help fulfill Darwin's ultimate vision by better integrating evolution into higher education.

Appendix 2.A
EvoS Membership Criteria as Established by the EvoS Consortium

In part to monitor the progress of the consortium for NSF, we have identified four levels of participation that make it easy for new members to become involved in an incremental fashion.[2]

Level 1: Joining the consortium as an incipient program.
Indices: One or more faculty express an interest in developing a program at their institution.
Benefits provided by grant:

1. incipient members added to email list and electronic journal;
2. are eligible for consulting services.

Level 2: Introductory course.
Indices: An introductory course comparable to "Evolution for Everyone" taught in coordination with courses at other member institutions.
Benefits provided by grant:

1. consulting services to establish the course;
2. using modules and contributing to the inventory of modules;
3. centralized course assessment services.

Level 3: Campus wide seminar series, including an associated "current topics" course for undergraduate students.
Indices: Number of seminars, audience size for each seminar, audience composition (e.g., undergraduate/graduate/faculty/member of community), ratings provided by participants.

---

[2] This material comes from the website of EvoS: The Evolutionary Studies Consortium at http://evostudies.org/members/membership-criteria/.

Benefits provided by grant:

1. consulting services and site visit to help establish the seminar series;
2. archiving service so that seminars at member institution can add their seminar to the archive maintained at Binghamton University.

Level 4: Establishment of formal interdisciplinary program.
Indices: Formal program designation by administration (e.g., as a minor or certificate program), number of faculty participants, number of student participants, number of courses, number of new courses and other courses relevant to undergraduate education.
Benefits provided by grant:

1. centralized whole-program assessment services;
2. participation in symposia and site visits promoting the EvoS consortium.

Appendix 2.B
University of Alabama's EvoS Minor Curriculum

COURSE IN THE MINOR (20 credits total)

The minor is designed to be taken over the course of your time at UA, beginning with "Evolution for Everyone." In this course, you will be introduced to evolutionary principles and a cross-section of applications of these principles. You will also outline an evolutionary project you can do while a UA undergraduate. In subsequent semesters, while taking other minor requirements and electives, you will take a 1-credit "Readings in ALLELE" course to stay involved with other students in the minor and a 1-credit "Independent Research" in conjunction with a research mentor in your primary discipline. In conjunction with this mentor, you will conduct the research you outlined during the first semester. In your final EvoS semester, you will take "Advanced Evolutionary Studies," which will again expose you to the principles and applications but which you will be able to integrate more deeply. Additionally, you will write-up your project results for submission to a student-level peer-reviewed publication. It is hoped that this course of study, while not as rigorous as a major, will enable you to be constructive critics of cultural use of evolutionary principles or prepare you for further studies.

REQUIRED "FOUNDATION" COURSES (12 credits)

ANT 150, Introduction to Evolutionary Studies

This team-taught course is the introduction to the minor. The objective is to expose students to the array of applications for evolutionary theory across the natural, social, and applied sciences and humanities. It features an array of guest lectures from across the university that changes each semester, as well as visits by visiting ALLELE lectures. Students design a project (based in the discipline of their major) that will be completed over the course of the minor and review a trade book for possible publication. Offered every spring semester.

BSC 220, Principles of Biological Evolution

This course is an introduction to the process and patterns of biological evolution. It is geared for non-biology majors. Offered approximately every two of three semesters.

ANT 480, Advanced Evolutionary Studies

This is the concluding course of the minor and should be taken in one's final spring semester. It meets concurrently with ANT 150 and features the same format. It's intention is to review basic mechanisms with the purpose of a deeper integration of principles and provide students exposure to a different set of instructors and ALLELE guests. Students compile a portfolio of their "EvoS Experience," present results from their minor project and, if relevant, submit it to a peer-reviewed journal for publication consideration. Offered every spring.

One of the following:

ANT 270, Physical Anthropology
GEO 102, Earth Thru Time
PHL 387, Philosophy & Evolution

INTEGRATIVE COURSES (2 credits)

ANT 431, Readings in ALLELE

This 1-credit course involves meetings to discuss readings of ALLELE guests scheduled for the semester.[3] Students attend ALLELE lectures and meeting as a group with ALLELE speakers for direct opportunity for in-depth discussion. The course is designed to ensure that students stay integrated in the EvoS program since, as an interdisciplinary minor, students are not taking evolution courses as part of a specific department or with a cohort. Can be arranged any semester between ANT 150 and ANT 450.

Independent Research

This 1-credit course is designed to provide an opportunity for student to collect data or complete their minor project. Students are mentored by a faculty member in their major. Students should arrange to take this between ANT 150 and ANT 450 and register for credit in their mentor's discipline. Credit will be manually assigned to the EvoS minor.

---

[3] This content is borrowed from the official website of the University of Alabama at https://evolutionarystudies.as.ua.edu/minor-requirements/.

Elective "Context" Areas (6 or more credits from at least TWO DIFFERENT departments that are not your major)

ANTHROPOLOGY ELECTIVES

ANT 270, Physical Anthropology*
ANT 208, Anthropology of Sex
ANT 275, Race, Ethnicity, & Variation
ANT 311, Population, Health, & Origins
ANT 312, Non-Human Primates
ANT 471, Fossil Humans and Evolution
ANT 473, Human Osteology
ANT 475, Biology, Culture, & Evolution
ANT 479, Human Paleopathology

ASTRONOMY ELECTIVES

AY 155, Life in the Universe

BIOLOGICAL SCIENCES ELECTIVES

BSC 315, Genetics
BSC 373, Vertebrate Zoology
BSC 400, Vertebrate Functional Morphology
BSC 420, Principles of Systematics
BSC 428, Biology of Fishes
BSC 441, Developmental Biology
BSC 442, Integrated Genomics
BSC 448, Animal Behavior
BSC 449, Endocrinology
BSC 434, Plant Systematics
BSC 464, Biology of Algae
BSC 470, Principles of Population Genetics
BSC 482, Conservation Biology
BSC 483, Evolution
BSC 487, Biogeography

GEOLOGICAL SCIENCES ELECTIVES

GEO 102, Earth Thru Time*
GEO 355, Invertebrate Paleontology
GEO 367, Sedimentology & Stratigraphy

GEO 401, Paleoclimatology
GEO 424, Dinosaur Paleobiology
GEO 462, Quaternary Climates & Environments

HISTORY ELECTIVES

HY 300-013, Race and Science
HY 400, Darwin, Evolution & Revolutions

NEW COLLEGE

NEW 490, Minds and Language

PHILOSOPHY ELECTIVES

PHL 387, Philosophy & Evolution*
PHL 367, Philosophy of Cognitive Science
PHL 390, Art & Human Nature

PSYCHOLOGY ELECTIVES

PY 313, Sensation & Perception
PY 352, Developmental Psychology
PY 413, Physiological Psychology

TELECOMMUNICATIONS AND FILM

TCF 420 Telecommunication Effects
TCF 433 Broadcast News Analysis

UNIVERSITY HONORS ELECTIVES

UH 300, Topics in Vertebrate Paleontology
UH 300, Primate Religion & Human Consciousness
UH 300, Language Evolution

* if not taken as Foundation course

References

Buller, D. (2005). Evolutionary psychology: The emperor's new paradigm. *Trends in Cognitive Science, 9*, 277–283.

Carmen, R. A., Geher, G., Glass, D. J., Guitar, A. E., Grandis, T. L., Johnsen, L., . . . Tauber, B. R. (2013). Evolution integrated across all islands of the human behavioral archipelago: All psychology as Evolutionary Psychology. *EvoS Journal, 5*(1), 108–126.

Carmen, R. A., Guitar, A. E., & Dillon, H. M. (2012). Ultimate answers to proximate questions: The evolutionary motivations behind tattoos and body piercings in popular culture. *Review of General Psychology, 16*, 134–143.

Galton, F. (1874). *English men of science: Their nature and nurture.* London, England: Macmillan.

Garcia, J. R., Geher, G., Crosier, B. C., Saad, G., Gambacorta, D., Johnsen, L., & Pranckitas, E. (2011). The interdisciplinary context of evolutionary approaches to human behavior: a key to survival in the ivory archipelago. *Futures, 43*, 749–761.

Geher, G., Crosier, B., Dillon, H. M., & Chang, R. (2011). Evolutionary psychology's place in evolutionary studies: A tale of promise and challenge. *Evolution: Education & Outreach, 4*, 11–16.

Glass, D. J., Wilson, D. S., & Geher, G. (2012). Evolutionary training in relation to human affairs is sorely lacking in higher education. *EvoS Journal, 4*(2), 16–22.

Locke, J. (1998). *An essay concerning human understanding.* New York, NY: Penguin Classics. (Original work published 1690)

Nelson, C. E. (2007). Teaching evolution effectively: A central dilemma and alternative strategies. *McGill Journal of Education, 42*, 265–283.

Peterson, A., Carmen, R., & Geher, G. (2013). Ovulatory shifts in mating intelligence. *Journal of Social, Evolutionary, and Cultural Psychology, 7*, 66–75.

Pinker, S. (2002). *The blank slate.* New York: Penguin.

Scott, E.C. (2008). *Evolution vs. creationism: An introduction.* Berkeley, CA: University of California Press.

Tooby, J., & Cosmides, L. (1992). The psychological foundations of culture. In J. Barkow, L. Cosmides, & J. Tooby (Eds.), *The adapted mind: Evolutionary psychology and the generation of culture* (pp. 19–136). New York, NY: Oxford University Press.

Walker, A. S. (2013). A missing link: Building STEAM with literary Darwinism. *EvoS Journal, 5*(1), 15–50.

Wilson, D. S. (2007). Evolution for everyone: How Darwin's theory can change the way we think about our lives. New York, NY: Delacorte.

CHAPTER 3 | Building an Evolutionary Studies Program at a Small Liberal Arts College

SUSAN M. HUGHES

SMALLER LIBERAL ARTS COLLEGES and institutions typically have overall fewer faculty members, course offerings, and resources in comparison to larger universities and this can pose some obstacles in developing an interdisciplinary program in Evolutionary Studies (EvoS). However, despite the fact that smaller colleges might have fewer faculty and resources, we have much to offer in terms of achieving a critical mass of faculty interactions across departments (Wilson, Geher, & Waldo, 2009) and sometimes have to deal with less bureaucratic oversights, which could possibly delay or impede program changes. With that in mind, the inception of an EvoS program at a small institution may be actually easier to accomplish than expected.

Similar to how Dr. David Sloan Wilson had begun the seminal EvoS program at Binghamton University (Wilson et al., 2009), I initiated the implementation of an EvoS program at my small liberal arts college, Albright College in Reading, Pennsylvania, by first reaching out to the few faculty members who were involved in teaching the recognized main courses within EvoS. Specifically, I contacted the faculty member in the Biology Department who teaches the course in "Evolution" and our college's only resident anthropologist who teaches a course in "Human Evolution" and other courses that focus specifically on evolutionary theory. My own training is in evolutionary psychology, and I teach a course specific in that area. Together, we represented the core faculty members who primarily teach evolution at my college and had brought a proposal for a minor degree in EvoS to our curriculum committee. The proposal was later presented to the faculty for review and approval. Within my first two years as a faculty member at Albright, we were granted formal approval by the faculty to implement this new special program in EvoS that bestows a minor degree.

Once word of the proposal to start this program was brought to the faculty for consideration, numerous other faculty members (such as those in the Philosophy and English departments) began contacting me as the director of the program, stating that they believed that some of the courses they currently teach would qualify under the curriculum for this program. I found that the faculty was generally curious and open-minded about contributing to the program, even if they did not regard themselves as necessarily qualified or trained evolutionists.

The impetus to create an EvoS program at Albright was prompted by my attendance at professional conference meetings where I became aware that my two colleagues, Dr. Glenn Geher and Dr. David Sloan Wilson, were in the process of submitting a National Science Foundation (NSF) grant for the development of a national consortium in EvoS. After requesting a curriculum addition at my college, I later applied for the NSF-granted seed funding to help grow our program, to which I was awarded. Albright College is listed on the NSF grant as one of the first members to endorse this national consortium, and with this funding in hand, I was able to further expand this EvoS program at my college. Since the program's inception, there have been several students who have graduated with an EvoS minor degree each year, and enrollments have increased with each progressive year. We have had several EvoS students enter graduate programs in EvoS at the master and doctoral levels and have reported that they feel their background and research experiences with EvoS placed them in a better position to pursue advanced training.

An Institution's Support

As with many small liberal arts colleges, Albright College is an institution that promotes and encourages an interdisciplinary program of education within a small liberal arts setting and provides a good foundation in which to sponsor an EvoS program. This is demonstrated in several ways throughout our curriculum. For instance, we offer special interdisciplinary capstone courses that students are required to take as part of their general studies requisites. These courses are often co-taught by faculty from two different disciplines and are an excellent opportunity to create a cross-disciplinary approach to the study of evolution. A few faculty members in the Religion and Chemistry Departments have been discussing ways to design and teach evolutionary principles in a cross-disciplinary fashion with those in the Psychology and Biology departments. Potentially, there could be courses created for a variety of subjects such as religion and evolution, evolutionary economics, and fashion design and evolutionary psychology (i.e., how fashion advertises sexual signaling).

Furthermore, our students are required to take a freshman seminar course during their first year, and these courses are designed with the intent to integrate a particular theme throughout the course. We have offered freshman seminars with an evolutionary theme and perhaps may be able to propose one that is similar to the "Evolution for Everyone" course that has been offered at Binghamton University. This introductory course could be taught as a module in a coordinated manner by different faculty members across different areas that each cover specific topics related to the study of evolution. These freshman forums may also be another way to inspire student interest in the EvoS program.

Both the involvement from other disciplines and a heightened student, faculty, and administrative awareness were needed to develop this program's strong potential at my college. Fortunately, our college's administration was already very receptive to and valued the idea of implementing a new special program of studies in evolution as it coincides with the emphasis on the interdisciplinary nature of education at our college. In particular, our administration seems to appreciate the value of curricular integration within our minors that can be taken in parallel with any major degree. From an admissions standpoint, having special program of studies sets our college apart from other similar schools, and are often "pitched" as attractive elements for enrollment in our school.

Implementing an EvoS Curriculum

The development of an EvoS program may involve both the creation of new, relevant courses and a minor degree (or its equivalent) in the curriculum. At Albright College, we have developed a curriculum that allows students to take courses across various disciplines that would qualify a degree on their academic record in EvoS. Similar to SUNY New Paltz (Wilson et al., 2009), Albright had elected to have a traditional minor but did so by allowing students to receive double credit for their major and/ or general studies courses so as to avoid undue additional course load.

Allowing students to simultaneously satisfy the requirements of this minor with their major/general studies became particularly attractive for enrollment into the program for students who were Psychology, Biology, Psychobiology, or Environmental Studies majors, as many of the course requirements overlapped. For those students, this minor simply provided a curricular pathway to follow to gain a cohesive course of study in evolution. For a sample curriculum, please visit the website: https://www.albright.edu/academic/undergraduate-programs/evolutionary-studies/

Despite our limited course offerings relative to a larger university, we did not seem to have difficulty in being able to satisfy the need for courses that qualified for this degree. There were several existing classes in our curriculum related to evolutionary theory, especially within biology and anthropology. One prominent area that was lacking was in evolutionary psychology. The idea to expand the EvoS education in the Psychology department began when I was given the opportunity to teach an upper-level "Special Topics" course in my specialty of Evolutionary Psychology the first semester that I arrived at Albright. Because of student interest and how well-received the course was, I submitted a proposal to the curriculum committee the following semester to have the course officially and permanently added to the curricula. This course currently qualifies not only toward the EvoS minor but is also an elected course for the Anthropology, Psychology, and Psychobiology majors. If your college is considering a new hire in the Psychology department, consider someone who could teach Evolutionary Psychology, as this is a growing field and graduate programs continue to produce very strong faculty candidates. If it is not feasible or affordable to hire a new faculty member to teach Evolutionary Psychology at your institution, consider that departments may already possess a resident faculty member who has decent knowledge, training, and/or a passion for evolutionary psychology and would be willing to teach such a course. This may even be the case within small departments. Many of the prominent scholars who present at professional evolutionary conferences and publish in evolutionary journals have backgrounds in a variety of psychological areas including social psychology, personality psychology, biopsychology, and even clinical psychology.

At smaller liberal arts colleges, students tend to have more freedom in the curriculum. Often smaller colleges are more flexible about requirements and give more leeway to construct programs that meet individual interests. Some institutions even allow students to design their own majors or do not have majors at all. For instance, at Albright, we have a built-in mechanism within the curriculum that makes it possible for students to create their own interdisciplinary concentration of study/major. Students could therefore combine areas of study across different disciplines through an evolutionary framework. Thus, in addition to the minor degree, becoming an EvoS major could also be an avenue that students can pursue.

Learning about evolution can be taken in so many directions and this should be reflected in the integrated curriculum. Therefore, instead of having a strict rubric of classes to take under the minor, consider the importance of having flexibility in the EvoS curriculum and allow students to have a menu of courses from which to formulate an integrated curriculum. This can be accomplished even within a narrower course selection at a small institution by offering a variety of courses that span different disciplines and by offering independent studies or special independent research opportunities to fulfill program requirements.

Another important consideration when developing the curriculum for an interdisciplinary program in EvoS are the prerequisite courses needed to be fulfilled before taking the upper level courses in evolutionary theory. It may be considerably easier for a Biology or Psychology student to already have taken the needed prerequisite courses because of their major requirements. But if an EvoS program is intended for students of all majors, assuring that all students have the opportunity to take prerequisite classes has to be taken into account when designing this curriculum for the minor degree. As such, both introductory psychology and biology courses are listed on our college's curriculum of required classes for the minor. Because it is often the case that students elect to take these courses for their general studies requirement, we found that having these prerequisites for our minor was not an impediment for students enrolling in the minor across different majors.

We also wanted to be certain that the introductory classes that serve as the program's prerequisites needed for upper-level evolution courses actually contained content relating to evolutionary theory. Our review of these courses concluded that they all did, but to varying degrees; the amount of content pertaining to evolutionary theory disseminated in these courses differed greatly depending upon the instructor. This variability appeared to be more common for introductory Psychology courses than introductory Biology courses. Studies have shown that while overall content and organization tend to be homogeneous across psychology introductory texts, substantial variance exists between texts on core concepts and key terms and that text selection by the instructor plays an important role in the type of topics that are covered (Griggs & Marek, 2001). Thus, the amount of evolutionary background offered in introductory prerequisite classes may vary within a college's offering of the same course depending upon the instructor. Nonetheless, the inclusion of evolutionary theory within introductory psychology texts seems to be expanding. As Gray (1996) highlighted, students will get a better education if they become familiar with evolutionary theory to help guide thought and research in the realm of psychology. Gray encourages teachers to think of the benefits of how to use evolutionary theory creatively in their courses to promote deeper understanding, critical thinking, and the advancement of the discipline. With an EvoS program in place, perhaps a greater consideration of teaching evolution within introductory psychology courses can be emphasized knowing the course is part of the program.

In addition to the practical curricular aspects, when designing an EvoS curriculum, there are also some challenging theoretical considerations. Sinatra, Brem, and Evans (2008) point out several ways instruction can foster conceptual change by teaching and learning about biological evolution. A curriculum in EvoS needs to consider that students may hold preconceived ideas related to the complex notion of evolution and should therefore develop instruction that allow students the opportunity to think deeply and critically about alternative perspectives. Instructional strategies should try to adopt ways to connect topics on evolution to students' everyday experiences and have students experience the phenomenon directly (for a review, see Sinatra et al., 2008). One way to address these ideas as part of the EvoS curriculum is to offer special seminars or special topics courses, independent study, or involvement in faculty research where students have more hands-on training in evolution.

Campus-Wide Events

Another instrumental component in building an EvoS program at a small liberal arts college has included sponsoring campus-wide speaker events. Speakers from various disciplines such as biology, psychology, and philosophy, have been invited to our campus to give talks that are related to evolutionary theory, whereby all students from all disciplines are invited to attend, and, in some cases, these invitations extend to the local community. These talks have raised further awareness of the study of evolutionary theory and demonstrate the value of evolutionary study in our academe.

An example of one such occasion at my college was an event held in honor of Charles Darwin's bicentennial in 2009 to celebrate his life, work, and legacy. There was a reception and a panel of speakers who discussed the impact of Darwin's theories across the various disciplines. A former alumnus and local retired medical doctor took part in this event by providing original Darwin artifacts that he has been avidly collecting for over the past 30 years to display as an exhibit in our college's library. The event was open to all students, faculty, and the general public. It was both advertised and featured in the local newspaper to allow the community to become more aware and engaged in evolutionary thought. It is events such as these that can serve as a good example of the type of activities that effectively advertise the EvoS program. Other such events could include informational meetings for students who want to pursue graduate work and/or careers in EvoS with researchers from the field serving as an advisement panel.

A major concern with being able to invite speakers to a small campus is often the funding needed to do so. A way in which this has been achieved at Albright College was to take advantage of another funding system already in place. Our college has an "outside the classroom" requirement called the Experience Program whose mission is to offer events and programs that highlight and emphasize linkages between the liberal arts curriculum and the scholarly, intellectual, cultural, and political life of the world outside the walls of the college. All students are required to attend certain designated events to earn a selected amount of "Experience" credits to graduate. This Experience Program provides funding to sponsor these different events, so I have utilized this program to fund invited EvoS speakers and as a method to more broadly reach out to the entire student population and educate students on evolutionary theory in an "experiential" manner. Therefore, thinking of other avenues in which to fund and sponsor campus-wide talks may be essential in smaller institutions where there may be less allocated funding within the EvoS program. If a school simply cannot fund such events, the other option is to take advantage of the seminars archived by the EvoS National Consortium from Binghamton University and SUNY New Paltz.

Student Involvement

An obvious key component in developing an interdisciplinary program of study is gaining student involvement. This can take form in several ways. In addition to enrolling students in the minor, there should be encouragement of student participation in attending and presenting at evolutionary conferences and workshops and facilitating student involvement in research projects and grants related to the field. At smaller schools, it is the undergraduates who are often called upon to aid in research (i.e., help conduct the experiments, help with the write-up, present at a conference the findings with the professor). In fact, part of the teaching mission at many small schools is to engage the students in the research of the faculty as a sort of a shared experience rather than a one-way communication of information. Smaller colleges are therefore an ideal forum to get students directly involved in evolutionary research.

I found that in a small institution, getting students to become more involved in scholarship related to evolutionary theory is primarily the result of direct faculty advisement and fostering of that student scholarship. Further, it is often through faculty communication that students become aware that they can take advantage of the benefits of being part of an institution that is affiliated with the EvoS Consortium. For instance, I encourage my students to become involved in either submitting articles to the e-journal associated with the consortium or even serving as student reviewers of articles submitted to the journal. Nonetheless, disseminating this information has proven to be somewhat difficult due to a general lack of funding and resources. My College Relations Department has recently tried to publicize student publications in the EvoS e-journal and participation in conferences in the field, but there needs to be more regular announcements of such opportunities for students.

There are other ways to encourage students to participate in the program even before arriving at the college. For instance, a school's College Admissions Department could allocate funding specific to advertising the EvoS program to prospective students. Another way to attract perspective students is to give research talks at community colleges or high schools within the area as a way to promote the program. For instance, I have been asked to give talks at local community colleges regarding my research in evolutionary psychology and found this to be an effective means to advertise and alert local college community students about the program we have at Albright College.

Our college, in particular, is also part of a local five-college consortium that holds annual local conferences to help advertise student research and further academic integration within the local community. These local conferences have also served as a medium through which we can advertise and alert students (such as those in our two-year local colleges) of our EvoS program, especially for the recruitment of underrepresented groups. Likewise, with only minimal funding for student research, students in our

EvoS program have been able to travel to present at these annual local conferences and disseminate information about EvoS to others.

Facing and Overcoming Barriers at a Small Institution

Smaller colleges might have fewer faculty and resources but we may have an easier time bridging faculty interactions across departments. Along with the assistance of just a few interested colleagues, I was able to develop a curriculum that allows students to take courses across various disciplines that would qualify for a degree on their academic records in EvoS. Moreover, membership in the consortium can supplement any missing facets of the program that are not immediately available at a smaller college (e.g., access to webinar speaker series).

Despite Albright's administrative backing and support of the academic and intellectual pursuit of an EvoS program, additional funding is needed for this program to flourish. Our program was fortunate to be awarded some seed funding to help foster its growth. However, as we move forward, our small liberal arts school often has to make budget cutbacks to contend with the financial state of the economy and the lack of large alumni endowments as compared to more affluent colleges. Due to these funding issues, not all aspects of the program have been fully implemented. For instance, our program could also profit from the development and maintenance of a more elaborate website that would permit information and announcements about the program to be more easily communicated and further encourage student interest.

It would be ideal to expand the program by holding a more regular, campus-wide seminar series as do the programs at SUNY Binghamton and New Paltz whereby speakers within the field could come to our campus to present on a variety of current topics related to EvoS. This could serve as an intellectual forum for not only our students and faculty but could also be extended to members of our local community as well as to a wider populace by providing podcasts of these seminars to be archived in the online library maintained by Binghamton University's EvoS program. Again, this takes financing, and one may have to be a bit creative in finding other methods of attaining financial support such as taking advantage of other facets of a college's budget to bring in speakers to campus.

Due to the faculty's demanding teaching load and service requirements, it is often difficult to coordinate the time to hold events related to EvoS (i.e., have reading groups, meetings) or to engage in collaborative research.

Our program does not appear to be as developed or impactful on the campus climate as compared to the more established EvoS programs at Binghamton University or SUNY New Paltz. Realizing that the demands of teaching and service at a small liberal arts college can place limitations on the faculty's time to contribute to the program should be considered when developing such a program. Colleges could encourage faculty participation by more heavily weighing the initiative, time, and effort it takes to facilitate a formal EvoS forum as an integral part of the faculty's service requirements in their performance evaluations.

Student enrollment numbers seem largely dependent upon the advertisement of the program by faculty advisors. At my institution, for instance, students seem generally unaware that such a minor degree exists unless a faculty member directly speaks of it to the student, so an important element of advisement is to encourage and share knowledge of the program with students. Our college offers over 15 different minors and, as with all minors, recruitment into these programs of study occurs through faculty advising, class announcements, faculty correspondence, and awareness via undergraduate research presentations. Campus-wide events also aid in the broadcasting of a program's existence, but there needs to be growth in this area. Personally, I have found great success in recruiting students into the program via class announcements; every time I teach Evolutionary Psychology (every other academic year), I am able to recruit approximately 25% of the class to sign into the minor degree. I also have had success enrolling most of my psychobiology advisees into the minor because of the overlap with their major requirements. Lastly, I often correspond with the Biology and Anthropology faculty regarding the recruitment of their students and encourage them to recommend the program to their students.

Outreach Beyond Your Institution

The process of starting an EvoS program at a smaller institution does not have to be achieved alone. Gaining support and resources by being affiliated with the EvoS National Consortium can prove to be very fruitful in the pursuits of establishing a program at one's institution. Particularly as a smaller institution with less resources and faculty, it would be beneficial to take advantage of the online seminar series, any centralized whole-program assessment services provided by the consortium, any assistance in developing participation in the symposia, and any site visits that could lobby for the EvoS program at an institution. It can also be very valuable to attend national workshop meetings and conferences that are associated with the EvoS Consortium (i.e., NorthEastern Evolutionary Society [NEEPS], Human Behavior and Evolution Society [HBES], American Society for the Study of Evolution) to interact with others who are also building similar programs, hear talks related to program development, and acquire ideas that will further aid in the development of such a program at one's school, as well as keep abreast of the cutting-edge research in this powerfully impactful field.

Overview of Goals for an EvoS Program at a Small College

As an outline, the following are some goals to consider when building an EvoS program at a smaller college/institution:

- Change and expand upon the curriculum with new courses and a minor/major degree.
- Encourage students to become involved in research within in the field and to submit articles to the EvoS Consortium e-journal.
- Hold campus-wide invited guest talks and other events sponsored by the program.
- Develop and maintain a website to encourage student interest.
- Advertise the program at local conferences and community colleges.
- Attend national workshop meetings and conferences that are associated with the EvoS Consortium.

References

Gray, P. (1996). Incorporating evolutionary theory into the teaching of psychology. *Teaching of Psychology, 23*, 207–214.

Griggs, R. A., & Marek, P. (2001). Similarity of introductory psychology textbooks: Reality or illusion? *Teaching of Psychology, 28*, 254–256.

Sinatra, G. M., Brem, S. K., & Evans, M. (2008). Changing minds? Implications of conceptual change for teaching and learning about biological evolution. *Evolution: Education and Outreach, 1*, 189–195.

Wilson, D. S., Geher, G., & Waldo, J. (2009). EvoS: Completing the evolutionary synthesis in higher education. *EvoS Journal, 1*(1), 3–10.

CHAPTER 4 | The Evolutionary Studies Program

Perspectives from the Flies on the Wall

NICOLE WEDBERG, KIAN BETANCOURT,
RICHARD H. HOLLER, AND VANIA ROLÓN

SINCE 2008, THE EVOLUTIONARY Studies (EvoS) program at SUNY New Paltz has held an annual speaker series—the EvoS Seminar Series. This series has included several renowned evolutionary scholars from across various disciplines (e.g., Niles Eldredge, Geology; Patricia Wright, Anthropology; Gordon Gallup, Psychology; Doug Emlen, Biology). This speaker series is connected with an academic undergraduate class (either the EvoS seminar class or the EvoS capstone)—a class that is designed to expose students to the breadth of applications of evolutionary principles in an intensive and exciting way. Part of the structure of the class includes graduate student teaching assistants (TAs) who help guide the undergraduate students along this academic journey. This chapter tells the story of EvoS at New Paltz from the perspective of these docents of evolution.

If the EvoS program at the State University of New York at New Paltz was to be summarized into one word, it would be *inspiring*. The EvoS program at SUNY New Paltz inspires individuals with different backgrounds and education levels. It inspires new ideas, and it inspires future directions for existing ideas. The ideas that bud from the EvoS program at New Paltz often become tangible research projects which help to inform our understanding of the world and our place in it.

From the talks, to the receptions, to the after-reception libations, all different minds connect through the EvoS program at New Paltz. Many academic programs can provide useful information and a worthwhile education, but not all *inspire* people to critically think for themselves in ways that are engaging. Part of what sets this program apart from others is that it inspires more questions and academic engagement than others.

Personal Perspectives on EvoS

Nicole Wedberg's Perspective on EvoS

Before I go any further, I think it's important to introduce myself and explain why my opinion might matter. I'm Nicole, and I recently graduated with my master's degree in Psychology from SUNY New Paltz. In 2013, I graduated with my bachelor's degree in Southern California (born and raised), and I knew I wanted to study Evolutionary Psychology. And that's *all* I knew. I started to Google where the best places to study this subject were, and I stumbled upon SUNY Albany and SUNY New Paltz. I saw that New Paltz had an amazing EvoS program that I desperately wanted to be a part of, but I needed some research experience first. So, in as little time as it takes to say "evolutionary studies," I packed up my little Honda Civic with whatever I could fit, and I drove my cat and my whole life across the country to upstate New York to pursue this field. I spent a year doing volunteer research with Dr. Gordon Gallup at SUNY Albany and was later accepted into the master's program at New Paltz. My passion for EvoS must have shone through because the head of the department, Dr. Glenn Geher, reached out to me and asked if I'd be interested in being the new assistant for the program. (*Hell yeah!*). I wore the coolest (nerdiest) evolution shirt I owned at my interview with Dr. Geher, and the position was mine that day. Two years later, I'm writing this chapter about the EvoS program at New Paltz from my perspective as the assistant for the program, the supervisor for the Evolutionary Psychology Lab, and as one of the TAs for the EvoS capstone course.

I have spent the last two years organizing the program, enrolling undergraduate students into the EvoS minor, advising students, offering my guidance as a TA, attending all lectures and receptions, designing all advertisements, helping with summer institutes in the program, writing about the program, conducting research on the program's benefits, and just about any other detail you can imagine. If anyone ever has a question about the program, I'm generally the first or second person they contact.

From my experience with this program, I have seen value and potential realized. I have not come across any other program that reaches as many people on such a positive level as the EvoS program. It can benefit all who encounter it, and its merit would be highly advantageous to any university.

The Format

There is a general format to the EvoS capstone course. It is a semester-long course with actual class time. Typically, the first few weeks of the course are spent solely in the classroom. It is a time for the professor to provide a couple of lectures on general evolutionary theory, while also establishing the goals of the course. Students spend these first few weeks learning about evolutionary theory, reading evolutionarily guided assignments, and providing questions and responses each week to consider with the class. The Speaker Series typically begins about a third of the way into the semester and invites about eight to nine speakers in total. Weeks that are not spent in a large lecture hall for one of the talks by an invited guest are spent back in the classroom to regroup and discuss ongoing talks and questions they might inspire. Once the Speaker Series portion of the semester ends, the students are left with final assignments designed to get students to think critically about all topics discussed throughout the semester and to generate their own research ideas that are evolutionarily guided.

The Speaker Series

The Speaker Series itself has its own agenda and format as well. On evenings when there is an invited guest speaker, the class meets first in a small lecture hall with the guest for an hour. This first hour is an opportunity for the guest to introduce himself or herself, describe his or her background, and explain what evolution means to him or her. Importantly, it is also an opportunity for the students to have a more personal encounter with the speaker. During this first hour, students are encouraged to ask the speaker questions about his or her work. Often, students report appreciating the value of this more intimate group conversation with the speaker even more than the main talk itself.

Following that first closed discussion with the class and speaker, everyone shuffles into a much larger lecture room for the main talk. This is the talk which has been advertised in the program and is open and free to the community. The guest speaker then gives about a 40-minute presentation of his or her work and its relevance to evolutionary theory. Following this, both the speaker and the host (in this case, the professor of the course) will then open up the room for questions. Without fail, there are always plenty.

Each and every time, countless hands shoot straight up to ask the speaker their question.

There is one final hour of each event—a reception, of course! Following the main talk, folks are invited to join the speaker in a separate room for pizza, cake (decorated with images typically relevant to the talk), and more questions. Anyone who wants to continue to pick the brain of the speaker can ask more questions. It's a final opportunity for people to gather together and share their thoughts on the talk, as well as converse with the speaker and generate ideas together. It really is a wonderful way to end the evening and tie everything together.

2016 Series

The 2016 series varied slightly from prior years. Since its inception at SUNY New Paltz, the EvoS program has always invited various speakers from various academic fields to come give a talk on their research. The topics presented were never related—they were just the speaker's main or current topic of study. In 2016, the program director and colleagues thought to try to connect each talk in a series around a particular theme. This would give students the opportunity to learn about a single topic from various evolutionarily guided academic fields. The team decided on *aggression* as the topic and invited speakers from around the country to give a talk on their research about aggression. We had experts in biology, psychology, primatology, and other fields all come to discuss aggression from an evolutionary standpoint. It was a truly unique and thorough way to learn about a single topic from so many different angles.

The Education It Provides

The EvoS program at SUNY New Paltz provides an extensive and interdisciplinary education. In fact, to complete a minor in EvoS students must complete courses from various fields including but not limited to Anthropology, Biology, Psychology, Geology, English, Art History, Black Studies, and Physics. The interdisciplinarity of the program makes it accessible and valuable for all students. Students who participate in the program often find that evolutionary theory can be applied just about everywhere. The interdisciplinarity of the EvoS Seminar is part of what sets it apart from other, similar programs. Again, upon discovering this, students often have ideas immediately and raise important and practical research questions that apply directly to their field of interest through an evolutionary framework. While several students who enroll in the capstone course major in psychology or biology, it is also true that several students who enroll in the course are majors in linguistics, math, and even art. The education provided from the program is tangible and applicable to our world today. As Dr. Glenn Geher puts it, once you have an understanding of evolutionary theory, everything thereafter is "like doing research with the light on." Students often have a more solid understanding of the big picture and of their place in the world after receiving an education through the EvoS program. It helps to inform and guide the way we think about everything!

EvoS in Action

Some examples of the EvoS program in action are seen in the work published by students. Undergraduate students at SUNY New Paltz have published manuscripts in *The Journal of the Evolutionary Studies Consortium* (see Muller, 2010). This is an accomplishment most students from other programs fail to even recognize as a possibility. Other examples include the numerous master's theses that have been sparked in some way by the EvoS program at New Paltz—many of which have been published in academic journals.

There are other ways to see the EvoS program in action, too. Several hikes inspired and hosted by evolutionary scholars take place every year. One hike in particular is designed to be both fun and educational—hikers will be able to travel along an evolutionary timeline through New Paltz's famous mountains. The trail itself has been marked with explanations along the way that guide hikers through the tree of life and point out when, chronologically, different species branched off from a common ancestor. It will be a very real and hands-on approach to EvoS.

Above all else, anyone who has been a part of the EvoS program at New Paltz before will be quick to tell you—if you really want to see EvoS in action, you have to check out the spring Speaker Series. That's where the really powerful stuff happens.

The Speaker Series

The EvoS Seminar Series is a lecture series like none other. Experts from different fields including Biology, Primatology, Psychology, and others are invited to SUNY New Paltz to give a guest lecture on their research and describe its connection to evolution. (See http://www.

newpaltz.edu/evos/seminar.html for streaming videos of any talk.) While the talks coincide with an EvoS capstone course each spring, they are also open and free to the general public. Announcements, posters, website event pages, and other various advertisements are made each year for upcoming talks, as the series is always a highly anticipated event. Experts have come far and wide to discuss topics like sexual competition, animal weaponry, family resemblance, lemur ecology, the psychology behind Facebook selfies, beauty, and one talk in particular had standing-room only and lines around the building to learn about the Paleo diet. Not only are the talks purely fascinating in their own right, but the speakers always do a superb job of connecting their subject matter to evolutionary principles.

The speaker series is the nucleus of inspirations. Far more often than not, people walk away from these talks with new questions and research ideas. Importantly, these ideas are typically applicable to the real world—students, faculty, and the general public alike will be inspired by the speakers and develop their own ideas and applications that are useful to the human condition. Instead of taking things at face value and leaving them as is, the speaker series often leaves listeners thinking, "What *else* can we do with this information?"

One great example comes from a recent talk in the 2016 seminar series on tonic immobility. Renowned researcher Dr. Gordon Gallup came to New Paltz to give a talk regarding his research on this subject—essentially, it's when animals play dead (however unwittingly) as a mechanism for survival. In true speaker series fashion, Dr. Gallup provided extensive examples of tonic immobility in several different species ranging from chickens to humans. He made a point to explain its potential adaptiveness in humans as in the case of sexual assault and how we might be able to include these findings into policymaking for a truer and more compassionate response to victims. As mentioned earlier, people often walk away from these talks thinking, "What *else?*" and this was no exception. While Dr. Gallup gave an amazing talk on his research and provided insight toward its applications, listeners began to do the same. Within minutes, attendees were asking questions and generating ideas about how we can implement programs to help victims of plane and car crashes when they are experiencing tonic immobility.

This is only one example drawn from the last nine years of the EvoS Seminar Series—many of which have ignited equally great and inspiring ideas in others.

2016 Series—Evolutionary Psychology and Aggression

For the first time since its conception, the EvoS program decided to have a central theme for the 2016 seminar series. In years past, speakers from all different fields have been invited to give a lecture on their research and expertise, and there was no central theme for the talks in a series other than their connection to evolutionary principles. Five of the masterminds who organize the EvoS program met over lunch in 2015 to discuss the upcoming series. The thought of centering all the talks for a series on a single theme was brought up, and we ran with it. We settled on the theme of aggression. So, all speakers invited for the 2016 seminar series gave talks about their respective research on aggression as well as the connection to evolutionary principles. We had lecturers from fields including Psychology, Biology, Anthropology, Biochemistry, and more. It really provided the students and the general audience with an opportunity to hear about and understand one topic from several different angles.

Among the 2016 speakers was Dr. Doug Emlen—the world's current leading researcher in animal weaponry. He flew to New Paltz from Montana to share his research and to talk about natural weaponry found in various animal species. It's every bit as cool as it sounds. He showed us pictures of the most incredible and elaborate natural animal weapons, he explained why animal weaponry may have been selected in some species and not in others and the adaptiveness behind animal weaponry for aggressive behaviors, and he concluded with human parallels that really unified all of his points. The human parallels were truly astonishing and relatable. For instance, he explained that caribou have the largest weapons relative to their body size out of any living vertebrate, yet more than 99.99% of their contests are actually settled without any violence. He pointed out that, generally, larger weapons actually work as deterrents. Often, the outcome of possessing very large and dangerous weaponry is peace. He paralleled this phenomenon to show that as much as it may be the case for humans, it also happens in the natural world among members of various other species as well. This is important because it directly spoke to individuals (myself included) in a way that allowed them to freely connect and reflect upon their real understanding of the world around them as it relates to human aggression and foreign warfare.

To hear about aggression from another angle, we also invited Dr. Rebecca Burch from SUNY Oswego to give a lecture from more of a psychological perspective. Using evolutionary psychology as a framework, Dr. Burch shared her research on the role family resemblance can play toward aggression among family members. She explained that the more similar a child looks to the father, the better the relationship is between father and child. As paternal resemblance increases, so does the positive treatment of that child, generally. This is another side of the same coin often explored in evolutionary psychology research that points out that stepchildren and unrelated children in the home are more frequently the victims of parental abuse (Daly & Wilson, 1996). Again, this talk offered a very tangible and applicable message about what might be going on in cases of familial abuse and aggression.

The EvoS program hosted seven different speakers for the 2016 Seminar Series from all different backgrounds to share their research and expertise as it relates to both evolutionary principles and aggression. Many attendees (myself included) will attest to the value of this series, and the interest it sparked in understanding the evolutionary psychology behind aggression.

Perspectives from the Teacher's Assistants

Richard H Holler's Perspective on EvoS

The evolutionary sciences demand teachers who both embrace multidimensional education and who have the ability to recognize the nuanced relationships between seemingly unrelated disciplines. The principles of evolutionary theory can appear so simplistic that even experienced professors can be warped by their simplicity. Successfully teaching a modern EvoS course to any student or audience may only be feasible when the curriculum includes both more than a one scholar teaching it and/or more than a one teaching philosophy employed. More specifically, evolutionary science is interdisciplinary, thus its lecturers should be too.

Charles Darwin's theory of evolution via natural selection is composed of principles outside of biology. Prior to publishing his famous book *On the Origin of Species* in 1859, Darwin had a surprisingly thorough background and understanding in geology aside from his initial devotion and understanding in medicine. A valuable merit in studying geology is learning to estimate the actual age of almost any geographical structure. In other words, a geologist could fathom how old the earth is. The year today is 2019 and two-thousand nineteen years is like a blink of an eye relative to the age of planet earth. Another characteristic of Darwin was being able to accurately calculate the trade-offs of a physical or mental attribute within a variety of species, which may have solidified his conception of *evolution* as *change over time* rather than *improvement over time*. Had Darwin not been a student of geology and zoology and had he not understood basic economics, then perhaps the phenomenon of evolution would have remained an enigma.

Applying evolutionary theory across the boundaries of biology requires the assistance of more than a single professor or area of expertise. In academia, it is now widely accepted that evolutionary theory is fact. It is also widely accepted, in academia, to appoint scholars who are highly specialized in a particular discipline as the professors or authority figures in that particular discipline. Although logical, employing specialized scholars only to their specialized field could still significantly deprive the student's educational welfare, especially when it comes to asking the biggest questions humankind has ever asked.

Professor Glenn Geher was wise to include speakers from various disciplines and career fields into the EvoS capstone course. After attending at least two or three talks, students could begin to identify recurring evolutionary themes, which would eventually serve as bridges from one discipline to the next. Geher's EvoS students were proudly exposed to speakers who presented varying behavioral and non-behavioral phenomena among different species while utilizing the similar scientific frameworks. Like myself, EvoS students come to realize that understanding the evolutionary sciences is almost too integrative to be constricted to only one classroom. Teaching within the evolutionary sciences is distinct from teaching within more traditional subjects like mathematics, chemistry, history, etc., in which a single textbook of any of the latter three classes or disciplines could sufficiently convey that discipline's intellectual philosophy in a single semester. A class like the EvoS capstone course probably could not survive in a parochial curriculum or school of thought.

As an EvoS capstone TA, I embraced Geher's philosophy to expose students to a variety of academic perspectives in learning the evolutionary lens, which led me to believe that a single question can be more valuable than a single answer. Each week students formulated questions from the assigned scholarly articles and presentations of each speaker, and each week I witnessed their questions and comments mature into inquiries that extended beyond biology.

Just how far did these students explore beyond biology? At the end of the semester, students developed PowerPoint presentations that integrated the evolutionary perspective into their own subjects of interest. Just to name a few, students presented their ideas about the application of evolutionary psychology to the motivations for nations to engage in war, to the underlying fears of mathematical anxiety and their relevance to biological evolution, and to the origins of the hammerhead shark's cephalofoil (otherwise known as the hammer). These presentations not only exemplified the students' ambition, critical thinking skills, and creativity but also their career interests.

Perhaps the most valuable reward in being a TA for EvoS capstone class is learning to perceive a scholarly presentation from an academic as a shared experience from a fellow human rather than just a presentation or collection of facts. The more diverse the shared presentations or experiences from each speaker, the better the experiences amass into a collage of stories, which also can be our best teachers. This approach inevitably teaches students not only new material to for them to learn, but also how to learn.

Inviting the general public to attend these colloquia can also enhance the intellect of the local community, which provides individuals who are unable to obtain a formal education with the opportunity to gain an advantage over individuals who do. The privilege of education, unfortunately, is not cheap, and there will always be individuals who will be deprived of any core of knowledge, but this should not dismiss an individual's right to learn novel disciplines. The ultimate goal of the EvoS capstone is not just to share the wonders of the condoned evolutionary perspective but also to equip humanity with a set of basic principles that can facilitate countless academic and real-world endeavors.

Perhaps there is no science like evolutionary science. An education firmly grounded in evolutionary theory is a virtue that is unfairly and disproportionally distributed throughout our university and likely the general population. Humanity accepting evolutionary theory as a fact stirs my optimism for the future of EvoS. In being a true interdisciplinarity, the general appreciation for EvoS may be delayed until individuals develop a thorough background in the multiple disciplines that gravitate around evolutionary science. Only after the introduction to and the acceptance of the negligence of evolutionary theory may then the practical applications of EvoS become obvious. The ability to perceive a discipline as a versatile perspective is a virtue that cannot be taught by a single teacher, class, or experience. Encouraging students and other individuals to seize opportunities like the EvoS capstone could be pivotal to both their and academia's future. As a *fly on the wall*, I am both honored to witness the practice of the philosophy of evolutionary science and am excited for the future of academia.

Kian Betancourt's Perspective on EvoS

Upon hearing that I would be selected as a TA for Dr. Glenn Geher's Evolutionary Seminar series, I was absolutely ecstatic. My real passions are for teaching and studying psychology, and being able to incorporate those into an EvoS class was an absolute dream. It made me very cognizant of my status as a graduate student. As a student who has just about completed his first year of graduate school, it took me a while to truly feel oriented, or feel as though there was a distinction between me and the undergraduate students. Importantly, I feel that I now have a different approach to critical thinking, understanding important psychological and evolutionary concepts, and an ability to reflectively look at other students' work and evaluate it as an instructor. These are all skills that developed relatively recently, and implementing them in such a way as that was helpful as a TA has been a phenomenal experience.

I particularly enjoy grading the students' discussions and biweekly responses to articles. Feeling as though I had the expertise to actually administer a grade to someone relatively close in age to me sparked feelings of the often-talked about "imposter syndrome." However, I simply did what I knew best and graded their responses as I would a professor with his or her own students—evaluating them for substance, critical thinking, questioning core concepts, and overall writing like a college student should. I genuinely believe that through assessing other students' work, I have become a better writer and thinker myself. Looking for nuances in logic and shortcomings in adequate comprehension of the subject matter, as well as simply an overall critique of writing style and fluidity, have all greatly contributed to the way I look at my own writing. When I write an essay now, I no longer look at it as simply grading my own work. Rather, I evaluate it as a third party, under the assumption that the work is that of a student and I had to assign a grade to. Doing this has resulted in a remarkable improvement in my own writing—an accomplishment I'd like to solely attribute to my experience as a TA for the EvoS capstone course.

Another reason I genuinely enjoyed my experience as a TA was because of how much I learned in the class, personally. While learning about how to teach and grade was a great experience, an even more worthwhile one was simply expanding my evolutionary knowledge throughout the course. Not only was I benefiting by thinking about evolution in new ways via the students' discussions and ideas, but I also benefited as an audience member who got to sit front row to some of the field's best. This had a major impact on me as a scholar—not only was I developing my teaching skills, but I was simultaneously becoming a better student. And to top it off, I was becoming a well-educated member of society by exposing myself to these ideas.

I am not the only person that felt that this class was a worthwhile experience—not by a long shot. Hearing students' reactions in class, as well as speaking to the other TAs has shown how much we've all grown in a short period of time. It's truly remarkable. It was a pleasure to take part in the first themed series for EvoS, with this semester focusing on aggression. Every speaker we had, it seemed, got better than the last. And through reading students' reactions to these talks, I know firsthand that they were just as blown away as I was.

Learning about concepts like animal weaponry, tonic immobility, and others really expanded my knowledge of not only EvoS but also biology and genetics, as well as overall scientific reasoning. This, in essence, is the core of the EvoS program—an interdisciplinary approach to understanding dimensions of human behavior as well as the world around us by understanding evolutionary theory. Being inclusive of all disciplines means a wider audience. This often leads to an increase in logical, empirical thinking regardless of one's discipline.

The benefits of an EvoS program like this speak for themselves. Students gain a wider understanding of science, as well as how to apply these skills to their own respective interests and perceptions of the world around them. The TAs also had the opportunity to refine and improve on their teaching ability, while simultaneously being a student of the class themselves. Professors teaching the classes get the satisfaction of knowing hundreds of students are gaining a more comprehensive, critical mindset of the world around them—something that anyone, irrespective of being in a science field or not, should possess. Further, making the events open to the public only increases exposure. It truly showcases one of SUNY New Paltz's prized possessions—an EvoS minor. Not only does it show the public that programs like these matter, but it simultaneously improves their own scientific literacy while exposing them to new and developing topics in the field of evolution.

Lastly, I'd argue that even the speakers gain experience in giving lectures. Generally, the speakers use feedback and questions from the audience to see perspectives and new angles on their own research that they may not yet have seen. New directions for research stem from the Seminar Series all the time.

Overall, both the minor and the EvoS Seminar are incredibly valuable components to SUNY New Paltz's overall academic reputation. They serve to expand EvoS as an interdisciplinary approach to understanding human behavior in our society.

Vania Rolon's Perspective on EvoS

To explain the benefits I believe come from a course like the EvoS capstone course, and from a minor in EvoS as a whole, I need to first describe the problem I have seen regarding the ambivalence of some professors when it comes to applying an evolutionary perspective in areas such as psychology. To do this, I mention some previous experiences I had even before becoming a TA for the EvoS capstone course, followed by how this course can address such problems.

During my first semester as a graduate student, I had the pleasure of being hired as a TA for the undergraduate psychology statistics course. Additionally, my mentor Glenn Geher suggested I help as a TA for his undergraduate evolutionary psychology course. As a result of both teaching assistantships, I saw several students often and had the chance to interact with them in both courses. I had a rather interesting exchange with one particular student taking both classes (whom I shall refer to as Sarah), and the professor for whom I was assisting with statistics (hereby referred to as Dr. X for lack of originality on my behalf).

Sarah had to write a one-page research proposal for the evolutionary psychology course, and since this was her first semester learning about statistics for the behavioral sciences, she approached Dr. X at the end of one of our class periods asking for advice on how to study her topic of interest. As I furtively overheard the conversation while packing my material, Sarah mentioned she wanted to look at attention deficit disorder and attention deficit/hyperactivity disorder from an evolutionary perspective and investigate whether these disorders were perhaps due to the mismatch between our environment of evolutionary adaptedness and our current lifestyle of locking children in classrooms for a significant portion of the day. Sarah's major problem was that she could not think of a good design for such a study. Dr. X mentioned she would be happy to assist Sarah, but that "evolutionary psychology didn't really do experimental studies." I pressed my lips together in an attempt to avoid a discussion. After all, Dr. X had years of experience and knowledge that I lacked, and any debate would have ended in a poor defense from my side. She must have noticed my reaction, as her next sentence was "I am sorry, Vania, but you know I'm right," followed by a slight giggle.

I must clarify that, despite this unusual episode, I truly enjoyed working for Dr. X, and I probably learned more about statistics by assisting her with her course than I did taking statistics as an undergraduate student. That being said, I would soon have to get accustomed to having similar encounters with other professors. The next semester, our Research Methods professor said she found the idea of evolutionary psychology interesting but didn't fully agree and felt the approach was biased and incorrect in many ways. Throughout our Psychology department, it seemed about a third of the faculty had either strong or mildly negative attitudes toward this approach that I so deeply loved. Opposite the faculty, students seemed to enjoy Glenn's evolutionary psychology course to the point that many of them registered for the EvoS Seminar Capstone course the following semester. Why were students eager to take these courses whereas some faculty members seemed slightly ambivalent about such a wonderful approach?

Turns out Dr. X did not agree with much of the literature on evolved sex differences and posited that many of these differences were due to society. Although I have heard this argument before, I do not believe it is enough to regard an entire field as "not suited for real experiments." Personally, it seemed to me that Dr. X, like many others, believed that evolutionary psychology focused almost exclusively on sex differences—a common misconception about the field. Our Research Methods professor, on the other hand, argued some ideas were simply wrong. "I do not have sex because I am thinking of reproducing and having babies." She mentioned. "I have sex because I like having sex." Hopefully readers who have received some evolutionary education will notice the fallacy in this argument. It was apparent to me that both of my professors held misconceptions and equivocal knowledge about evolutionary psychology. It was through these exchanges with both professors that I realized the importance of the EvoS minor and the difference a course such as the EvoS Capstone could have for students.

The first apparent benefit of such a course is the increase in evolutionary knowledge and the provision of tools that makes students think openly and with more flexibility. Because of this course and the EvoS minor students come to understand that not all studies that use an evolutionary approach focus on evolved sex differences. To rest my point, one only needs to look at the wide gamut of proposals that students submitted for their final presentations. Topics included improving social connections in the classroom, the co-evolution between humans and canines, an evolutionary theory of emotion, locus of control and lying behavior in children, and so on. The only limit for student presentations was their own creativity, and I am sure these students will not hold the same idea as Dr. X when it comes to understanding all the topics that can be examined with an evolutionary lens.

More importantly, students come to understand the misconceptions behind evolutionary theory. A common (and perhaps erroneous) critique of this approach in psychology is that, by arguing in favor of evolved sex differences, the field somehow justifies sexism and inequality across genders. Evolutionary psychology is seen by some as evil; promoting misogyny, prejudice, and selfishness. A talk that clearly debunked this malevolent view came from none other than Dr. Gordon Gallup. Not only did his talk open our eyes to how studies on animal behavior can shed light into human behavior, but he also made a strong argument on legal implications with regards to allowing rape victims to sue their attackers. By enlightening us on tonic immobility as an adaptive mechanism in situations in which an organism can no longer escape a predator and thus has few options left but to "play dead," Dr. Gallup went on to describe how this immobility can instinctively kick in—even in humans. Several rape victims, he reported, describe feeling paralyzed during the attack. This paralysis is a response from the body, and yet some laws demand the victim attempts to fight the attacker to be able to file a lawsuit. Demanding such an action is demanding the victim to not experience a common biological reaction to threats. Dr. Gallup ended his talk noting that perhaps these laws should change and that rape victims should learn about the mechanisms of tonic immobility to perhaps help alleviate part of the burden they carry. Clearly, such points are far from evil, and hopefully students found this talk fascinating and highly educative on how understanding our evolved mechanisms can help us create laws that serve to better protect people.

One last advantage of the capstone course and the EvoS minor that cannot go unmentioned is its vast interdisciplinarity. I had the chance to take an evolutionary psychology course as an undergraduate, and I knew in theory that the field is highly linked to evolutionary biology. Being a TA for the capstone course allowed me to see this in practice. From aggressive signaling in competing male crickets to the evolution of extreme animal weapons and its parallels to human weapons, the seminar series constantly challenged my conceptions of how and where evolutionary theory can be applied. Additionally, such interdisciplinarity provides more than a new lens with which to view the world. It also provides something that, in my opinion, is lacking in today's academia: cooperation. Academia today is a competitive world, with scholars aiming to publish as many results as possible before someone else comes up with the same idea. In addition, departments typically exist as worlds of their own, rarely intermingling with one another. EvoS, however, brings great minds from several fields into one room, where each expert can add his or her own knowledge to create something bigger than any area could fathom alone. Again, one only need to look at the EvoS capstone course to realize talks range from anthropologists to biologists to even cross fit experts in past years. This interdisciplinary approach is one that should strongly be considered in the world of academia. After all, have we not evolved as a social, cooperative species? Cooperation has allowed us to move out of the African savannah and into the metropolises of today. Cooperation has helped us send a man to the moon, and only cooperation can help us understand the ultimate causes of human behavior.

An EvoS minor would prepare students for arguments such as the one I had with Dr. X. For instance, a knowledgeable student could cite Ketelaar and Ellis (2000) to argue that Dr. X's accusations of evolutionary psychology as unfalsifiable are unwarranted and rooted in a misguided Popperian interpretation of how science operates, while, in fact, evolutionary theory has a unique ability to explain what seem to be anomalies and generate novel predictions and explanations that can actually be tested through experiments. Similarly, to argue Dr. X's claim that much of evolutionary theory focuses on sex differences that could result from cultural rather than biological differences, one could cite studies with evidence on how sex differences in characteristics such as sociosexuality (i.e., engaging in sexual behavior without an emotional attachment) can be found cross-culturally in societies ranging from Argentina to Zimbabwe (Schmitt, 2005) or studies on how sex differences can be found even among primates (Alexander & Hines, 2002). Getting involved with evolutionary theory helped me prepare proper rebuttals for future encounters with anti-EvoS faculty.

Kian Betancourt: What EvoS Does for Us

The EvoS program does more than inspire. Typically, it teaches students and faculty alike to critically think about the world in a way that makes sense. It arouses within people a drive to better understand the human condition, the world, and whatever other topic interests them. The program often gives people a sense of direction. I know it has for me.

For example, I mentioned that Dr. Burch gave a very well-received talk about family resemblance and aggression in humans. In typical EvoS fashion, several students raised questions about future directions and implications for this research. Dr. Burch's response to nearly every question about a future direction or implication was something to the effect of "Great question! That's your homework. Go find out the answer, publish it, and that'll be your thing." It would not be an exaggeration to say that seven different potential master's theses could have stemmed from that hour alone with Dr. Burch.

The seminar series takes place within an EvoS course. The students in the course are required to generate a research proposal by the end of the semester, and many of the proposals are truly worthy of pursuit. It always amazes me how much students are able to understand and apply evolutionary concepts by the end of the seminar series to generate their own research ideas. One student developed a new model that could help to predict when a nation should become involved in warfare, and he developed this model using his understanding of male dominance hierarchies and altruism. Another student developed a research idea designed to measure the extent to which an environment may influence disordered eating in women. Talk about a couple of projects that would have a real world application!

The program is extremely valuable to students. It extends and expands students' understanding of evolution, it helps students understand their world, it encourages students to apply evolutionary principles to the world around them, and it often generates opportunities for students to publish research projects and other scientific manuscripts.

For how valuable the program is for students, it offers just as much to faculty, staff, guest speakers, graduate students, TAs, community, and others exposed to it. The faculty and TAs (often graduate students who are also members of the Evolutionary Psychology Lab on campus) are exposed to just as many research opportunities. The EvoS program truly enlightens us to understand our place in the world, and it has generated much good and service.

References

Alexander, G. M. & Hines, M. (2002). Sex differences in response to children's toys in nonhuman primates. *Evolution and Human Behavior, 23*, 467–479.

Daly, M., & Wilson, M. I. (1996). Violence against Stepchildren. *Current Directions in Psychological Science, 5*(3), 77–81.

Ketelaar, T., & Ellis, B. J. (2000). Are evolutionary explanations unfalsifiable? Evolutionary psychology and the Lakatosian philosophy of science. *Psychological Inquiry, 11*, 1–21.

Muller, K. (2010). Evolutionary educational psychology: The disparity between how children want to learn and how they are being taught. *EvoS Journal, 2*(1), 12–23.

Schmitt, D. P. (2005). Sociosexuality from Argentina to Zimbabwe: A 48-nation study of sex, culture, and strategies of human mating. *Behavioral and Brain Sciences, 28*, 247–311.

SECTION 2 | Evolutionary Studies Embodied within Disciplines

CHAPTER 5 | Extraordinary Claims, Extraordinary Proof

THOMAS G. NOLEN

An extraordinary claim requires extraordinary proof.
—MARCELLO TRUZZI (1978, p. 11)

WHEN DARWIN PROPOSED HIS theory of the origin of species, he was making an extraordinary claim that not only could natural selection account for the occurrence of new species but that it could account for why organisms were so well suited to their environments—that is, it explained the origin of adaptations. In the introduction to *On the Origin of Species*, Darwin (1859) remarks how by that time in the 19th century it was not an extraordinary observation to make that species are not fixed and that modern species appear to have descended with modification from earlier forms. But he admitted: "Nevertheless, such a conclusion, even if well founded, would be unsatisfactory, until it could be shown how the innumerable species inhabiting this world have been modified so as to acquire that perfection of structure and co-adaptation which most justly excites our admiration" (Darwin, 1859, p. 3).

While the quote by the professional skeptic Marcello Truzzi with regard to parapsychological claims would be popularized by Carl Sagan a few years later (Sagan, 1995), the notion has been around for a long time. Darwin clearly understood the implications of his grand idea: Despite the overwhelming evidence suggesting that modern species descended from earlier forms by a process of "descent with modification" (what we now call *evolution*) before Darwin and Wallace (1858) no mechanism other than special creation could explain the well-known adaptive features of organisms (Paley, 1809). Well versed in William Paley's (1809) *Natural Theology*, Darwin knew that his claims needed extraordinary "proof," that is, a mechanism to account for biological adaptations. He was also aware of the enormous *consequences* of his theory for religion and for how we view ourselves in this world: "Light will be thrown on the origin of man and his history" (Darwin, 1859, Chapter 14, pg488). Those are the reasons he waited so long before publishing his little "sketch" (Darwin, 2005) as he was compelled to collect sufficient evidence of natural selection and then elaborate on its extraordinary explanatory power by compiling copious examples of its application (Darwin, 2005). Only later did he publish a more in-depth discussion of just what his theory meant for modern humans (see especially Darwin 1879/2004).

The Meaning of "Proof" in Science

> *The criterion of the scientific status of a theory is its falsifiability, or refutability, or testability.*
> —KARL POPPER (2002, p. 48)

Although not specifically reviewed by Thomas Kuhn in his *Structure of Scientific Revolutions* (Kuhn, 1962/2012), Darwin's theory was indeed a radical *paradigm shift* and therefore required extraordinary evidence if it was to displace the long-held paradigm of *special creation* (embodied in natural theology). It is not an overstatement to say that the scientific field of Biology (as distinct from Natural History) started with the publication of *On the Origin of Species* in 1859. Darwin's theory provided a predictive, theoretical framework that replaced mere description of the natural world and extended the explanation beyond introspection (i.e., interpreting nature from a purely human experience). Unlike the theory of natural theology (i.e., special creation), descent with modification via selection was testable. In fact, it was both testable and falsifiable, as formulated by the 20th-century philosopher science, Sir Karl R. Popper in his 1935 work, *The Logic of Scientific Discovery* (Popper, 1935/2005). Thus Darwin's theory was constructed such that some observations or data (testability) could prove fatal to its acceptance (falsifiability). Darwin (1859, pg189) himself said, "If it could be demonstrated that any complex organ existed, which could not possibly have been formed by numerous, successive, slight modifications, my theory would absolutely break down" (p. 189). Tellingly, even by the sixth edition of the *Origin*, he concluded, "But I can find out no such case" (Darwin, 1872, Chapter 6, p. 146). And we have yet to do so.

Alternative Hypotheses and Popper's Critical Rationalism

> *According to my proposal, what characterizes the empirical method is its manner of exposing to falsification, in every conceivable way, the system to be tested. Its aim is not to save the lives of untenable systems but, on the contrary, to select the one which is by comparison the fittest, by exposing them all to the fiercest struggle for survival.*
>
> —KARL POPPER (2005, p. 454)

Simply being *able* to "falsify" a hypothesis or theory was not enough for Popper.[1] He recognized that since science employs the empirical method, it was necessary to rigorously test and modify the hypothesis or theory, by comparing alternate hypotheses or assumptions with empirical observations. In this, Darwin anticipated Popper and later philosophers of science. Darwin's many publications with regard to his theory provided many empirical tests, modified assumptions and modified hypotheses[2] to support the basic idea of selection as a cause of "descent with modification."

Biological Evolution, an Extraordinary Claim No Longer

Darwin and Wallace's (1858) theory of biological evolution is a prime example of a claim that was extraordinary and, indeed, fantastic at the time. Since then it has been tested and supported and, most important, has survived numerous falsification attempts by an extraordinary set of evidence compiled over 160 years—starting with Darwin's *On the Origin of Species* and now extending to thousands upon thousands of papers not only providing more support for Darwin's origins "sketch"

---

[1] "In point of fact, no conclusive disproof of a theory can ever be produced; for it is always possible to say that the experimental results are not reliable, or that the discrepancies which are asserted to exist between the experimental results and the theory are only apparent and that they will disappear with the advance of our understanding." (Popper, 2005, pp. 691–692).

[2] "Scientific theories are perpetually changing. This is not due to mere chance but might well be expected, according to our characterization of empirical science. . . . for a severe test of a system presupposes that it is at the time sufficiently definite and final in form to make it impossible for new assumptions to be smuggled in. In other words, the system must be formulated sufficiently clearly and definitely to make every new assumption easily recognizable for what it is: a modification and therefore a revision of the system." (Popper, 2005, pp. 454–455).

but also allowing us to make ordinary predictions and even incorporate the modern theory into new bodies of knowledge (classical and population genetics, molecular biology, biochemistry, animal behavior, paleontology, anthropology, predator–prey interactions, development of public health policy, solutions for the emergence of drug resistance in pathogens, etc). Moreover, the modern synthesis of evolutionary theory itself has seen applications in psychology and other social sciences (e.g., Daly & Wilson, 1988; Gough et al., 2008; Schmidt, 2015) and in other traditionally nonbiological fields as well (e.g., economics: see Waltman et al., 2011; also, peruse any of the articles in the *Journal of Evolutionary Economics*; music: see Dostal, 2005; Matić, 2010; mathematics: see Clark, 2013).

Now, most evolutionary claims fit into a continuum of biological possibilities. Evolutionary theory is often said to be a "fact"—but what that means is that evolutionary hypotheses for biological phenomena are ordinary, testable by straightforward experiments and observations. Modern evolutionary theory now falls into the category of practice that Thomas Kuhn called *ordinary science* (Kuhn, 2012). Building on its empirical and theoretical foundation, we can propose ordinary explanations with very simple (but still solid) supporting evidence.

Although evolutionary claims within the realm of biology, biomedical applications and biological anthropology are no longer considered extraordinary, adaptive hypotheses about humans are often subjected to extraordinary scrutiny. The mere claim that evolutionary theory says anything important about the human condition still is met with resistance, and some consider such claims to be extraordinary, at least within the social sciences, cultural anthropology, and even education. Requiring that claims for the evolution of human adaptations meet a higher level of evidence than all other evolutionary questions misses the point of what is meant by extraordinary "proof." Acceptance of an evolutionary foundation of the human animal no longer represents a paradigm shift (Kuhn, 2012). A real paradigm shift would be to hypothesize that humans have been uncoupled from our biological, evolutionary past by a unique, modern environment (social, technological, political, economical, etc.)—that we are no longer evolving by natural selection and nothing about the biological past informs us about ourselves. That claim, although falsifiable, would be extraordinary. Still, even "ordinary" adaptive hypotheses need to be falsifiable and tested.

Strong Inference: A Different Kind of Extraordinary "Proof"

In the lead up to this, the 210th anniversary year of Darwin's birth, I've attended my fair share of evolutionary research seminars. Surprisingly, few presented alternative hypotheses or, better yet, multiple alternative hypotheses for the subject of the seminar, even though that may be characteristic of their published works. In fact, rarely was a *specific* evolutionary hypothesis enunciated, and most of the time a mechanistic explanation was offered but not an evolutionary one. And when one was, it was often just an untested hypothesis. The speaker usually failed to point out what critical experiment or observation could falsify it. Admittedly, these talks were directed toward a general, nonspecialist audience. But many of those in attendance were students, and this "omission" was a missed didactic opportunity. Moreover, modern evolutionary theory is championed (all too often in courthouses in the United States) as a true science (as opposed to creation "science") because its hypotheses are falsifiable (Popper, 2005). So, where are all these falsifiable hypotheses?

My undergraduate Invertebrate Zoology professor at University of California–Santa Barbara, Demorest Davenport emphasized (i.e., drummed it into our skulls) that adaptive questions can be addressed using *strong inference*. He had us all read the 1964 *Science* article of that name by Platt (1964). At the time I was not especially impressed because it sounded like what we had been taught in General Biology and General Chemistry and had already accepted as the standard operating procedure in science. Professor Davenport believed that at that time (mid-1970s) evolutionary biology had lost its empirical rigor. As Massimo Pigliucci (2009) suggests, the main point of Platt's article was to explain why the "soft sciences" (viz. social sciences, animal behavior, and even sociobiology) were less successful than the new (at the time) "hard sciences" like molecular biology and modern physics.

This "hard versus soft" characterization was based on the contemporary perception that molecular biology of the time had experienced extraordinary success (think of the double helix, the decoding of the genetic code, protein sequencing, the "central dogma") largely because of its systematic use of rigorous hypothesis testing. The still common distinction between "hard" and "soft" emphasizes different approaches of providing "proof." But generally, strong inference has been adopted to varying extents, if not routinely practiced by most "hard" and "soft" sciences today. (For a recent re-evaluation of the relationship of biology and sociology, see Schutt & Wilson, 2016).

In any case, as a young scientist, my take-home message was that strong inference could be applied to all kinds of questions and that it should have been applied more often in evolutionary biology. This is still true today, even within traditional biology as well as anthropology, psychology, medicine, and public health. When it is employed, it is not always obvious when researchers present their work to a general audience.

Specifically, the elements of strong inference are straightforward (Platt, 1964, p. 347):

> The steps are familiar to every college student and are practiced, off and on, by every scientist. The difference comes in their systematic application. Strong inference consists of applying the following steps to every problem in science, formally and explicitly and regularly:
> 1) Devising alternative hypotheses;
> 2) Devising a crucial experiment (or several of them), with alternative possible outcomes, each of which will, as nearly as possible, exclude one or more of the hypotheses;
> 3) Carrying out the experiment so as to get a clean result;
> 1') Recycling the procedure, making sub-hypotheses or sequential hypotheses to refine the possibilities that remain; and so on. (p. 347)

An important part of this approach is the systematic probing, subhypothesis development, and recycling (with modification) of the procedure. For an extraordinary claim, more bits of evidence from many different directions is better than one, seeming extraordinary experimental outcome that necessarily addresses very specific aspects of the problem in question and therefore is not as strong. But why should it matter? Isn't a satisfactory hypothesis good enough to account for "what may have happened in the unobservable past?" No, according to Popper, it is not, for an empirical science. Moreover, a further advantage of testing a main hypothesis against multiple, alternative hypotheses is that it protects the scientist against what T. C. Chamberlin called "[over]affection for his intellectual child," which could include "non-biological vs biological" biases (cited in Platt, 1964, p. 350).[3]

---

[3] "The moment one has offered an original explanation for a phenomenon which seems satisfactory, that moment affection for his intellectual child springs into existence and as the explanation grows into a definite theory his parental affections cluster about his offspring and grows more and more dear to him. . . . There springs up also unwittingly a pressing of the theory to make it fit the facts and a pressing of the facts to make them fit the theory" (Chamberlin, cited in Platt, 1964).

TABLE 5.1. Levels of Explanation

EXPLANATION	TINBERGEN'S FOUR QUESTIONS
1. Proximate	Mechanistic questions about how some character is produced in the organism now.
	How does it develop? → Ontogeny
	How is it produced? → Physiology, biochemistry, hormones, etc.
2. Ultimate	Evolutionary questions about why some character came to be.
	Why did it evolve? What is its biological function? → adaptation value; role of natural selection, drift, mutation, etc.
	What is its evolutionary history → phylogeny

Formulating Multiple Testable Adaptive Hypotheses

Although, like most concepts in Biology, there are differences of opinions on details (e.g., Burian, 1992; West-Eberhardt, 1992), we can define an adaptation as a simple or complex character that has a biological function—it enhances fitness and has evolved by natural selection as a consequence of that function (West-Eberhardt, 1992). The Nobel Prize–winning Classical Ethologist, Nikko Tinbergen (1967) provided an evolutionary context for testing interesting characteristics of organisms (see Table 5.1). Out of that formulation comes the idea of different types of explanations in Biology: proximate versus ultimate explanations (see Table 5.1).

Proximate explanations come from a mechanistic level of analysis that would include how the trait develops; how it is produced in response to the environment, what may trigger it, etc. (invoking neurophysiology, physiology, development, hormones, biomechanics, etc.). *Ultimate explanations* are at an evolutionary level of analysis that provides an explanation for the trait in terms of its evolutionary history, mechanisms by which it evolved, why it may have been adaptive and/or still is adaptive, etc.

Elaborating Tinbergen's (1967) approach, adaptations are *proximate mechanisms* of ultimate, evolutionary solutions for fitness challenges in the organism's adaptive environment. Examining adaptations and evaluating their function requires determining what alternative forms of the character existed in the evolutionary past and why one or more characters survived while alternatives have gone extinct (via negative selection). Once reasonable alternatives have been identified, then direct fitness comparisons can be made between all the variants (Tables 5.1 and 5.2; Williams, 1996, Chapter 9; also see Dawkins, 1986). Alternatively, more indirect tests of adaptive hypotheses involve either comparing the specific functional features of the purported adaptation with an optimal analog (or human design parallel) to see if it meets minimal functional specifications (Table 5.2). According to Tinbergen, comparative studies of functional characters in different adaptive environments are also considered indirect tests (Table 5.2).

Many aspects of an organism may have come under extreme selection for the same reason (e.g., auditory system, vocalization apparatus, bat echolocation: Kick & Simmons, 1984; passive chemical defenses in monarch butterflies: Brower, 1984; Parsons, 1965). This is no less likely to have been the case for humans as it was for bats (e.g., skeletal adaptations for bipedalism in hominins: Crompton et al., 2010). So, which *one* adaptive feature is the best "adaptive character" to test to support a particular adaptive hypothesis?

TABLE 5.2. Testing Adaptive Hypotheses

TYPE OF TEST	KINDS OF TESTS	EXAMPLE
1. Direct	**Tests of fitness differences between alternative variants**	
	a. Direct (experimental or observational) comparisons of the fitness of existing variants	Direct measures of the survival value of cryptic coloration (melanic vs. peppered varieties) in the peppered moth, *Biston betularia*, in response to bird predation
	b. Direct (experimental or observational) comparisons of the fitness of artificially produced (model) variants	Direct measures of the survival of normal and deafened *Noctuid* moths, in response to bat predation
2. Indirect	**Tests of Predictions of Adaptive Hypotheses**	
	a. Observational studies: evaluation of the design features of the purported adaptation	Comparing the frequency range and sensitivity of the *noctuid* moth's ear and sensitivity to bat-like ultrasonic biosonar and the ear's detection distance of hunting bats
	b. Comparative studies of characters in different selection environments (i.e., different areas)	The relative frequency of the different color morphs of *biston betularia* in different environments where only one or the other would cryptic

Selection leads to the evolution of characters that solve specific biological problems (e.g., finding the right mate) in a particular selection environment, what we call the environment of evolutionary adaptiveness (EEA; Tooby & Devore, 1987). A better way to think of adaptation is to consider that the whole organism has been adapted under the hypothesized EEA so that the organism's characters work together as an adaptive set of proximate mechanisms. But these individual characters experience a variety of developmental/evolutionary constraints. So, an ultimate question might be: Have humans adapted to a certain kind of diet (e.g., hunter-gatherer)? If so, which of many specific characteristics (proximate mechanisms) of modern humans are most consistent with that diet enhancing fitness in our evolutionary past (digestive anatomy and physiology, dentition, detoxification mechanisms, food preferences, etc.)? Which aspects are inconsistent with that hypothesis? What other evolutionary hypotheses are consistent with alternatives to our (favored) hunter-gatherer diet hypothesis?

This approach needs to deal with the complication that our dietary EEA has changed over our evolutionary history and different populations of humans experienced different local EEAs. So, just *which* EEA do we use to formulate a test of our adaptive hypothesis? For example, after our hunter-gatherer EEA some populations entered an agricultural EEA with varying degrees of dairy as a major source of protein and fat (Gluckman et al., 2016). Other early agrarian populations did not co-evolve with a dairy diet. The agrarian diet is hypothesized to be lower in antioxidants than a hunter-gatherer diet, so we might also expect selection on antioxidant metabolism (if we were adapted in the previous EEA to a diet rich in antioxidants). What are modern humans "normally" adapted to eat then? (And what does that mean?) What proximate mechanisms are consistent with our ultimate hypothesis? The complication is there is no one early-agriculture EEA for modern humans. And the way a hunter-gatherer adaptive set might have been modified in a subsequent agrarian EEA will not be the same for all populations of humans.

A Beautiful Hypothesis

> *The great tragedy of Science—the slaying of a beautiful hypothesis by an ugly fact.*
> —THOMAS H. HUXLEY *(1870, p. 244)*

Evolutionary theory can generate some great gut-worthy hypotheses that simply "feel" right (see "Truthiness," 2016). Here is an example you all know: Why does the giraffe have a long neck? As long ago as Lamarck (see the preface to Darwin, 1872), the explanation has been "to get to the top of the acacia tree to reach the tender, most nutritious leaves." Darwin and Lamarck may have differed in their notion of how the giraffe acquired its long neck (ontogeny) but they would have agreed that it was advantageous in competing for food (an adaptation). We'll call this the interspecific foraging competition hypothesis (IFCH). Soon after Darwin and Wallace proposed natural selection theory, the IFCH had become the accepted *explanation* for the giraffe's long neck. This hypothesis was so convincing that even Huxley (Huxley, 1942) refers to it in *Evolution: The Modern Synthesis*, as if it was well tested and well supported by facts.

An Ugly Fact

Now you have to admit that the IFCH is a beautiful hypothesis. It just feels right. It just makes sense (*sensu* "truthiness"). It fits within the paradigm of historical and modern Darwinian natural selection. It explains *so* much and is easy to explain to others. It has become an iconic illustration of Darwinian natural selection. Why ruin it by testing it?

Why? Because there might be a *better* explanation. One obvious test of the IFCH is to determine how giraffes actually use their neck (i.e., behavior as a proximate mechanism). In 1996, Robert Simmons and Lue Scheepers decided to do just that, and in their review of the literature (Simmons & Scheepers, 1996), they found that giraffes do not use their neck in a way consistent with the IFCH—they tend to spend most of their time foraging at about shoulder height even when food is scarce and competition high (Young & Isbell, 1991). Now you could attempt to save the beautiful hypothesis by special pleading, or by suggesting that all of the many studies cited by Simmons and Scheepers "missed" something (invoking Popper's lamentation about there being "no conclusive disproof' " of a theory or hypothesis; Popper, 2005). Of course, then you would be reduced to simply refuting ugly facts.

However, if there were plausible *alternative* hypotheses to the IFCH, then those could be explored and perhaps we can reject the IFCH without feeling empty-handed. The irony of the story of the giraffe's long neck is that Darwin (1859, 1879/2004) had developed another theory (sexual selection theory) that could have been used to generate plausible alternative hypotheses to IFCH. Although he didn't know it at the time, he came close when he recognized that male giraffes use their long necks and their heavy skulls and stubby horns as weapons and therefore could be seen as intrasexually selected adaptations (Darwin, 1879/2004).

Simmons and Scheepers (1996) considered what sexual selection predicts about the giraffe's neck if it is an intrasexually selected (Darwin, 1879/2004) weapon used in physical combat for access to mates. As seen in Table 5.3, observations tend to support not only an intrasexual selection history of the male giraffe's large/long neck but also a possible female choice component (intersexual selection). Less support was found for the IFCH (Table 5.3). Pratt and Anderson (1982, 1985) looked at mating success and with male–male aggression as a function of size and their results (see Table 5.4 and Table 5.5) anticipated Simmons and Scheepers (1996; see Table 5.3).

The story is more interesting than this, however. More recently, the challenge to the standard explanation was accepted, and an alternative to the IFCH hypothesis was proposed and supported (Cameron & Toit, 2007; Mitchell et al., 2009), in contradiction to Simmons and Scheepers (1996). Their hypothesis is that intraspecies competition for food drove the evolution of longer necks. The point is that with more refined alternative hypothesis testing the closer we get to a better understanding—which may result in a conclusion that the long neck used in male–male sexual competition could be an *exaptation* (co-opted from a prior adaptation; see Gould & Vrba, 1982) of a feeding adaptation that now functions in more than one way. It may also benefit females' foraging efficiency and be, for them, an exaptation maintained by positive direct natural (survival) selection.

Chemical Defense in Anaspidea

Here is another example of a beautiful hypothesis that was long taken a "proof" and that, when rigorously tested, lead in unexpected, fruitful directions: It had been known for a long time that many species of marine snails in the gastropod order Anaspidea (called sea hares) have an ink gland (Eales, 1921) and secrete a thick, dark purple "ink" when disturbed (Figure 5.1). It has been widely believed that the ink acted as a defense against predators, but a specific functional hypothesis remained untested (reviewed by Johnson & Willows, 1999). Variation in just what stimulus triggers this behavior and the absence of dramatic, anecdotal observations of inking as a defense against a specific predator led researchers to suggest other functions, but none of these were ever systematically tested as alternatives. Then, in 1994 three undergraduate research students in my lab were challenged to answer the question, "What is ink *for*?" First, they

TABLE 5.3. Comparison of Multiple Hypotheses for Giraffe's Long Neck

HYPOTHESIS	PREDICTIONS	OBSERVATIONS
Intrasexual selection	Sexual size dimorphism in body mass and in neck length and size (Simmons & Scheepers, 1996).	Males are larger than females; although females have a long neck, the male's neck is disproportionately larger than the female's. The thickness and mass of the male's neck continues to increase after maturity, unlike the female. Males, but not females, have disproportionately thick skulls and ossicones (Simmons & Scheepers, 1996).
	Uses neck and associate structures as a weapon against conspecific males unlike related ungulates (Simmons & Scheepers, 1996).	Males engage in neck-to-neck contact, pushing and shoving other males. Males swing their heads, striking other males on the head, neck or back, sometimes inflicting fatal injuries with the mass of the skull or the ossicones (Simmons & Scheepers, 1996).
	Larger males with larger necks gain more mates (Pratt & Anderson, 1982, 1985).	Surveys of mating success show that larger males "win" more contests and get more matings. Males that are size matched tend to split their contests and get an equal share of mates (Pratt & Anderson, 1982, 1985).
	Neck is disproportionate in size compared to that of related, nonsexually selected species.	The male giraffe's neck is longer, thicker and their skulls ticker than those of other related browsers that do not appear to engage in male-male contest for mates (Simmons & Scheepers, 1996).
	Neck demonstrates positive allometry (Simmons & Scheepers, 1996).	Larger males have disproportionately lager necks, skulls and ossicones (Simmons & Scheepers, 1996).
	Long, large neck is costly to survival.	Various adaptations suggest that maintaining the elongated neck is costly (special valves in neck veins help maintain blood pressure to the brain when drinking and raising the head; the heart must pump at a much higher rate to maintain blood pressure at the brain, compared to related mammals of about the same size). The animals are more vulnerable while drinking. Some contradictory evidence of higher male mortality compared to females.

Hypothesis	Prediction	Findings
Intersexual selection	Large neck used by females to assess male quality as a mate (i.e., females choose males with larger/longer necks)	Females also engage in necking with males, not in an aggressive way; could assess a male's size. No evidence females are free to choose among all suitors, but let larger (and not smaller males) "urine test" them (sign of receptivity; Pratt & Anderson, 1982, 1985). Like many females in species with male-male contests for mates, females could avoid injury by consorting with larger, winning males who are not likely to be challenged by smaller males.
	Larger neck indicates male's offspring will have higher fitness	No indication (yet) that female choice results in higher fitness for her offspring. A possible Runaway Selection or Sexy Sons effect has not been tested.
Interspecific foraging competition hypothesis (IFCH)	Use maximum reach of the neck to get food during times of food scarcity. There should be no sex difference in neck length relative to body size (Mitchell et al., 2009).	Neither males or females forage at the highest levels, even during food shortage and in competition with other browsers. Generally, forage at shoulder height like the okapi (Simmons & Scheepers, 1996). But contrary results found both sexes forage at about the height of shorter browsers (Cameron & DuToi, 2007).
	In the evolutionary history of the giraffe's neck, heterospecific competitors in the EEA would have favored a costly solution such as the long neck (Simmons & Scheepers, 1996).	Giraffes appeared to be the tallest browsers in their EEA and so the advantage of the longer neck for foraging would appear too low to balance its costs (Simmons & Scheepers, 1996).
Intraspecific foraging competition hypothesis (IntraFCH)	Use maximum reach of the neck to get food during times of food scarcity.	Taller giraffes forage higher than shorter browsers in competition for food on the same trees (Cameron & DuToi, 2007).
	In the evolutionary history of the giraffe's neck, intra-specific competitors in the EEA would have favored a costly solution such as the long neck (Simmons & Scheepers, 1996).	Giraffes appeared to be the tallest browsers in their EEA. But an interspecific arms race could drive directional selection for longer necks.

*Simmons and Scheepers (1996); †Pratt and Anderson (1982, 1985); ¥ Cameron and DuToi (2007); § Mitchell et al. (2009)

NOTE: EEA = environment of evolutionary adaptiveness.

TABLE 5.4. Male–Male Encounters during Mating Season

LARGER DISPLACES SMALLER	EQUAL SIZE	SMALLER DISPLACES LARGER
82/127 (64.6%)	39/127 (30.1%)	6/127 (4.7%)

TABLE 5.5. Male Size and Mating Success

RELATIVE MALE SIZE	SUCCESSFUL	UNSUCCESSFUL	% SUCCESSFUL
Large	34	22	60.7
Medium	76	61	55.5
Small	45	89	33.6

did an exhaustive search of the literature and found that much was known about the neurophysiology of the trigger mechanism for ink release (Carew & Kandel, 1977a, 1977b, 1977c), except exactly what the natural stimulus was. Ink could be directed toward the eliciting stimulus as expected if

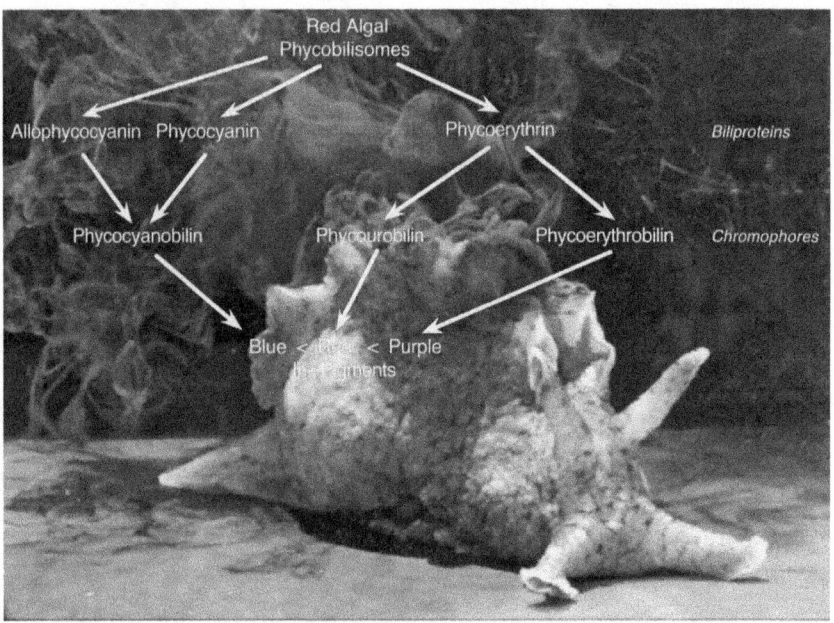

FIGURE 5.1. Purple Ink Release from the sea hare *Aplysia californica*. Modified from Yang et al. (2005; pigment pathway added). Ink pigments are derived from phycobilisomes (accessory photosynthetic chromoproteins) of red seaweed. Three biliproteins are broken down to their chromophores (blue, red and purple ink pigments). The major ink pigment is phycoerytthrobilin.

it was deployed as an active weapon (Walters & Erickson, 1986). They also found that some potential natural predators were sea anemones, crabs, and lobsters (reviewed in Johnson & Willows, 1999; Nolen et al., 1995). They also had to consider that in this larger phylogenetic group (order Anaspidea) antipredator defenses would have evolved over a long evolutionary history (Johnson & Willows, 1999). Then, they formulated a series of testable hypotheses and put them to the test. The empirical results along the way required modifying assumptions and hypotheses or throwing some out altogether.

I summarize the major alternative hypotheses, predictions, results, and conclusions in Table 5.6, but the big picture is that ink secretion is a part of an elaborate chemical defensive arsenal—an adaptive set of proximate mechanisms—that protects the animal from a variety of predators over different parts of its life history (Figure 5.2). An ink gland is found in all 35 species of *Aplysia,* and purple ink is known to be secreted by all but five *Aplysia* species that secrete ink[4] and by all other members of the larger order (Anaspidea). The purple color is a photosynthetic pigment, r-phycoerythrobilin derived from the breakdown of the biliprotein, r-phycoerythrin, that the snail obtains from its red seaweed diet. Although not particularly toxic, ink is aversive to crabs and is an antifeedant for sea anemones, crabs, and lobsters. Moreover, purified r-phycoerythrin strongly activates the taste chemoreceptors used by crabs to find food (Carlson & Nolen, 1997). Other active defensive chemicals are found in another mantle gland secretion; opaline works synergistically with ink, and it may provide protection for larger animals (who may no longer eat red seaweed and make purple ink) and species that have lost the ability to secrete ink. Amino acids (e.g., taurine) are also active components of purple ink, and all stimulate lobster olfactory chemoreceptors creating a novel antipredator effect by acting as a phagomimetic super stimulus—with other ink components overstimulating chemoreceptors and mechanoreceptors, causing predators to drop their prey (Johnson et al, 2005; Kicklighter et al, 2005). A flavin-containing L-amino acid oxidase (LAAO) called escapin was isolated from ink (Yang et al., 2006). It acts with separately packaged L-lysine and L-arginine as substrates to produce hydrogen peroxide, explaining its very powerful antimicrobial action. Escapin is stored and released from distinct amber colored ink gland vesicles (Johnson et al, 2005). L-lysine is a major component of opaline so the ink and opaline glands act synergistically to

---

[4] White ink is secreted by two species of *Aplysia*. Three others have not been studied (Johnson & Willows, 1999).

TABLE 5.6. Comparison of Multiple Hypotheses for Purple Ink Secretion by *Aplysia sp* (Order Anaspidea)

HYPOTHESIS	PREDICTIONS	OBSERVATIONS
Source of purple ink is red seaweed	Ink pigments are diet derived.	Ink is produced only when snails eat a diet of red, but not green or brown seaweed. However, a snail derived enzyme, escapin with strong antibiotic properties recently has been found to be an active component of ink in *Aplysia californica* (Coelho et al., 1998; Kicklighter et al., 2005; Nolen & Johnson, 2001; Nolen et al., 1995; Yang et al., 2005).
		Ink is a complex mixture of chemicals, not all necessarily waste products and not all derived from the diet. Some (escapin) found only in the ink gland (Yang et al., 2005). Prediction partially confirmed.
	Red seaweed, but not other potential foods contain purple (and other) ink pigments.	The purple color is primarily from r-phycoerythrobilin (RPEB) along with a few other phycobilisome pigments found only in a red seaweed diet. Anaspideans that don't eat red seaweed don't produce purple ink (Coelho et al., 1998; Johnson & Willows, 1999; Nolen & Johnson, 2001).
		Prediction confirmed.
Ink release is bile excretion (removal of bile pigments following digestion of red seaweed accessory photosynthetic phycobilisomes)	Red seaweed phycobilisomes are processed in the snail's digestive system where proteins are processed and the biliprotein component digested away from the chromophore (e.g., RPEB).	Antibody labeling visualized under transmission electron microscopy suggested that RPEB is cleaved from the biliprotein of phycobilisomes in the digestive "gland" and transported in the hemolymph to the ink gland where special cells filter out large proteins and allow smaller RPEB to be concentrated in special ink granulate cells via endocytosis (Coelho et al., 1998; Prince et al., 1998). Prediction confirmed.

HYPOTHESIS	PREDICTIONS	OBSERVATIONS
	Waste bile pigments (e.g., RPEB) would would be collected and stored in a special gland. The simplest way to remove the wastes is to expel them into the excurrent stream from the mantle cavity whenever the storage gland is full.	Ink secretion is not a function of fullness or length of waste storage. Ink glands of snails kept in the laboratory for up to three months remained full of old, oxidized ink, evident as many large red-brown ink vesicles. Snails with freshly sequestered ink have a low density of smaller dark purple ink vesicles. Snails forced to release their ink take several days to partially replenish their stores (Nolen & Johnson, 2001; Prince et al., 1998). Prediction not supported.
	Waste processing of bile materials should be a simple process if ink has no function.	Ink pigment collection, packaging, storage, and release is quite complicated. Granulate cells mature into muscled secretory ink vesicles, with a capped secretory canal that empties vesicle content into the mantel cavity when the muscles contract. Ink vesicles are innervated by cholinergic neurons under control of cells in the abdominal ganglion in response to high threshold external stimuli (Coelho et al., 1998; Prince et al., 1998). Prediction not supported.
	There is no reason to expel the waste in response to a particular external sensory stimulus. It should be more likely to be expelled when the gland is full.	Ink is released by a specific stimulus configuration (tactile stimulation of the body as the foot is lifted off the substrate). It is directed by the parapodia covering the mantle in to the of stimulation (=directed response). The red seaweed derived ink material is in limited supply and it is judicially meted out based on supply and potential danger (Nolen & Johnson, 2001; Walters & Erickson, 1986). Prediction not supported.

(continued)

TABLE 5.6. Continued

HYPOTHESIS	PREDICTIONS	OBSERVATIONS
Ink is an active chemical defense	Ink should be released in response to danger associated with particular, sympatric predators. Specific releasers should be detected by the sensory system that are consistent with danger.	Ink is released in response to anemone tentacles, and the prickly poking of crab and lobster appendages when they pick the snail up off the substrate. Anecdotally, poking the animal is supposed to trigger inking, but it is hard to get the snail to ink by just poking it, even with sharp forceps to the tail (Carlson & Nolen, 1997; Nolen & Johnson, 2001; Nolen et al., 1995). Prediction confirmed.
	Ink should prevent attack, capture or consumption by sympatric predators.	Ink secretion allows the snail to escape after capture by anemones (prevents consumption) and by crustaceans (but not specialized predators such as the gastropod *Navanax sp* (Johnson & Willows, 1999; Nolen & Johnson, 2001; Nolen et al., 1995; Walters & Erickson, 1986)). Prediction confirmed.
Antifeedant	Ink should act as a feeding deterrent (antifeedant) or should prevent the predator from capturing and/or consuming the snail.	The ink acts as an anti-feedant for a variety of predators (crabs, lobsters, sea anemones, and possibly sea stars) and allows the snail to escape capture or avoid getting eaten (in sea anemones ink causes them to regurgitate the snail). In interactions with anemones, and crabs and lobsters, ink is released after capture but allows the animal to escape consumption. Ink is aversive to crabs and lobsters, although lobsters can use their tail to blow away ink that has engulfed them and the snail (then the snails passive chemical defense comes into play; Carlson & Nolen, 1997; Johnson et al., 2006; Kicklighter et al., 2005; Nolen et al., 1995; Walters & Erickson, 1986). Prediction mostly confirmed (ink prevents consumption, if not capture).

TABLE 5.6. Continued

HYPOTHESIS	PREDICTIONS	OBSERVATIONS
	Ink would act on the chemosensory systems (olfaction and taste) that mediates prey localization and consumption.	Fresh ink is chemoreceptively active on crustacean chemoreceptors. Purified r-phycoerythrin (RPE) from red seaweed activates amino acid chemorecepotrs in crabs; but opaline contains amino acids that also activates olfactory chemoreceptors in lobster (Carlson & Nolen, 1997; Johnson et al., 2006; Kicklighter et al., 2005). Prediction confirmed.
	Ink should be effective as an anti-predator weapon independently of the snail's passive chemical defense (unpalatability due to incorporation of red seaweed derived antiherbivory chemicals in its skin).	Snails raised on green seaweed do not make, store, or release (purple) ink. They are not (as) unpalatable as red seaweed fed snails. Green fed snails are eaten readily by sea anemones but can be saved by application of fresh ink just as they are being engulfed. De-inked, but red seaweed fed snails have no ink but are distasteful and are eaten as readily as chemically unprotected green fed snails by anemones, who can digest the outside of the snails but then regurgitate the most toxic viscera (Nolen & Johnson, 2001; Nolen et al., 1995). Prediction confirmed.
Subhypothesis: Mechanism of antifeedant action is conventional	Based on comparative studies, expect conventional activation of nociceptive, antagonistic sensory receptors that prevent or suppress appetitive responses to food (analogous to the reaction to bitter, distasteful food).	Unconventionally, RPE (and probably RPEB) and fresh ink activates appetitive amino acid chemoreceptors in crabs. Ink components do the same to olfactory receptors in lobsters producing phagomimicry (mimicry of the sensory stimulation associate with food) and disrupting appetitive behavior. Opaline acts synergistically as an antagonist of appetitive behavior.

(continued)

TABLE 5.6. Continued

HYPOTHESIS	PREDICTIONS	OBSERVATIONS
		Surprisingly escapin acts with separately packaged L-lysine and L-arginine as substrates but its activation seems to be too slow to account for the phagomimicry effect of other ink products. Its antimicrobial function is uncertain (Johnson et al., 2006; Kicklighter et al., 2005; Yang et al., 2005). Prediction partially confirmed but a new, unexpected mechanism was discovered.
Camouflage (visual or chemical smoke screen)	As an active chemical defense, ink could act as a chemical smoke screen, allowing the snail to escape while the predator is disoriented in the ink cloud.	While creating a cloud of ink and mucus (which helps ink adhere to the snail as well as predators) ink does not seem capable of creating a visual smokescreen. It could act as an obscuring chemical smoke screen. Except, the snails don't move all that fast away from predators after releasing their ink (Johnson & Willows, 1999; Nolen et al., 1995) Prediction confirmed for swimming species.
	It would not be effective and should not be deployed in response to sessile predators.	Sea anemone tentacles readily trigger ink release if the animal has been picked up off the substrate (Nolen et al., 1995). Prediction not supported.
Startle	For the nine species that can swim, ink could startle a predators such as a fish, while the snail swims away.	In *Aplysia basiliana* stimuli that trigger ink release also trigger swimming (Nolen & Johnson, 2001). Prediction Confirmed for swimming species. Ink's origin for this function in the family depends on whether swimming is the ancestral condition.
Ink is a Chemical Signal	Signals, whether species specific pheromones or general alarm signals, must benefit both the sender as well as the receiver.	Conclusion: Mixed, but prediction not supported in general.

TABLE 5.6. Continued

HYPOTHESIS	PREDICTIONS	OBSERVATIONS
Aposematism	A signal directed at predators of the snail's unprofitability as prey (because of the distasteful/toxic chemicals incorporated into tis tissues). Ink signals the animal's toxicity.	In this case both sender (the snail) and receiver (crab, lobster, fish, sea star) may benefit. A pure signal of unprofitability would not need to be an antifeedant, but the two functions would be possible together. But only predators that find red seaweed-feeding snail unpalatable would respond to ink as an aposematic signal. This hypothesis would not work for sea anemones, crabs, and lobsters who are not deterred by the snail's distastefulness (because of special ways of dealing with the toxic flesh) nor for some fish (Johnson & Willows, 1999; Nolen et al., 1995). Prediction: viable in some cases but not in general.
Alarm Signal	Alarm signals must cause the receiver to behave in a way that benefits the signaler; this could be via (a) the confusion effect or (b) warning a close genetic relative that then can escape even if the signaler does not.	In *Aplysia californica*, ink triggers escape galloping. It is not clear that that would create enough of a confusion effect in a group of snails to make it less likely that the predator would take the signaler. Ink triggers swimming is *A. basiliana* so for swimming species, a confusion effect is a possibility (Johnson & Willows, 1999; Nolen et al., 1995). Prediction viable for swimming species. However, given *Aplysia*'s extended larval phase, it is unlikely that the signaler would be near any kin (Johnson & Willows, 1999; Nolen et al., 1995). Prediction not supported.

(*continued*)

TABLE 5.6. Continued

HYPOTHESIS	PREDICTIONS	OBSERVATIONS
Pheromone	Ink specifically communicates with conspecifics, about food, mates, etc.	Given *Aplysia's* extended larval phase, it is unlikely that the signaler would be near any kin, so indirect fitness effects seem to be ruled out. The context of ink's antipredator use is not consistent with ink serving as a mating or aggregation pheromone (Johnson & Willows, 1999; Nolen et al., 1995). Prediction not supported.
Ink is an adventitious cue of danger	If ink serves a defensive function, its detection by other snails could be used to avoid danger of ongoing predation. But it would not have evolved for that function.	Juvenile *A. californica* do avoid clouds of fresh ink and do engage in galloping escape locomotion. If engulfed in the ink cloud (not able to localize it), they tend to stop locomotion (and don't then encounter a sessile predator such as a sea anemone). In this way, ink could be used as an adventitious cue of danger, and its reception would be adaptive. It does not benefit the inking snail (Johnson & Willows, 1999; Nolen & Johnson, 2001; Nolen et al., 1995). Prediction supported.

NOTE: Some sources are primary, and some are reviews of earlier primary sources.

facilitate escapin function. But its role in antipredator chemical defense is unclear. One hypothesis is that it may act to prevent infection of injuries after attacks by predators (Johnson et al., 2005; Yang et al., 2006). Finally, the snail employs a passive chemical defense by repurposing the red seaweed's antiherbivory chemicals and incorporating them into its eggs and its skin, thus making it distasteful to some predators (Figure 5.2; for reviews, see Johnson & Willows, 1999; Nolen et al., 1995).

The combination of active (ink, opaline) and passive (distasteful) chemical defenses apparently has had implications for the evolution of this successful group. The heavy calcareous shell characteristic of gastropod molluscs is a major co-evolved defense against crabs and their heavy claws (Vermeij, 1971, 1976). In Anaspidea, the shell has become vestigial as its function was subsumed by a diverse set of active and passive chemical weapons. Because of their reduced (internalized) shell, some anaspidean

Sea Hare Defenses Through Ontogeny

Egg Mass	Free Swimming Larva	Metamorphosis	Crawling Snail with Reduced/Internalized Shell

			Size
Odor		Odor	
		Escape Locomotion: Gallop and Swim	
Coloration		Coloration	
		Opaline	
		Ink	
	Withdrawal and Drop		
		Withdrawal/Local Contraction	
	Shell		
Distastefulness		Distastefulness	

Egg	Veliger	Recruit	Juvenile	Small Adult	Adult

FIGURE 5.2. An adaptive set of proximate mechanisms protect sea hares (*Aplysia sp*) from predation throughout ontogeny. Egg masses are laid on red seaweed and include red seaweed pigments and anti-herbivory compounds the parent acquired from their diet. Egg masses are distasteful, and visually and chemically camouflaged. But simply by the association with the red seaweed, many predators avoid egg masses due to the odor and distastefulness of the red seaweed. Veliger: After hatching, the veliger larva swims in the water column for several months. As a typical gastropod veliger, their main defense is their shell into which they can pull their head and ciliated swimming apparatus. Doing so causes them to drop/sink in water column. Veligers will only settle down and metamorphose into a tiny snail if they settle on red seaweed. Recruits are protected by their shell for a short time until they outgrow it. Juvenile snails feed on the red seaweed and grow to small adults while accumulating distasteful chemicals and ink pigments. They are protected by their objectionable odor of red seaweed and their distastefulness. They are camouflaged by pigments from the red seaweed and can withdraw vulnerable structures (gill, siphon) into their mantel cavity if mildly disturbed. If attacked they may engage in "fast" galloping or swimming. Ink and Opaline can be deployed as active chemical defenses. Adults: As they grow larger (some species can get as large as 4 kg) they become too large for some predators. For larger predators their combined passive and active defenses suffice. Modified from Johnson and Willows (1999).

species are capable of swimming as adults and therefore are better able to disperse over a larger area (mitigating local extinction and facilitating speciation) and adding another predator escape mechanism.

A phylogenetic analysis (Johnson & Willows, 1999; Medina & Walsh, 2000; Figure 5.3) supports the hypothesis that ink secretion is a by-product of bile excretion, resulting from the processing of the photosynthetic biliproteins from the snail's diet of red seaweed. Eventually, it is hypothesized, the neural control of bile excretion from the waste storage gland was adapted to respond to predators grabbing and picking the snail off the substrate (i.e., best way to trigger ink release). In short, our working

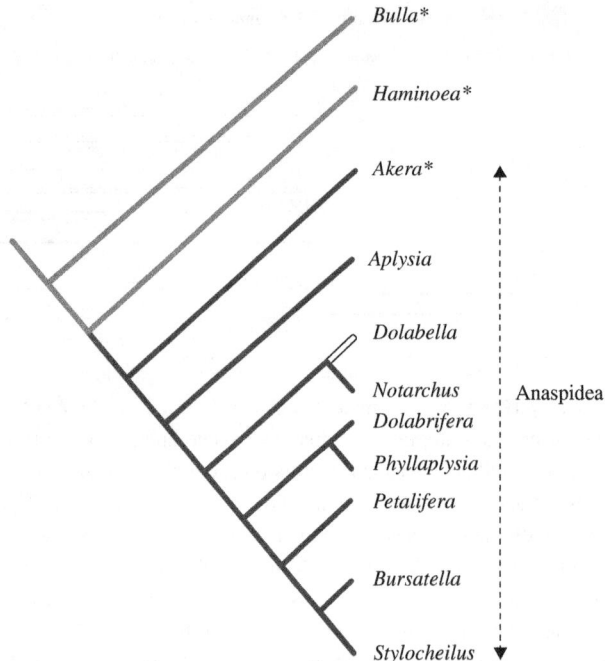

FIGURE 5.3. Phylogeny of the order Anaspidea. Tree based on rDNA (Medina and Walsh, 2000) with additional overlay data from Johnson and Willows (1999) and Nolen et al. (1995). The outgroups are the Cephalaspideans in the genera *Bulla* and *Haminoea*, neither of which have an ink gland. Inking has been verified for eight of the nine genera in the order Anaspidea. The ability to release purple ink has been lost in few species. For example of the 35 species of *Aplysia*, two produce a white ink and three have not been studied. Ink release in the genus *Dolabella* has not been verified so it is equivocal as to whether the ability has been lost.

Bulla, Haminoea and *Akera* have an external shell; all the rest have an internal shell or it has been lost altogether in adults. Swimming species are found in the genera *Akera, Aplysia,* and *Notarchus*.

(ultimate) hypothesis is that an active chemical defense (directed, triggered ink secretion) was co-opted from a physiological mechanism to dispose of dietary waste products. Opaline may have evolved in parallel, maybe several times independently, and it has replaced ink in some species and certainly adds to the active chemical defense of adults (Johnson & Willows, 1999). Each of my three undergraduate students went on to win a best Honors Senior Thesis, and all went on to get PhDs. Their work itself lead to in a number of publications, a PhD thesis on the topic and to two postdocs on the topic. All because a student asked their General Biology instructor: "What is *that* for?"

Evolution and Human Health

One problem with the evolutionary research on humans can be a lack of multiple hypothesis testing, a tenet of Platt's strong inference as well as Popper's (2005) critical rationalism. In the most powerful studies, alternative hypotheses are systematically revised and the process recycled. An early goal of the systematic study of human behavior has been to establish "what makes us different from the other animals?" That is, we have been obsessed with human exceptionalism. But, time and time again, a purportedly uniquely human characteristic is found to be only *quantitatively* and not *qualitatively* different from that of our closest (animal) relatives, that is, on an evolutionary continuum and very much explained within an evolutionary context of our "ordinary science" (Tooby & DeVore, 1987). As we found in the study of inking in the sea snail, a review of the evolution literature can provide insights and alternative hypotheses even for applications to new problems. One generalization that arises is that humans are not so exceptional that we cannot apply comparative methods commonly used to study adaptive questions in other organisms. But testing human adaptation is not always easy and it is often controversial.

Why Is Testing Human Adaptation Controversial?

On the one hand, the controversy is largely due to misperceptions about the "fixity" of our characters, with the argument often made that if a character is "biological" or (what they often mean) "genetic," and if we cannot change our genes, then we cannot change our behavior or our society. If true, this could have implications for social justice and eliminating social inequities in the human condition. Luckily, this is not the case (see Herron & Freeman, 2013; Gough et al., 2008).

On the other hand, another reason testing human adaptation is controversial is that it challenges some older traditions in the social sciences, which emphasizes cultural influences and interactions between individuals at a different level of interaction than genes (Gough et al., 2008). In fact, most biological characters of interest are not the result of a single gene (allele) but of continuously varying, multigene traits whose final developmental outcome depends on a complex interaction between genes, the environment, and the developmental program (Figure 5.4). Both genetic variation and variation in the developmental environment play a role in natural selection of the final product. Without heritable variation (Zimmer & Emlen,

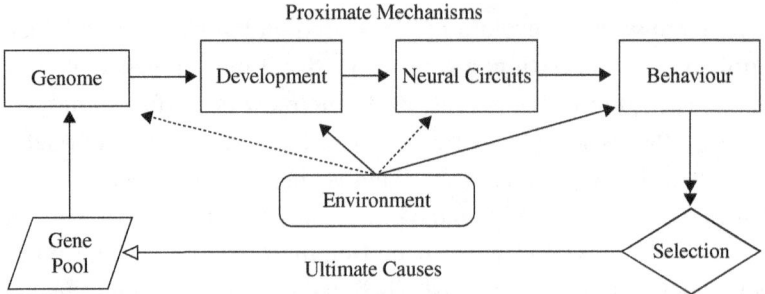

FIGURE 5.4. Role of genes and environment on development and selection of behavior. Selection operates on the consequences of behavior for the organism's fitness. But behavior is a complicated phenotype that results from a developmental program which includes interactions between genes and environment.

2016), there will be no evolution via natural selection.[5] Finally, because of the inherent complexity of a contingent evolutionary process, there has always been criticism that an adaptationist approach is too simplistic, although its success has been shown time and again[6] (for an insightful review of the current state of Evolutionary Biology, see Welch, 2016).

Why Is Testing Human Adaptation Especially Difficult?

Partly this is because it is hard to determine what is heritable and capable of having evolved by natural selection (for relatively accessible introductions to the concept of heritability and selection, see Herron & Freeman, 2013; Zimmer & Emlen, 2016). Moreover, it is hard to test the adaptive value of a character that might no longer exist in its original selective environment (EEA), which itself might not be well understood. In addition, a problem with explaining adaption is not that we cannot define what an adaptation is or how to test its adaptive value. The problem is deciding what we mean by a phenotypic *character* (see Darden, 1992; Fristrup,

---

[5] An interesting thought experiment, useful in the classroom is Olivia Judson's question: "What would happen if all mutations stopped being produced here on earth?" (Judson, 2008).

[6] "It is then argued that special discontent stems from misunderstandings and dislike of one well-known but atypical research programme: the study of adaptive function, in the tradition of behavioural ecology. To achieve its goals, this research needs distinct tools, often including imaginary agency, and a partial description of the evolutionary process. This invites mistaken charges of narrowness and over-simplification (which come, not least, from researchers in other subfields), and these chime with anxieties about human agency and overall purpose" (Welch, 2016, p. 1).

1992) and determining alternative, variant characters that presumably lost the evolutionary battle for genetic representation in the next generation. And we have trouble nailing down just which EEA is the relevant adaptive context for testing the adaptiveness of a character that evolved long ago.

By neglecting to consider our evolutionary history, we tend to miss the big picture. As an example, consider lactose intolerance: Most populations of humans cannot digest lactose as adults because the production of the enzyme lactase has been shut down after weaning (Gluckman et al., 2016). Still, the World Health Organization (2003) classifies lactose intolerance as a metabolic "disease" or disorder. Because of the recent cultural dominance of northern Europeans, what is considered the "normal" condition is a high degree of tolerance of dairy in the adult diet. In fact, the inability to digest the milk sugar lactose as an adult represents a norm for most of our evolutionary history and still is in the majority of modern human populations around the world. The so-called normal condition, retaining lactase activity as an adult (after weaning) is linked to a single nucleotide polymorphism, C/T(−13910), in a regulatory element upstream of the lactase gene which emerged about 10,000 years ago in northern Europeans (Tishkoff et al., 2007). In fact, lactase persistence in northern European populations is a recent evolutionary adaptation (6,000–7,000 ya) to a new EEA—the introduction of dairy cattle to Europe in which dairy products were a common major source of protein and fat and the disaccharide lactose. This adaptation evolved independently several times and involved selection on different genetic loci: G/C(−14010) in Kenya and Tanzania and another lactase-persistence allele in Arab populations, associated with consumption of camels' milk (Enattah et al., 2008). Convergent selection on lactase persistence suggests it was an especially important adaptation in analogous dietary EEAs, which differed from those of other contemporary populations (Gluckman et al., 2016).

Public Health and Strong Inference

In a recent survey published in the *Journal of the American Medical Association* (Aguilar et al., 2015, p. 1974), it is reported that "nearly 35% of all adults and 50% of those aged 60 years or older" in the United States suffer from a set of metabolic diseases called metabolic syndrome (MetS): The four health conditions of MetS are obesity, insulin resistance, dyslipidaemia, and hypertension (i.e., too much fat in adipose tissue, blood glucose levels consistent with type 2 diabetes, high cholesterol, and high blood pressure). So, obesity in modern humans (well, the US population) is indicative of several metabolic disorders (American Heart Association, National Cholesterol Education Program Adult Treatment Panel III, cited in Aguilar et al., 2015). In addition, obesity is associated with fatty liver disease and atherosclerosis, which with the others are all thought to be diet related. We will call this collection of poor health conditions: obesity-associated metabolic disorders (OAMD).

One hallmark of biology is natural variation, so it should not be surprising that about 10% to 20% of obese individuals do *not* suffer from OAMD (Fabbrini et al., 2015). And the degree to which other obese individuals suffer from OAMD is quite variable. There are 59 different "types" of obesity (proximate causes), and as many as 300 genes have been identified that may play a role in small weight gain (Corey & Kaplan, 2014); a mutation in any 1 of 25 of those genetic sites has been strongly associated with the development of obesity (O'Rahilly & Farooqi, 2013). Even though the heritability of obesity is high (an estimated 40% to 70% genetic contribution to interindividual variation), identifying specific genes has been difficult (Loos et al., 2008), which is consistent with obesity being a multigene, quantitative trait.

A recent set of studies on individuals with a genetic disorder called lipodystrophy (serious lack of fat cells) and OAMD victims has revealed how medical science and public health policy may suffer when strong inference is not employed (Huang-Dorin et al., 2010). Medical and public health officials (and many in the medical community) may have misconstrued cause and effect and concluded that the association of obesity with metabolic disorders is indicative of causation (i.e., "Obesity causes OAMD"). Following the dictates of strong inference, this "conclusion" should have been a subsequent, or *subhypothesis,* and rigorously tested. For example, "proximate aspects of obesity trigger diabetes, heart disease, HBP, atherosclerosis, etc." Alternatively, perhaps whatever causes some people to overeat (and get fat) also causes metabolic disorders. *Or* perhaps metabolic disorders cause obesity. Admittedly, medical researchers did not stop at the demonstration of the association between obesity and OAMD, but public health policy was prematurely developed with medical recommendations and other actions purportedly because the potential consequences of the association were too important to ignore. The message was "reduce obesity and we will reduce the expensive consequences of metabolic diseases." But the "proof" was not extraordinary and major public policy—advocating for a change in human behavior (lifestyle)—has been enacted on the basis of an association.

In fact, recent studies have hinted that the proximate cause of some, if not all, of the OAMDs is the storage of fat in nonfat storage cells (i.e., visceral adipose tissue; Virtue & Vidal-Puig, 2010). Sometimes, OAMD is not associated with obesity. The nonobese sufferers of OAMD (with lipodystrophy) have little subcutaneous adipose fat tissue (and, therefore, are "skinny"). They also cannot produce much of the hormone leptin, which is produced by subcutaneous fat cells. Leptin signals the brain when fat stores are high, so low levels of leptin in the blood are indicative of starvation, even if there is a sufficient intake of calories. Individuals with lipodystrophy had voracious appetites but no place to safely store the excess fat their bodies were producing by overeating. This fat then accumulated inappropriately in liver, pancreas, heart, and other muscles apparently causing the OAMD illnesses. (Misplaced fat is metabolized in different ways, and one product is diacylglycerol, which interferes with the action of insulin, resulting in insulin insensitivity and type 2 diabetes). Interestingly, some obese sufferers of OAMD had a reduced amount of subcutaneous adipose tissue fats cells but more than the lipodystrophy OAMD sufferers. (It is not clear if this deficiency is heritable, environmental, and/or epigenic.) The result was the same: relatively lower levels of fat storage cells —> low leptin levels in the blood/brain —> elevated appetites —> excess caloric intake —>improper fat storage —> tissue damage and OAMD with obesity. So obesity is not necessary for OAMD to develop, nor is it likely to be sufficient. OAMD can result from a low level of leptin-producing subcutaneous fat storage cells, which results in elevated appetites that outstrip the body's ability to safely store fat. That leads to a disease state along with obesity.

Does this change the conclusion that obesity is something to avoid? Not necessarily, as the risks are still high that excess caloric intake causes other problems. But not everyone who is overweight is unhealthy (and not everyone who is skinny is healthy). The tremendous genetic variation inherent in obesity and the recognition of multiple proximate causes harkens back to the evolutionary perspective that over a long evolutionary history modern humans collectively have experienced a diverse set of dietary EEAs and perhaps exhibit a diverse adaptive set of proximate mechanisms of calorie storage and metabolism. It re-emphasizes the idea that as far as public health policy goes, the concept of "normal" is not particularly useful.

Evolutionary Hypotheses for Obesity

What are the evolutionary implications for all of this? Here is one theory: In our previous hunter-gatherer or early agrarian EEAs, food was scarce at certain times of the year, and our ancestors experienced times of feast and famine. Hominins evolved an adaptive set of proximate mechanisms that helped deal with regular cycles of feast and famine. When in periods of feast, as body fat accumulated in adipose tissue, the fat cells there released leptin, which caused the brain to suppress the appetite and therefore reduce foraging (food searching) activities. When in periods of famine, calories stored in fat were used up, and so body fat was low and fat cells released little leptin into the blood. The brain is programmed to search for food naturally unless suppressed by leptin. With little leptin to suppress appetite, these people tended to overeat and fill all their fat cells, ostensibly to be used during the next famine. Populations living under different cycles of feast and famine (equatorial, temperate, subarctic) may have been adapted differently to a variety of food availability fluctuations.

So our first hypothesis is that obesity is a consequence of an adaptive foraging mechanism to deal with the feast and famine aspects of our distant EEAs. It is no longer adaptive in the modern EEA because feast and famine cycles in first-world countries have ceased with an abundance of inexpensive calories year round, mostly in the form of carbohydrates and fat. OAMD may or may not result depending on individual genetic (evolutionary) background. We also have to take into account the diverse prior EEAs modern human populations may have. Is the "modern" diet unhealthy? Yes, and no. It depends on your evolutionary history and family genetic background, but probably yes for most of us—but not because overeating is "unnatural." Does simply reducing caloric intake solve the obesity problem? No. A lot of this depends on how many and what type of fat cells you have. Despite years of public health policy and medical school education, and due to our diverse evolutionary histories, it is not as simple as just "eat less and move more" (Sacks et al., 2009; for a popular science introduction, see Kolata, 2016).

Conclusion

At public lectures at most colleges in the United States, you are likely to hear from critics in the audience who insistent that an adaptive claim for a human trait requires extraordinary proof, but a claim of a cultural origin does not. Is this a double standard? Why would evolutionary explanations of human behavior (especially) require more extraordinary evidence than the nonbiological claim? What does that mean? Do we need to provide an overabundance of evidence or unusually powerful evidence? Generally the justification goes like this: *If the potential consequence of misapplication of a conclusion is dire, then the evidence must be strong.* Think of the repercussions of eugenics laws in the United States and Europe in the mid-20th century (e.g., *Buck v. Bell,* 1927; Krisch, 2014) or the implementation of educational policies based on a discovery of sex differences in, say, math abilities (for an insightful discussion, see Zuk, 2002, Chapter 12).

But the alternative also holds for the social sciences: If social or public health policy is going to be made regarding human behavior, we must ensure we understand the phenomenon, regardless of where that understanding takes us. Think of the repercussions to education of behaviorism or intelligence testing or perhaps misguided or misdirected efforts to minimize the development of perceived sex differences in academic performance in our school system. In the end, both aspects of the false dichotomy, biological versus cultural, needs to be addressed for the best understanding of the human animal.

It is generally assumed by biologists, evolutionary psychologists, and biological anthropologists (Gough et al., 2008; Tooby & DeVore, 1987) that many aspects of humans were adaptations in the past and that they evolved by natural selection from prior adaptations or as exaptations (Schmitt, 2015). Unlike pre-Darwinians, it is not a huge factual jump to propose that a particular human character *might* have evolved from a prior form, because Darwin and many others have shown that not only is that possible but that it is quite an ordinary prediction. Dismissing a biological explanation for, say, behavior simply because we don't like the implications is not only short-sighted, it hinders a deeper understanding of the human animal, and it is intellectually dishonest. Likewise, accepting a beautiful hypothesis as explanation leaves us ignorant to what has happened.

Still, labeling a character as "biological" doesn't necessarily mean it is "adaptive." Best practices dictate that multiple adaptive hypotheses be tested—preferably by the application of strong inference. In some cases, biological remnants of an adaptive past may be so obscured by more recent evolutionary history that the underlying adaptive value is hard to pinpoint, because our EEA has changed throughout our long evolutionary history. Employing multiple hypothesis testing can resolve some of these difficulties, especially if the researcher is not allowed to become too enamored with his or her pet hypothesis. Dealing with alternative hypotheses allows one to let go of those that don't stand up to scrutiny, because an alternative can replace it (until it, too, is modified, retested, and perhaps rejected in favor of another).

Finally, when testing the possible adaptive value of human behavior, the standard fall back—or null hypothesis—should no longer be that the character is "cultural" (or a result of "nurture" or "environmental"). Applying strong inference, we also test the alternative, nonbiological hypothesis and do not assume it to be the default explanation if we reject a particular biological one. Moreover, when comparative studies suggest biological homologies between a human character and another in a close ancestor, the onus is on us to show it is not biological rather than having to prove it is biological (of course, if the biological hypothesis is rejected and a nonbiological one accepted as the best explanation so far, then we have learned something).

Essentially reflecting on Tooby and DeVore (1987), the null hypothesis is based on the fact that we are animals, we evolved, and much of what makes us "unique" is at one extreme end of an evolutionary/developmental continuum. It is at the extremes that we may find evidence of extraordinary abilities, and then extraordinary "proof" would be needed to show that human uniqueness cannot be explained as an elaboration of our biology. And that will be a paradigm shift.

References

Aguilar, M., Bhuket, T., Torres, S., Liu, B., & Wong, R. J. (2015). Prevalence of the metabolic syndrome in the United States, 2003–2012. *JAMA, 313*, 1973–1974.

Brower, L. P. (1984). Chemical defence in butterflies. *Symposia of the Royal Entomological Society of London, 11*, 109–134.

Burian, R. M. (1992). Adaptation: Historical perspectives. In E. F. Keller & E. A. Lloyd (Eds.), *Keywords in evolutionary biology* (pp. 7–12). Cambridge, MA: Harvard University Press.

Cameron, E., & du Toit, J. T. (2007). Winning by a neck: Tall giraffes avoid competing with shorter browsers. *American Naturalist, 169*, 130–135.

Carew, T. J., & Kandel, E. R. (1977a). Inking in *Aplysia californica*: I. Neural circuit of an all-or-none behavioural response. *Journal of Neurophysiology, 40*, 692–707.

Carew, T. J., & Kandel, E. R. (1977b). Inking in *Aplysia californica*: II. Central program for inking. *Journal of Neurophysiology, 40*, 708–720.

Carew, T. J., & Kandel, E. R. (1977c). Inking in *Aplysia californica*: III. Two different synaptic conductance mechanisms if trigger central program for inking. *Journal of Neurophysiology, 40*, 721–734.

Carlson, B., & Nolen, T. G. (1997). The effect of *Aplysia's* defensive chemical ink on the dactyl chemoreceptors of predatory crabs (*Cancer anntenarius*). *Society for Neuroscience: Abstracts, 23,* 188.

Clark, D. (2013). Evolution of algebraic terms: 1. Term to term operation continuity. *International Journal of Algebra and Computation, 23,* 1175–1205.

Coelho, L., Prince, J., & Nolen, T. (1998). Processing of defensive pigment in *Aplysia californica:* Acquisition, modification and mobilization of the red algal pigment, r-phycoerythrin by the digestive gland. *Journal of Experimental Biology, 201,* 425–438.

Corey, K. E., & Kaplan, L. M. (2014). Obesity and liver disease: the epidemic of the twenty-first century. *Clinical Liver Disease, 18,* 1–18.

Crompton, R. H., Sellers, W. I., & Thorpe, S. K. S. (2010). Arboreality, terrestriality and bipedalism. *Philosophical Transactions of the Royal Society B: Biological Sciences, 365,* 3301–3314.

Darden, L. (1992). Character: Historical perspectives. In E. F. Keller & E. A. Lloyd (Eds.), *Keywords in evolutionary biology* (pp. 41–44). Cambridge, MA: Harvard University Press.

Darwin, C. (1859). *On the origin of species, by means of natural selection* (1st ed.). London, England: John Murray.

Darwin, C. (1872). *On the origin of species, by means of natural selection* (6th ed.). London, England: John Murray.

Darwin, C. (1899). *The expression of emotions in man and animals.* New York, NY: D. Appleton. (Original work published 1872)

Darwin, C. (2004). *Descent of man, and selection in relation to sex* (2nd ed.). New York, NY: Penguin. (Original work published 1879)

Darwin, C. (2005). *The autobiography of Charles Darwin, with an introduction by Brain Regal.* New York, NY: Barnes and Noble.

Darwin, C., & Wallace, A. (1858). On the tendency of species to form varieties; and on the perpetuation of varieties and species by natural means of selection. *Journal of the Proceedings of the Linnean Society of London: Zoology, 3,* 45–50.

Daly, M., & Wilson, M. (1988). *Homicide.* New York, NY: Aldine de Gruyter.

Dawkins, M. (1986). *Unraveling animal behaviour.* Essex, England: Longman.

Dostal, M. (2005). Genetic algorithms as a model of musical creativity—on generating a human-like rhythmic accompaniment. *Computing and Informatics, 22,* 321–340.

Eales, N. B. (1921). Aplysia. *Liverpool Marine Biological Committee. Proc. Trans. Liverpool Biol. Soc. L.M.B.C. Mem. Vol.* 35 (24), 183–266.

Enattah, N. S., Jensen, T. G., Nielsen, M., Lewinski, R., Kuokkanen, M., Rasinpera, H., ... Peltonen, L. (2008). Independent introduction of two lactase-persistence alleles into human populations reflects different history of adaptation to milk culture. *American Journal of Human Genetics, 82,* 57–72.

Fabbrini, E., Yoshino, J., Yoshino, M., Magkos, F., Tiemann, C., Samovski, D., . . . Klein, S. (2015). Metabolically normal obese people are protected from adverse effects following weight gain. *Journal of Clinical Investigation, 125,* 787–795.

Fristrup, K. (1992). Character: Current usages. In E. F. Keller & E. A. Lloyd (Eds.), *Keywords in evolutionary biology* (pp. 45–51). Cambridge, MA: Harvard University Press.

Gough, I., Runciman, G., Mace, R., Hodgson, G., & Rustin, M. (2008). Darwinian evolutionary theory and the social sciences. *Twenty-First Century Society, 3*, 65–86.

Gould, S. J. (1984). Carrie Buck's daughter. *Natural History Magazine, 7*(84), 14–18.

Gould, S. J., & Vrba, E. S. (1982). Exaptation: A missing term in the science of form. *Paleobiology, 8*, 4–15.

Gluckman, P., Beedle, A., Buklijas, T., Low, F., & Hanson, M. (2016). *Principles of evolutionary medicine* (2nd ed.). Oxford, England: Oxford University Press.

Herron, J. C., & Freeman, S. (2013). *Evolutionary analysis* (5th ed.). Boston: Pearson.

Huang-Dorin, I., Sleigh, A., Rochford, J. J., O'Rahilly, S., & Savage, D. (2010). Lipodystrophy: Metabolic insights from a rare disorder. *Journal of Endocrinology, 207*, 245–255.

Huxley, J. (1942). *Evolution: The modern synthesis.* New York, NY: Harper.

Huxley, T. H. (1870). *Biogenesis and abiogenesis: Critiques and addresses. Collected essays VIII.* Retrieved from http://aleph0.clarku.edu/huxley/CE8/B-Ab.html

Johnson, P. M., Kicklighter, C. E., Manfred Schmidt, M., Kamio, M., Yang, H., . . . Derby, C. D. (2006). Packaging of chemicals in the defensive secretory glands of the sea hare *Aplysia californica*. *Journal of Experimental Biology, 209*, 78–88.

Johnson, P. M., & Willows, A. O. D. (1999). Defense in sea hares (Gastropoda, Opisthobranchia, Anaspidea): Multiple layers of protection from egg to adult. *Marine and Freshwater Behaviour and Physiology, 32*, 147–180.

Judson, O. (2008). Stop the mutants! *New York Times Science Blogs.* March 11, 2008. http://judson.blogs.nytimes.com/

Kick, S. A., & Simmons, J. A. (1984). Automatic gain control in the bat's sonar receiver and the neuroethology of echolocation. *Journal of Neuroscience, 4*, 2725–2737.

Kicklighter, C. E., Shabani, S., Johnson, P. M., & Derby, C. M. (2005). Sea hares use novel antipredatory chemical defenses. *Current Biology, 15*, 549–554.

Kolata, G. (2016, December 12). One weight-loss approach works for all? No, not even close. *New York Times.* Retrieved from http://nyti.ms/2hEgtax

Krisch, J. (2014). When racism was a science. *New York Times.* http://nyti.ms/1tpl30k

Kuhn, T. S. (2012). *The structure of scientific revolutions* (4th ed.). Chicago, IL: University of Chicago Press. (Original work published 1962)

Loos, R. J. F. et al. (2008) Common variants near MC4R are associated with fat mass, weight and risk of obesity. *Nature Genetics, 40*(6), 768–775.

Malcom, S. B., & Brower, L. P. (1989). Evolutionary and ecological implications of cardenolide sequestration in the monarch butterfly. *Experientia, 45*, 284–295.

Matić, D. (2010). A genetic algorithm for composing music. *Yugoslav Journal of Operations Research, 20*, 157–177.

Medina, M., & Walsh, P. J. (2000). Molecular systematics of the order Anaspidea based on mitochondrial DNA sequence (12s, 16s, and CO1). *Molecular Phylogenetics and Evolution, 15*, 41–58.

Mitchell, G., van Stittert, S. J., & Skinner, J. D. (2009). Sexual selection is not the origin of long necks in giraffes. *Journal of Zoology, 278*, 281–286.

Nolen, T. G., & Johnson, P. M. (2001). Defensive inking in *Aplysia spp*: Multiple episodes of ink secretion and the adaptive use of a limited chemical resource. *Journal of Experimental Biology, 204*, 1257–1268.

Nolen, T. G., Johnson, P. M., Kicklighter, C. E., & Capo, T. (1995). Ink secretion by the marine snail *Aplysia californica* enhances its ability to escape from a natural predator. *Journal of Comparative Physiology A, 176*, 239–254.

O'Rahilly, S., & Farooqi, I. S. (2013). The genetics of obesity in humans. In L. J. De Groot, G. Chrousos, K. Dungan, K. R. Feingold, A. Grossman, J. M. Hershman. . . A. Vinik (Eds.), Endotext. South Dartmouth, MA: MDText.com. https://www.ncbi.nlm.nih.gov/books/NBK279064/

Paley, W. (1809). *Natural theology: or, evidences of the existence and attributes of the deity* (12th ed.). London, England: Printed for J. Faulder.

Parons, J. A. (1965). A digitalis-like toxin in the monarch butterfly, *Danaus plexippus* L. *Journal of Physiology, 178*, 290–304.

Pigliucci, M (2009, May 31). Strong inference and the distinction between soft and hard science [Web log post]. Retrieved from http://www.scientificblogging.com/rationallyspeaking/strong

Platt, J. R. (1964). Strong inference: Certain systematic methods of scientific thinking may produce much more rapid progress than others. *Science, 146*, 347–353.

Popper, K. (2002). *Conjectures and refutations.* Oxford, England: Routledge Classics.

Popper, K. (2005). *The logic of scientific discovery.* London, England: Taylor and Francis. (Original work published 1935)

Pratt, D. M., & Anderson, V. H. (1982). Population, distribution, and behavior of giraffe in the Arusha National Park, Tanzania. *Journal of Natural History, 16*, 481–489.

Pratt, D. M., & Anderson, V. H. (1985). Giraffe social behavior. *Journal of Natural History, 19*, 771–781.

Prince, J., Nolen, T., & Coelho, L. (1998). Defensive ink pigment processing and secretion in *Aplysia californica:* Concentration and storage of phycoerythrobilin in the ink gland. *Journal of Experimental Biology*, 201, 1595–1613.

Sacks, F. M., Bray, G. A. Carey, V. J. Smith S. R., Ryan, D. H. Anton S. D., . . . Williamson, D. A. (2009). Comparison of weight-loss diets with different compositions of fat, protein, and carbohydrates *New England Journal of Medicine, 360*, 859–873.

Sagan, C. (1995). *The demon-haunted world: Science as a candle in the dark* (1st ed.). New York, NY: Random House.

Schmitt, D. (2015). "Yes, but . . ." Answers to ten common criticisms of evolutionary psychology. In Evolution Institue (Ed.), *What's wrong (and right) about evolutionary psychology?* (pp. 25–35). San Antonio, FL: Evolution Institute. https://evolution-institute.org/wp-content/uploads/2016/03/20160307_evopsych_ebook.pdf

Schutt, R., & Wilson, D. S. (2016, January 4). Why did sociology declare independence from biology (and can they be reunited)? An interview with Russell Schutt. *This View of Life.* Retrieved from https://evolution-institute.org/article/why-did-sociology-declare-independence-from-biology-and-can-they-be-reunited-an-interview-with-russell-schutt/

Simmons, R., & Scheepers, L. (1996). Winning by a neck: Sexual selection in the evolution of giraffe. *American Naturalist, 148*, 771–786.

Tinbergen, N. (1967). Adaptive features of the black-headed gull *Larus ridibundus*. In D. Snow (Ed.), *Proceedings of the 14th International Ornithological Congress* (pp. 43–59). Oxford, England: Blackwell Scientific.

Tishkoff, S. A., Reed, F. A., Ranciaro A., Voight, B. F., Babbitt, C. C., Silverman, J. S., . . . Deloukas, P. (2007). Convergent adaptation of human lactase persistence in Africa and Europe. *Nature Genetics, 39*, 31–40.

Tooby J., & DeVore, I. (1987). The reconstruction of hominid behavioral evolution through strategic modeling. In W. Kinzey (Ed.), *Primate models of hominid behavior* (pp. 183–237). Albany, NY: SUNY Press.

Truthiness. (n.d.). *Dictionary.com*. Retrieved from http://www.dictionary.com/browse/truthiness

Truzzi, M. (1978). On the extraordinary: An attempt at clarification. *Zetetic Scholar, 1*(1), 11.

Vermeij, G. J. (1971). Gastropod evolution and morphological diversity in relation to shell geometry. *Journal of Zoology, 163*, 15–23.

Vermeij, G. J. (1976). Inter-oceanic differences in vulnerability of shelled prey to crab predation. *Nature, 260*, 135–136.

Virtue, S., & Vidal-Puig, A. (2010). Adipose tissue expandability, lipotoxicity and the metabolic syndrome: An allostatic perspective. *Biochimica et Biophysica Acta: Molecular and Cell Biology of Lipids, 1801*, 338–349.

Walters, E. T., & Erickson, M. T. (1986). Directional control and functional organization of defensive responses in *Aplysia*. *Journal of Comparative Physiology A, 159*, 339–351.

Waltman, L., van Eck, N.J., Dekker, R., & Kaymak, U. (2011). Economic modeling using evolutionary algorithms: the effect of a binary encoding of strategies. *Journal of Evolutionary Economics, 21*, 737–756.

Welch, J. J. (2016) What's wrong with evolutionary biology? *Biology & Philosophy, 32*, 263–279.

West-Eberhrdt, M. J. (1992). Adaptation: Current usages. In E. F. Keller & E. A. Lloyd (Eds.), *Keywords in evolutionary biology* (pp. 13–18). Cambridge, MA: Harvard University Press.

Williams, G. C. (1996). *Adaptation and natural selection: A critique on some current evolutionary thought*. Princeton, NJ: Princeton University Press.

World Health Organization. (2003). Endocrine, nutritional and metabolic diseases (E00-E90) Metabolic disorders (E70-E90). E73 Lactose Intolerance. In *International Statistical Classification of Diseases and Related Health Problems* (10th rev.). Geneva, Switzerland: World Health Organization. http://apps.who.int/classifications/apps/icd/icd10online2003/ge70.htm

Yang, H., Johnson, P. M., Ko, K., Kamio, M., Germann, M. W., Derby, C. D., & Tai, P. C. (2005). Cloning, characterization and expression of escapin, a broadly antimicrobial FAD-containing L-amino acid oxidase from ink of the sea hare *Aplysia californica*. *Journal of Experimental Biology, 208*, 3609–3622.

Young, T. P., & Isbell, L. A. (1991). Sex differences in giraffe feeding ecology: Energetics and social constraints. *Ethology, 87*, 79–89.

Zimmer, C., & Emlen, D. J. (2016). *Evolution: Making sense of life* (2nd ed.). New York, NY: W. H. Freeman.

Zuk, M. (2002). *Sexual selections: What we can and can't learn about sex from animals*. Berkeley, CA: University of California Press.

CHAPTER 6 | Charles Darwin and Selection in Relation to Sex in the Colors of Monkeys

SANDRA WINTERS, MEGAN PETERSDORF,
AND JAMES P. HIGHAM

MANY ANIMALS EXHIBIT STRIKING colors and patterns that have long fascinated naturalists and stimulated them to ask the question: Why are animals often so colorful? Darwin provided a roadmap for understanding how we might investigate these questions through the lens of natural and sexual selection. Thanks to Darwin, Wallace, and other pioneering evolutionary biologists, we now approach this question by asking: How and why might such conspicuous colors have evolved? In some cases, conspicuous coloration evolves via natural selection, such as warning colors that advertise the toxicity or unpalatability of an animal to potential predators (i.e., aposematic colors; Poulton 1890; Ruxton, Sherratt, & Speed, 2004). Many conspicuous colors, however, appear to have evolved via sexual selection, facilitating an individual's access to a greater number or higher quality of mates. Darwin was fascinated by questions regarding animal coloration; his collected works include many descriptions of animals exhibiting brilliantly colored ornaments, and he used many examples of animal color patterns to support his theories. Darwin's conceptualization of sexual selection, relayed primarily in his 1871 book *The Descent of Man*, was in part an attempt to understand and explain how animals evolved these types of conspicuous ornamentation.

In his paper "Sexual Selection in Relation to Monkeys," published in the journal *Nature* in 1876, Darwin wrote, "No case interested and perplexed me so much as the brightly-coloured hinder ends and adjoining parts of certain monkeys" (p. 18; Figure 6.1). It is no wonder that Darwin

FIGURE 6.1. Images of male mandrills. According to Darwin (1871, p. 292), "no other member in the whole class of mammals is coloured in so extraordinary a manner as the adult male mandrill". (Color figure in online version.)
L: Public domain image licensed under Creative Commons CCO 1.0 Universal: http://creativecommons.org/publicdomain/zero/1.0/legalcode; R: Image by Amit Patel (2017). Image licensed under Creative Commons CCO BY 2.0 Attribution: http://creativecommons.org/licenses/by/2.0/legalcode

emphasized the brilliant colors of primates. Comfortably the most colorful group of mammals, primates exhibit extensive variation in color and patterning across species (Bradley & Mundy, 2008; Caro, 2005). This is precisely the type of variation that captivates biologists, from early evolutionists like Darwin to modern practitioners using his theories to understand animal diversity. Like other animals, the colors displayed by primates are influenced by both natural and sexual selection as primates seek to both avoid predators and obtain mates. Among primates, natural selection tends to produce cryptic colors that render them difficult to spot against the background, whereas conspicuous primate colors tend to be selected through reproductive contexts and result from sexual selection (Bradley & Mundy, 2008). In this chapter, we explore how Darwin's theory of sexual selection has influenced and structured research focusing on the evolution of conspicuous coloration in primates (Figure 6.1).

Sexual Selection: Darwin's Roadmap to Understanding the Extravagant Colors of Monkeys

Darwin presented the theory of sexual selection in *The Origin of Species (1859)* and greatly expanded it in *The Descent of Man and Selection in*

Relation to Sex (1871) to explain the evolution of elaborate traits that are often differentiated by sex and appear to hinder survival. Sexual selection, he wrote in *The Origin of Species*, "depends, not on a struggle for existence . . . but on a struggle between the individuals of one sex . . . for the possession of the other sex. The result is not death to the unsuccessful competitor but few or no offspring" (p. 88). Traits evolve under sexual selection because they increase an individual's lifetime reproductive success through a competitive advantage in reproduction. Darwin believed this competition took two distinct forms: intrasexual selection, which selects for traits that assist in competition among individuals of the same sex for reproductive opportunities with the opposite sex, and intersexual selection, which selects for traits that improve an individual's attractiveness to the opposite sex. While sexual selection acts on both sexes, albeit to different degrees depending on which sex is the limiting resource to reproduction (Bateman, 1948; Gowaty, 2004; Trivers, 1972), intrasexual and intersexual selection are commonly referred to as male–male competition and female mate choice, respectively.

Intrasexual selection often leads to the evolution of armaments to assist in competition (e.g., the horns, tusks, and canines of many mammals), either directly in fights or indirectly by advertising competitive ability (Darwin, 1871). In comparison, intersexual selection leads to the evolution of ornaments that are attractive to potential mates (Clutton-Brock & McAuliffe, 2009; Darwin, 1871) such as the elaborate plumage of many birds (Hill & McGraw, 2006). Due to the tendency of mammalian females to group together to avoid predation in terrestrial habitats where they can be corralled and monopolized by males, mammals are often subject to high degrees of direct male–male contest competition. This results in a high degree of sexual dimorphism in body size, in addition to the evolution of weaponry in males. Primates have evolved classic intrasexually selected traits such as sexual dimorphism in body and canine size, but unusually among mammals, primate males also exhibit brightly colored visual signals (Figure 6.2). Similarly, unlike other mammals, females of many primate species also exhibit colorful ornamentation.

History of Primate Coloration Research

Scientists and philosophers have long been interested in animal coloration, with texts on this topic tracing back to ancient times (e.g., Aristotle, 400 BCE/1910). With the development of his theory of sexual selection in

FIGURE 6.2. Primates exhibit a wide variety of conspicuous colors that may have evolved under sexual selection. Composite image from Allen and Higham (2013). Top row: bald uakari *Cacajao calvus* (oIpaat), ring-tailed lemur *Lemur catta* (Yves Picq), emperor tamarin *Saguinus imperator* (B. Inaglory), mantled guereza *Colobus guereza* (C. Burnett). Middle row: mandril *Mandrillus sphinx* (R. Young), De Brazza's monkey *Cercopithecus neglectus* (W. Allen), Diana monkey *Cercopithecus neglectus* (W. Allen). Bottom row: common squirrel monkey *Saimiri sciureus* (T. Montfort), crowned guenon *Cercopithecus pogonias* (W. Allen), gelada *Theropithecus gelada* (Kolumbusjogger), collared mangabey *Cercocebus torquatus* (W. Allen). (Color figure in online version.)

the 19th century, Darwin (1871) gave biologists a framework by which to investigate how and why mating traits have evolved. The modern study of the evolution of conspicuous animal coloration in mating contexts, including in primates, has its roots in Darwinian thinking.

In providing support for his new theory of sexual selection, Darwin drew examples from a wide variety of taxa, including primates. He generally considered conspicuous primate colors to have evolved in males via female mate choice. These conclusions were based on observations of primate morphology (e.g., color patch location and color change across time) and behavior (e.g., the presentation of colorful regions during sociosexual interactions; Darwin, 1876). Although Darwin brought attention to primate colors as important examples of sexual selection, it would be over a century before much research was conducted in this area. Early Darwinists and their opponents used animal coloration extensively as support for their views on the process of evolution (Blaisdell, 1992); however, this work focused primarily on natural selection. The theory of sexual selection was less well received than that of natural selection, with many early evolutionists expressing doubts about its validity (e.g., Wallace 1870, 1889) and seeking alternative explanations. In an extreme example, Thayer (1909) considered all animal colors to have evolved under natural selection for crypsis; even the peacock's tail, now a textbook example of a sexually selected trait, was considered by Thayer to have evolved for concealment in flowering trees. Research on coloration in nonhuman primates was also rare during this time, with the majority of this work focused on other taxa such as birds and insects. Information about primate colors is notably absent from Poulton's volume *The Colours of Animals*, published in 1890, as well as Cott's impressive tome *Adaptive Coloration in Animals*, published half a century later in 1940. Both of these books were heralded by contemporaries as impressive and comprehensive works, and the lack of references to primates within their pages is representative of the general lack of research at the time.

Historically, the majority of research on conspicuous coloration and sexual selection in primates that did occur was focused on the sexual skin of Old World monkeys and apes. Primate sexual skin is a region of largely hairless skin, often in the hindquarters, genitals, and/or face, that can vary in color across time (Dixson 1983, 2012). Females of some species also exhibit varying degrees of swelling of perineal sexual skin, called sexual swellings (Zinner, van Schaik, Nunn, & Kappeler, 2004).

Many studies analyzing the coloration of primate females' sexual skin were published throughout the 20th century. Much of the earlier work in this area was physiological in nature, focusing primarily on how changes in sexual skin color are tied to hormones and the reproductive cycle. These studies did not generally emphasize an evolutionary perspective or attempt to understand the signaling function of sexual skin based on sexual selection, but they did provide useful information for later researchers. For instance, researchers demonstrated that sexual skin color changes are linked to female reproductive cycling (Baulu, 1976; Corner, 1932; Czaja, Eisele, & Goy, 1975; Elder & Yerkes, 1936; Gillman & Gilbert, 1946; Heape, 1896; Matthews, 1956; Zuckerman, van Wagenen, & Gardiner, 1938), pregnancy status (Altmann, 1973; Baulu, 1976; Bielert et al., 1976; Gilbert & Gillman, 1945, 1951; Hartman, 1928; Tinklepaugh & Hartman,

1930), and breeding seasonality (Baulu, 1976; Koford, 1965). They also showed that sexual skin color changes across development (Zuckerman et al., 1938), that red sexual skin is caused by increased vascularization of the relevant tissue (Collings, 1926), that color saturation is linked to estrogen and progesterone levels (Corner, 1935; Czaja, Robinson, Eisele, Scheffler, & Roy, 1977), and that hormone receptors mediating these changes are found in the sexual skin (Kato, Onouchi, & Oshima, 1980; Onouchi & Kato, 1983; Ozasa & Gould, 1982). Some early investigators, however, did explicitly tie their research back to Darwin and sexual selection theory. For instance, Pocock (1906) concluded that conspicuous female sexual skin may "reveal to the males the sex and condition of adult non-pregnant females . . . act[ing] as an aphrodisiac impelling them to pair with the females in which the characters are produced rather than with those in which they are poorly developed or absent" (p. 560).

Darwin's sexual selection framework emphasized colorful ornamentations used to obtain mates by males, however, and the use of similar ornamentations by females was originally not well understood. Explicit links between female sexual skin and sexual selection were initially rare but became more common later in the 20th century as these ideas became more developed. For instance, a number of hypotheses were put forth to explain how colorful sexual swellings could have evolved as sexually selected traits whose attractiveness to males increases female fitness (e.g., Clutton-Brock & Harvey 1976; Nunn, 1999; Pagel, 1994). Throughout the 20th century, sexual skin color changes in females were described and analyzed in a variety of species, including macaques (e.g., Baulu, 1976; Bielert et al., 1976; Collings, 1926; Corner, 1932, 1935; Czaja et al., 1975; Hartman, 1928; Heape, 1896; Pocock, 1925; Tinklepaugh & Hartman, 1930; Zuckerman et al., 1938), gelada (Garrod, 1879; Matthews, 1956; Pocock, 1925), baboons (e.g., Altmann, 1973; Gillman & Gilbert, 1945, 1946, 1951; Pocock, 1925), mangabeys (e.g., Pocock, 1906), colobines (e.g., Pocock, 1935), and chimpanzees (e.g., Elder & Yerkes, 1936).

The 20th century also saw research on male sexual skin. As in research on females, most studies initially focused on physiology and ontogeny. Researchers linked the expression of darker sexual skin color with reproductive opportunities and the mating season (Gordon & Bernstein, 1973; Sade, 1964), described the development of colorful secondary sexual characteristics (Wickings & Dixson, 1992; Zuckerman et al., 1938), and linked the expression of these traits to hormone levels (both estrogen and testosterone; Vandenbergh, 1965; Wickings & Dixson, 1992; Zuckerman et al., 1938; Zuckerman & Parkes, 1939). Noting the conspicuousness of male sexual skin color, researchers reasoned that it may have a signaling function (Guthrie, 1970; Rowland, 1979; Wickler, 1967). Some suggested functions related to male rivalry and that male signals were products of intrasexual selection (e.g., Bercovitch, 1996), while others suggested more general social functions such as appeasement (e.g., Wickler, 1967). Descriptions of male sexual skin focused on a variety of species during the 20th century, including macaques (e.g., Zuckerman et al., 1938), baboons (Gordon & Bernstein, 1973; Kummer, 1968; Sade, 1964; Vandenbergh, 1965; Zuckerman & Parkes, 1939), mandrills (e.g., Hill, 1970; Wickings & Dixson, 1992), drills (e.g., Zuckerman & Parkes, 1939), guenons (e.g., Bercovitch, 1996; Kingdon, 1980; Wickler, 1967), and langurs (e.g., Zuckerman & Parkes, 1939).

Overall, much of the early research on primate sexual skin coloration focused on proximate mechanisms, such as the role of hormones and development on color expression. Increasingly throughout the 20th century, however, researchers started to focus more on the ultimate function and comparative evolution of colorful primate traits and to undertake investigations of how sexual selection theory can explain the evolution of conspicuous coloration in primates.

Conceptual and Methodological Transitions

Over time, primate coloration has been studied using a variety of different approaches, with recent developments ushering in a new era of research on conspicuous coloration and sexual selection in primates. In Darwin's time, biologists generally amassed detailed descriptions of animal behaviors and morphologies, which were then used to infer their function and identify evolutionary trends. Early naturalists amassed wonderful descriptions of animal colors that contributed to the knowledge of their contemporaries and continue to captivate those interested in natural phenomena to this day.

In the early 20th century, biologists began to focus on experimental research and more precise measurements of animal behavior and morphology. While qualitative descriptions still occurred, there was a growing trend to quantify descriptions of animals and to test explicit predictions. Previous research, including that of Darwin, was criticized as being overly anthropomorphic, and backing up conclusions with numbers became paramount. Some researchers studying primate coloration began quantifying color change based on visual categorizations (e.g., Sade, 1964) or via comparisons with color cards (e.g., Bercovitch, 1996; Corner, 1932).

While these are relatively crude methods by modern standards, the transition to analyzing primate coloration based on quantifiable changes and comparisons was an important advancement in the field.

Eventually, biologists also began to conduct experiments analyzing primate colors that had been manipulated, allowing researchers to determine the effect of variations in a signal while controlling for all other potential factors. The first experimental assessment of primate coloration and sexual selection focused on the color of primate sexual swellings. Bielert, Girolami, and Jowell (1989) fitted female baboons with a model of a sexual swelling that was painted a variety of colors and measured resulting male sexual arousal. The red swelling yielded the highest male arousal, leading to the conclusion that sexual skin color is an important factor in female attractiveness to males when all other factors are held constant. Later studies have used experimental methods to investigate the use of a variety of colors exhibited by primates in both intrasexual competition (e.g., vervet monkey scrotal color; Gerald, 2001) and intersexual mate choice (e.g., rhesus macaque facial skin color; Dubuc et al., 2016; Higham et al., 2011; Waitt, Gerald, Little, & Kraiselburd, 2006; Waitt et al., 2003). Such experiments that directly test subjects' responses to visual stimuli are essential for understanding the functions of primate colors.

Recent times have seen an increase in research focused on conspicuous coloration and sexual selection in primates. The advent of digital photography provided researchers with a more accurate method of quantitatively measuring primate colors in a variety of settings (Stevens, Párraga, Cuthill, Partridge, & Troscianko, 2007), and information about the visual systems of primate species has allowed researchers to model colors as seen by their subjects (Stevens, Stoddard, & Higham, 2009) and take perceptually relevant measurements of color distances (e.g., units of "just noticeable differences"; Stern & Johnson, 2010). These new methods, along with a renewed interest in sexual selection throughout evolutionary biology, have spurred much additional research in modern times.

Recent Studies of Primate Coloration

Darwin's theory of sexual selection created the template for much of the modern research on primate colors. Current research questions are generally framed to assess whether conspicuous coloration primarily evolved under intrasexual or intersexual selection. This is often established by identifying the primary receivers of the signal and determining how they respond to variation in the signal's expression. If the receivers are the same sex as the signaler, the signal has likely evolved under stronger intrasexual selection, whereas if the receivers are the opposite sex, the signal has likely evolved under stronger intersexual selection. It is probable that many signals evolved under both selective mechanisms, given that this distinction is Darwin's rather than the animals', but empirical studies typically separate the two. There have, however, been recent calls to integrate selective mechanisms when evaluating the form and strength of selection on sexual signals, as intrasexual and intersexual selection may reinforce or oppose one another (Hunt, Breuker, Sadowski, & Moore, 2009).

Recent methodological advances in digital photography have allowed coloration to be quantified (Stevens et al., 2009) and compared to social variables in captive, wild, and free-ranging primate populations. To test the alternative hypotheses that a colorful primate signal is under intrasexual versus intersexual selection, both observational and experimental approaches are undertaken and often complement one another. Observational approaches include assessing how variation in the signal relates to social status, mediates intrasexual agonistic interactions, and/or influences intersexual mating behavior. Experimental approaches include the looking-time paradigm, which uses photographic stimuli to assess attentional biases between the sexes (Winters, Dubuc, & Higham, 2015). Attentional biases can be due to both intimidation and attraction, making this a relevant paradigm for uncovering bias that results from male–male competition or female mate choice. Some studies have used both experimental and observational approaches by pairing captive individuals together in same-sex or opposite-sex pairs that express different signal intensities to understand how signal variation mediates dyadic social interactions (e.g., Gerald, 2003; Gerald, Ayala, Ruíz-Lambides, Waitt, & Weiss, 2010).

Males of many primate taxa exhibit bright coloration on their face, hindquarters, genitals, or chest, which is highly unusual for mammals. Given the similarities in form to the colorful and elaborate ornaments of many male birds, these bright primate colors have been hypothesized to be a product of female mate choice (Clutton-Brock & McAuliffe, 2009). In contrast, recent research has shown that they generally function as "badges of status," correlating strongly with social status, similar to most other male mammalian traits that have evolved under male–male competition (e.g., Emlen, 2014). Male mandrills (*Mandrillus sphinx*) are characterized by multicolored faces, rumps, and genitals; drills (*M. leucophaeus*) exhibit multicolored rumps and genitals and a red lower lip; crested macaques

(*Macaca nigra*) exhibit red scrotal coloration; gelada (*Theropithecus gelada*) exhibit red chest patches; vervet monkeys exhibit red, white, and blue genital coloration; and black-and-white snub-nosed monkeys (*Rhinopithecus bieti*) exhibit red lips. In all of these species, both observational and experimental evidence show that coloration generally correlates with social status, such that males of higher dominance rank or those who hold one-male units express more intense coloration (mandrills: Renoult, Schaefer, Sallé, & Charpentier, 2011; Setchell & Wickings, 2005, drills: Marty, Higham, Gadsby, & Ross, 2009; crested macaques: Engelhardt, Neumann, Heistermann, & Perwitasari-Farajallah, 2008; gelada: Bergman, Ho, & Beehner, 2009; vervet monkeys: Cramer, 2012; snub-nosed monkeys: Grueter et al., 2015). This coloration is also responsive to changes in social status, as males rapidly change in color after the take-over of a high-ranking position (Bergman et al., 2009; Setchell & Dixson, 2001). These conspicuously colored signals appear to be important in mediating male–male interactions (e.g., Gerald, 2001), and there is limited evidence that they are attractive to females and influence female mating behaviors after controlling for male dominance rank (e.g., drills: Marty et al., 2009; vervet monkeys: Gerald et al., 2010). When a female preference for male coloration has been found (e.g., mandrills: Setchell, 2005), it is difficult to separate the effects of coloration and dominance on male attractiveness since dominance is so strongly correlated with coloration in these taxa. Therefore, it appears that intrasexual selection is generally the stronger selective pressure on the evolution of colorful male sexual signals in many primate species.

An interesting case of male coloration in primates that differs from this typical pattern is the rhesus macaque (*Macaca mulatta*). Rhesus macaques express co-varying red coloration on their face, rump, and genitals (Higham, Pfefferle, Heistermann, Maestripieri, & Stevens, 2013) that intensifies in the mating season due to seasonal changes in testosterone (Baulu, 1976). Unlike in other primates studied, facial coloration does not correlate with dominance rank (Dubuc, Allen, Maestripieri, & Higham, 2014; Higham et al., 2013), and instead there is observational and experimental evidence that coloration is attractive to females (Dubuc, Allen et al., 2014; Dubuc et al., 2016; Waitt et al., 2003). Additionally, red ornamentation is heritable and influences fecundity; darker red males experience higher reproductive success when combined with high dominance rank (Dubuc, Winters et al., 2014). This evidence suggests that red ornamentation in male rhesus macaques is primarily under intersexual selection, which is unusual for a mammalian sexual signal. Despite this, male coloration may still mediate male–male interactions. Adult males experimentally show an attentional bias toward dark red compared to pale pink males (Dubuc et al., 2016), and there is observational evidence that coloration differences may influence male dyadic agonism (Petersdorf et al., 2017), similar to the pattern seen in vervet monkeys (Gerald, 2001).

Conspicuous color signals are more common in primate males, but uniquely among mammals, females of many species also exhibit elaborate visual ornamentation. Significantly less work has focused on coloration in female sexual signals, particularly on the intrasexual and intersexual selective mechanisms. Most research on female color signals has been in relation to fertility advertisement. In some catarrhine primate taxa, intraindividual variation in facial coloration over time signals the fertile phase (e.g., rhesus macaques: Dubuc et al., 2009, Higham et al., 2010; Japanese macaques, *Macaca fuscata*: Fujita, Sugiura, Mitsunaga, & Shimizu, 2004). Whether this variation influences responses of conspecifics has only been assessed in rhesus macaques: males can discern intraindividual fertility differences in female facial coloration, although this ability increases with social familiarity (Higham et al., 2011). Fertility does not seem to be advertised in the coloration of other sexual signals examined, such as the sexual swellings of olive baboons, with swelling size cues seemingly more important in indicating the timing of the fertile phase (*Papio anubis*: Higham, MacLarnon, Ross, Heistermann, & Semple, 2008).

In addition to within individual variation, female coloration also varies between individuals. In rhesus macaques, female facial coloration is heritable and positively influences fecundity (Dubuc, Winters et al., 2014). Males show attentional biases toward images of redder females (Waitt et al., 2006), and females with redder faces reproduce at higher rates (Dubuc, Winters et al., 2014), suggesting that female facial coloration evolved under intersexual selection via male mate choice. More work on other species is needed to confirm how different sexual selection pressures have led to the evolution of female primate color ornamentation.

It is not yet clear what quality coloration may advertise in primates that opposite-sex individuals may find attractive. Sexual skin coloration is under proximate control of sex steroid hormones, and its expression is related to the degree of blood flow and oxygenation in the sexual skin (Changizi, Zhang, & Shimojo, 2006; Rhodes et al., 1997; Vandenbergh, 1965). Coloration might therefore be an honest signal of condition (Folstad & Karter, 1992) or current health (Hamilton & Zuk, 1982); however, female facial coloration in some species does not appear to reflect body condition (mandrills: Setchell, Wickings, & Knapp, 2006) or parasite load (Japanese macaques: Rigaill et al., 2017).

On the whole, recent research on conspicuous coloration in primates has progressed extensively in recent years, and we now have a much greater understanding of how intrasexual and intersexual selection function in primates, the ways in which these selective dynamics make primates unique among mammals, and the underlying characteristics that may be signaled by primate colors. The Darwinian sexual selection framework has been essential for structuring research in this area and for understanding how and why primate colors are related to reproductive competition.

Future Directions

While the past decade has seen a surge of research in conspicuous coloration and sexual selection in primates, much work remains to be done, including assessments of specific color patterns in individual primate groups as well as broader trends across species. Primates are an extraordinarily colorful group, yet relatively few primate taxa have been the subject of quantitative analyses of color patterns that may be relevant in a mating context. For example, in a review of the evidence for mate choice based on color in primates, Higham and Winters (2016) identified 13 studies investigating the phenomenon in nine species: three lemurs, five cercopithecines (Old World monkeys), and one ape. Some entire branches of the primate phylogenetic tree have yet to be studied for such questions, including New World monkeys and the Asian colobines. This is particularly noteworthy because these groups contain some primates with extremely interesting appearances, such as the emperor tamarin (*Saguinus imperator*) and the golden snub-nosed monkey (*Rhinopithecus roxellana*). Studies investigating how sexual selection has influenced the colors and patterns displayed by species in these groups are ripe for the taking.

In addition to studies seeking to explain species-specific coloration, additional work should focus on identifying evolutionary trends based on comparisons across species. In 2008, Bradley and Mundy published an excellent review of primate coloration in which they evaluated the selective forces likely to have helped shape primate diversity, including sexual selection. They pointed to many unanswered questions that could be addressed using comparative analyses assessing traits and species characteristics across primate species. These include the fascinating phenomena of sexual dichromatism—whereby females and males of the same species are differently colored, and of natal coat colors—whereby the infants of primate species show coats that are conspicuous orange and yellow colors that contrast strikingly with the darker fur of their parents. However, to date few of these interspecific studies have been undertaken using modern methods with detailed measurements of primate colors and appropriate statistical controls for the phylogenetic relationships between species. In part, the lack of broader comparative analyses of the role of sexual selection in primate coloration is due to the lack of available data from many species. Many primate colors likely to be involved in sexual signaling are located on the skin. These colors do not preserve post mortem and therefore cannot be analyzed based on museum specimens. It is notable that recent comparative analyses of primate coloration (e.g., Kamilar & Bradley, 2011a, 2011b; Winters, Kamilar, Webster, Bradley, & Higham, 2014) have focused on the evolution of pelage colors with respect to crypsis and predator avoidance (i.e., natural selection). To understand why and how primate colors have evolved by sexual selection, we need to amass more information about the diversity of primate coloration, including quantitative descriptions (e.g., spectral measurements) of primate skin colors as well as analyses of functionality (e.g., determining that a color patch in a given species has evolved via mate choice).

Rather than always studying a single color patch in isolation, research should also focus on the structures and patterns associated with primate colors, as well as how different patches combine to create a total visual phenotype (Allen & Higham, 2013). Contrasting colors, patterns, and optical effects created by primate colors could be under intrasexual and/or intersexual selection, but this has not yet been tested. New methods for analyzing animal patterns (e.g., Stoddard, Kilner, & Town, 2014; Troscianko & Stevens, 2015; Tylor, Gilbert, & Reader, 2013) could be applied to the analysis of primate color patterns to test this hypothesis.

Conclusion: Darwin's Roadmap to Primate Coloration

Darwin was fascinated by primate coloration. He explicitly began asking questions about how and why colors had evolved, posing questions about their functions and underlying mechanism. His insights into sexual selection gave researchers a roadmap to follow for investigating primate colors, which researchers of the topic have used to frame their studies and research questions. As such, the study of primate coloration can be held up as an exemplar for how Darwinian thinking framed and continues to shape academic study in the 21st century. Colors that appear to have evolved via intrasexual and intersexual selection have been identified in a variety of primate species, and it is clear that primates are an excellent group in which to study sexual selection in mammals. In many ways, research on conspicuous coloration and sexual selection in primates has come full circle. Darwin used colorful primate traits as key examples when initially articulating his theory of sexual selection, but research on primate coloration and sexual selection was rare for decades afterward. Renewed interest in this topic as well as developments in methods and technology making it easier to quantify primate colors has led to a surge of research in this area in recent times. This reflects sexual selection research returning to its roots, as researchers in the 21st century provide the required evidence that Darwin's ideas about the colors of monkeys were right all along.

References

Allen, W. L., & Higham, J. P. (2013). Analyzing visual signals as visual scenes. *American Journal of Primatology, 75*, 664–682.

Altmann, S. A. (1973). The pregnancy sign in savannah baboons. *Journal of Zoo Animal Medicine, 4*, 8–12.

Aristotle. (1910). Historia animalium. (D. W. Thompson, Trans.) In J. A. Smith & W. D. Ross (Eds), *The works of Aristotle translated into English* (Vol. 4). Oxford, England: Clarendon. (Original work published 400 BCE)

Bateman, A. J. (1948). Intrasexual selection in Drosophila. *Heredity, 2*, 349–368.

Baulu, J. (1976). Seasonal sex skin coloration and hormonal fluctuations in free-ranging and captive monkeys. *Hormones and Behavior, 7*, 481–494.

Bercovitch, F. B. (1996). Testicular function and scrotal coloration in patas monkeys. *Journal of the Zoological Society of London, 239*, 93–100.

Bergman, T. J., Ho, L., & Beehner, J. C. (2009). Chest color and social status in male geladas (*Theropithecus gelada*). *International Journal of Primatology, 30*, 791–806.

Bielert, C., Czaja, J. A., Eisele, S., Scheffler, G., Robinson, J.A., & Goy, R. W. (1976). Mating in the rhesus monkey (*Macaca mulatta*) after conception and its relation to oestradiol and progesterone levels throughout pregnancy. *Journal of Reproductive Fertility, 46*, 179–187.

Bielert, C., Girolami, L., & Jowell, S. (1989). An experimental examination of the colour component in visually mediated sexual arousal of the male chacma baboon (*Papio ursinus*). *Journal of Zoology London, 219*, 569–579.

Blaisdell, M. L. (1992). *Darwinism and its data: The adaptive coloration of animals*. New York, NY: Garland.

Bradley, B. J., & Mundy, N. I. (2008). The primate palette: The evolution of primate coloration. *Evolutionary Anthropology, 17*, 97–111.

Caro, T. (2005). The adaptive significance of coloration in mammals. *BioScience, 55*, 125–136.

Changizi, M. A., Zhang, Q., & Shimojo, S. (2006). Bare skin, blood and the evolution of primate colour vision. *Biology Letters*, *2*, 217–221.

Clutton-Brock, T. H., & Harvey, P. H. (1976). Evolutionary rules and primate societies. In P. O. G. Bateson & R. A. Hinde (Eds.), *Growing points in ethology* (pp. 195–237). Cambridge, England: Cambridge University Press.

Clutton-Brock, T., & McAuliffe, K. (2009). Female mate choice in mammals. *Quarterly Review of Biology*, *84*, 3–27.

Collings, M. R. (1926). A study of the cutaneous reddening and swelling about the genitalia of the monkey, *Macacus rhesus*. *Anatomical Record*, *33*, 271–278.

Corner, G. W. (1932). The menstrual cycle of the Malayan monkey, *Macaca irus*. *Anatomical Record*, *52*, 401–410.

Corner, G. W. (1935). Influence of the ovarian hormones, oestrin and progestin, upon the menstrual cycle of the monkey. *American Journal of Physiology*, *113*, 238–250.

Cott, H. B. (1940). *Adaptive coloration in animals*. London, Methuen.

Cramer, J. D. (2012). *Scrotal color and signal content among South African vervet monkeys* (Chlorocebus aethiops pygerythrus). Unpublished doctoral dissertation.

Czaja, J. A., Eisele, S. G., & Goy, R. W. (1975). Cyclical changes in the sexual skin of female rhesus: Relationships to mating behavior and successful artificial insemination. In W. J. Goodwin & J. Augustine (Eds.), *Primate research* (pp. 83–92). New York, NY: Plenum.

Czaja, J. A., Robinson, J. A., Eisele, S. G., Scheffler, G., & Roy, R. W. (1977). Relationship between sexual skin colour of female rhesus monkeys and midcycle plasma levels of oestradiol and progesterone. *Journal of Reproductive Fertility*, *49*, 147–150.

Darwin, C. (1871). *The descent of man, and selection in relation to sex*. London, England: John Murray.

Darwin, C. (1876). Sexual selection in relation to monkeys. *Nature*, *15*, 18–19.

Dixson, A. F. (1983). Observations on the evolution and behavioral significance of "sexual skin" in female primates. *Advances in the Study of Behavior*, *13*, 63–106.

Dixson, A. F. (2012). *Primate sexuality: Comparative studies of the prosimians, monkeys, apes, and humans*. Oxford, England: Oxford University Press.

Dubuc, C., Allen, W. L., Cascio, J., Lee, D. S., Maestripieri, D., Petersdorf, M., . . . Higham, J. P. (2016). Who cares? Experimental attention biases provide new insights into a mammalian sexual signal. *Behavioral Ecology*, *27*, 68–74.

Dubuc, C., Allen, W. L., Maestripieri, D., & Higham, J. P. (2014). Is male rhesus macaque red color ornamentation attractive to females? *Behavioral Ecology and Sociobiology*, *68*, 1215–1224.

Dubuc, C., Brent, L. J. N., Accamando, A. K., Gerald, M. S., MacLarnon, A., Semple, S., . . . Engelhardt, A. (2009). Sexual skin color contains information about the timing of the fertile phase in free-ranging macaca mulatta. *International Journal of Primatology*, *30*, 777–789.

Dubuc, C., Winters, S., Allen, W. L., Brent, L. J. N., Cascio, J., Maestripieri, D., . . . Higham, J. P. (2014). Sexually selected skin colour is heritable and related to fecundity in a non-human primate. *Proceedings of the Royal Society B: Biological Sciences*, *281*, 20141602.

Elder, J. H., & Yerkes, R. M. (1936). The sexual cycle of the chimpanzee. *Anatomical Record*, *67*, 119–143.

Emlen, D. J. (2014). *Animal weapons: the evolution of battle*. New York, NY: Henry Holt.

Engelhardt, A., Neumann, C., Heistermann, M., & Perwitasari-Farajallah, D. (2008). Sex skin coloration in male Sulawesi crested black macaques (*Macaca nigra*). *Primate Eye, 96*, 337.

Folstad, I., & Karter, A. J. (1992). Parasites, bright males, and immunocompetence handicap. *American Naturalist, 139*, 603–622.

Fujita, S., Sugiura, H., Mitsunaga, F., & Shimizu, K. (2004). Hormone profiles and reproductive characteristics in wild female Japanese macaques (Macaca fuscata). *American Journal of Primatology, 64*, 367–375.

Garrod, A. H. (1879). Notes on the anatomy of *Gelada rueppelli*. *Proceedings of the Zoological Society of London, 47*, 451–457.

Gerald, M. S. (2001). Primate colour predicts social status and aggressive outcome. *Animal Behaviour, 61*, 559–566.

Gerald, M. S. (2003). How color may guide the primate world: Possible relationships between sexual selection and dichromatism. In C. B. Jones (Ed.), *Sexual selection and reproductive competition in primates: new perspectives and directions*. Norman, OK: American Society of Primatologists.

Gerald, M. S., Ayala, J., Ruíz-Lambides, A., Waitt, C., & Weiss, A. (2010). Do females pay attention to secondary sexual coloration in vervet monkeys (*Chlorocebus aethiops*)? *Die Naturwissenschaften, 97*, 89–96.

Gilbert, C., & Gillman, J. (1945). The reactions of the perineum of the pregnant baboon with special reference to some aspects of the hormonal regulations of pregnancy. *South African Journal of Medical Science, 10*, 51–55.

Gilbert, C., & Gillman, J. (1951). Pregnancy in the baboon (*Papio ursinus*). *South African Journal of Medical Sciences, 16*, 115–124.

Gillman, J., & Gilbert, C. (1946). The reproductive cycle of the chacma baboon (*Papio ursinus*) with special reference to the problem of menstrual irregularities as assessed by the behavior of the sex skin. *South African Journal of Medical Sciences, 11*, 1–54.

Gordon, T. P., & Bernstein, I. S. (1973). Seasonal variation in sexual behavior in all-male rhesus troops. *American Journal of Physical Anthropology, 38*, 221–227.

Gowaty, P.A. (2004). Sex roles, contests for the control of reproduction, and sexual selection. In P. M. Kappeler & C. P. van Schaik (Eds.), *Sexual selection in primates: New and comparative perspectives* (pp. 37–54). Cambridge, England: Cambridge University Press.

Grueter, C. C., Zhu, P., Allen, W. L., Higham, J. P., Ren, B., & Li, M. (2015). Sexually selected lip colour indicates male group-holding status in the mating season in a multi-level primate society. *Royal Society Open Science, 1*, 150490.

Guthrie, R. D. (1970). Evolution of human threat display organs. *Evolutionary Biology, 4*, 257–302.

Hamilton, W. D., & Zuk, M. (1982). Heritable true fitness and bright birds: A role for parasites? *Science, 218*, 384–387.

Hartman, C. (1928). The period of gestation in the monkey, *Macacus rhesus*, first description of parturition in monkeys, size and behavior of the young. *Journal of Mammalogy, 9*, 181–194.

Heape, W. (1896). The menstruation and ovulation of Macacus rhesus. *Proceedings of the Royal Society B: Biological Sciences, 60*, 202–205.

Higham, J. P., Brent, L. J. N., Dubuc, C., Accamando, A. K., Engelhardt, A., Gerald, M. S., . . . Stevens, M. (2010). Color signal information content and the eye of the beholder: a case study in the rhesus macaque. *Behavioral Ecology, 21*, 739–746.

Higham, J. P., Hughes, K. D., Brent, L. J. N., Dubuc, C., Engelhardt, A., Heistermann, M., . . . Stevens, M. (2011). Familiarity affects the assessment of female facial signals of fertility by free-ranging male rhesus macaques. *Proceedings of the Royal Society B: Biological Sciences, 278*, 3452–3458.

Higham, J. P., MacLarnon, A., Ross, C., Heistermann, M., & Semple, S. (2008). Baboon sexual swellings: information content of size and color. *Hormones and Behavior, 53*, 452–462.

Higham, J. P., Pfefferle, D., Heistermann, M., Maestripieri, D., & Stevens, M. (2013). Signaling in multiple modalities in male rhesus macaques: Sex skin coloration and barks in relation to androgen levels, social status, and mating behavior. *Behavioral Ecology and Sociobiology, 67*, 1457–1469.

Higham, J. P., & Winters, S. (2016). Color and mate choice in non-human animals. In A. J. Elliot, M. D. Fairchild, & A. Franklin (Eds), *Handbook of color psychology* (pp. 502–530). Cambridge, England: Cambridge University Press.

Hill, C. O. (1970). *Primates, comparative anatomy and taxonomy. Vol. 8: Cynopithecinae, Papio, Mandrillus, Theropithecus.* Edinburgh, England: Edinburgh University Press.

Hill, G. E., & McGraw, K. J. (2006). *Bird coloration. Vol. 2: Function and evolution.* Cambridge, MA: Harvard University Press.

Hunt, J., Breuker, C. J., Sadowski, J. A., & Moore, A. J. (2009). Male–male competition, female mate choice and their interaction: determining total sexual selection. *Journal of Evolutionary Biology, 22*, 13–26.

Kamilar, J. M., & Bradley, B. J. (2011a). Countershading is related to positional behavior in primates. *Journal of Zoology, 283*, 227–233.

Kamilar, J. M, & Bradley, B. J. (2011b). Interspecific variation in primate coat colour supports Gloger's rule. *Journal of Biogeography, 38*, 2270–2277.

Kato, J., Onouchi, T., & Oshima, K. (1980). The presence of progesterone receptors in the sexual skin of the monkey. *Steroids, 46*, 743–749.

Kingdon, J. S. (1980). The role of visual signals and face patterns in African forest monkeys (guenons) of the genus Cercopithecus. *Transactions of the Zoological Society of London, 35*, 425–475.

Koford, C. B. (1965). Population dynamics of rhesus monkeys on Cayo Santiago. In I. DeVore (Ed.), *Primate behavior* (pp. 160–174). New York, NY: Holt, Rinehart & Winston.

Kummer, H. (1968). Social organization of hamadryas baboons: A field study. *Bibliotheca Primatologia, 6*, 1–189.

Marty, J. S., Higham, J. P., Gadsby, E. L., & Ross, C. (2009). Dominance, coloration, and social and sexual behavior in male drills *Mandrillus leucophaeus*. *International Journal of Primatology, 30*, 807–823.

Matthews, L. H. (1956). The sexual skin of the Gelada baboon (*Theropithecus gelada*). *Transactions of the Zoological Society of London, 28*, 543–552.

Nunn, C. L. (1999). The evolution of exaggerated sexual swellings in primates and the graded-signal hypothesis. *Animal Behaviour, 58*, 229–246.

Onouchi, T., & Kato, J. (1983). Estrogen receptors and estrogen-inducible progestin receptors in the sexual skin of the monkey. *Journal of Steroid Biochemistry, 18,* 145–151.

Ozasa, H., & Gould, K. G. (1982). Demonstration and characterization of estrogen receptor in chimpanzee sex skin: Correlation between nuclear receptor levels and degree of swelling. *Endocrinology, 111,* 231–248.

Pagel, M. (1994). The evolution of conspicuous oestrous advertisement in Old World monkeys. *Animal Behaviour, 47,* 1333–1341.

Petersdorf, M, Dubuc, C., Georgiev, A. V., Winters, S., & Higham, J. P. (2017). Is male rhesus macaque facial coloration under intrasexual selection? *Behavioral Ecology, 28,* 1472–1481.

Pocock, R. I. (1906). Notes upon menstruation, gestation, and parturition of some monkeys that have lived in the society's gardens. *Proceedings of the Zoological Society of London, 2,* 558–570.

Pocock, R. I. (1925). The external characters of the catarrhine monkeys and apes. *Proceedings of the Zoological Society of London, 95,* 1479–1579.

Pocock, R. I. (1935). The external characters of a female red colobus monkey (*Procolobus badius waldroni*). *Proceedings of the Zoological Society of London, 105,* 939–944.

Poulton, E. B. (1890). *The colours of animals, their meaning and use, especially considered in the case of insects.* New York, NY: D. Appleton.

Renoult, J. P., Schaefer, H. M., Sallé, B., & Charpentier, M. J. E. (2011). The evolution of the multicoloured face of mandrills: insights from the perceptual space of colour vision. *PLoS One, 6,* e29117.

Rhodes, L., Argersinger, M. E., Gantert, L. T., Friscino, B. H., Hom, G., Pikounis, B., ... Rhodes, W. L. (1997). Effects of administration of testosterone, dihydrotestosterone, oestrogen and fadrozole, an aromatase inhibitor, on sex skin colour in intact male rhesus macaques. *Journal of Reproduction and Fertility, 111,* 51–57.

Rigaill, L., MacIntosh, A. J. J., Higham, J. P., Winters, S., Shimizu, K., Mouri, K., ... Garcia, C. (2017). Testing for links between face color and age, dominance status, parity, weight, and intestinal nematode infection in a sample of female Japanese macaques. *Primates, 58,* 83–91.

Rowland, W. J. (1979). The use of color in intraspecific communication. In E. H. Burtt (Ed.), *The behavioral significance of color* (pp. 380–421). New York, NY: Garland.

Ruxton, G. D., Sherratt T. N., & Speed, M. P. (2004). *Avoiding attack: The evolutionary ecology of crypsis, warning signals, and mimicry.* Oxford, England: Oxford University Press.

Sade, D. S. (1964). Seasonal cycle in size of testes of free-ranging *Macaca mulatta. Folia Primatologica, 2,* 171–180.

Setchell, J. M. (2005). Do female mandrills prefer brightly colored males? *International Journal of Primatology, 26,* 715–735.

Setchell, J. M., & Dixson, A. F. (2001). Changes in the secondary sexual adornments of male mandrills (*Mandrillus sphinx*) are associated with gain and loss of alpha status. *Hormones and Behavior, 39,* 177–184.

Setchell, J. M., & Wickings, E. J. (2005). Dominance, status signals, and coloration in male mandrills (Mandrillus sphinx). *Ethology, 111,* 25–50.

Setchell, J. M., Wickings, E. J., & Knapp, L. A. (2006). Signal content of red facial coloration in female mandrills (Mandrillus sphinx). *Proceedings of the Royal Society B: Biological Sciences*, *273*, 2395–2400.

Stern, M. K., & Johnson, J. H. (2010). Just noticeable difference. In I. B. Weiner & W. E. Craighead (Eds.), *The Corsini encyclopedia of psychology* (pp. 886–887). Hoboken, NJ: Wiley.

Stevens, M., Párraga, C. A., Cuthill, I. C., Partridge, J. C., & Troscianko, T. S. (2007). Using digital photography to study animal coloration. *Biological Journal of the Linnean Society*, *90*, 211–237.

Stevens, M., Stoddard, M. C., & Higham, J. P. (2009). Studying primate color: Towards visual system-dependent methods. *International Journal of Primatology*, *30*, 893–917.

Stoddard, M. C., Kilner, R. M., & Town, C. (2014). Pattern recognition algorithm reveals how birds evolve individual egg pattern signatures. *Nature Communications*, *5*, 4117.

Thayer, A. (1909). *Concealing coloration in the animal kingdom: An exposition of the laws of disguise through color and pattern: Being a summary of Abbott H. Thayer's discoveries*. New York, NY: MacMillan.

Tinklepaugh, O. L., & Hartman, C. G. (1930). Behavioral aspects of parturition in the monkey (*Macacus rhesus*). *Comparative Psychologist*, *9*, 63–98.

Trivers, R.L (1972) Parental investment and sexual selection, 1871–1971. In B. Campbell (Ed.), *Sexual selection and the descent of man* (pp 136–179). London, England: Heinemann.

Troscianko, J., & Stevens, M. (2015). Image calibration and analysis toolbox: A free software suite for objectively measuring reflectance, colour, and pattern. *Methods in Ecology and Evolution*, *6*, 1320–1331.

Tylor, C. H., Gilbert, F., & Reader, T. (2013). Distance transform: A tool for the study of animal colour patterns. *Methods in Ecology and Evolution*, *4*, 771–781.

Vandenbergh, J. G. (1965). Hormonal basis of sex skin in male rhesus monkeys. *General and Comparative Endocrinology*, *5*, 31–34.

Waitt, C., Gerald, M. S., Little, A. C., & Kraiselburd, E. (2006). Selective attention toward female secondary sexual color in male rhesus macaques. *American Journal of Primatology*, *68*, 738–744.

Waitt, C., Little, A. C., Wolfensohn, S., Honess, P., Brown, A. P., Buchanan-Smith, H. M., & Perrett, D. I. (2003). Evidence from rhesus macaques suggests that male coloration plays a role in female primate mate choice. *Proceedings of the Royal Society B: Biological Sciences*, *270*(Suppl), S144–S146.

Wallace, A. R. (1870). *Contributions to the theory of natural selection*. London, England: Macmillan.

Wallace, A. R. (1889). *Darwinism: An exposition of the theory of natural selection with some of its applications*. London, England: Macmillan.

Wickings, E. J., & Dixson, A. F. (1992). Testicular function, secondary sexual development, and social status in male mandrills (*Mandrillus sphinx*). *Physiology and Behavior*, *52*, 909–916.

Wickler, W. (1967). Socio-sexual signals and their intrasexual specific imitation among primates. In D Morris (Ed.), *Primate ethology* (pp. 69–147). Chicago, IL: Aldine.

Winters, S., Dubuc, C., & Higham, J. P. (2015). Perspectives: The looking time experimental paradigm in studies of animal visual perception and cognition. *Ethology, 121*, 625–640.

Winters, S., Kamilar, J. M., Webster, T. H., Bradley, B. J., & Higham, J. P. (2014). Primate camouflage as seen by felids, raptors, and conspecifics. *American Journal of Physical Anthropology, 153*, 275.

Zinner, D. P., van Schaik, C. P., Nunn, C. L., & Kappeler, P. M. (2004). Sexual selection and exaggerated sexual swellings of female primates. In P. M. Kappeler & C. P. van Schaik (Eds.), *Sexual selection in primates: New and comparative perspectives* (pp. 71–89). Cambridge, England: Cambridge University Press.

Zuckerman, S., & Parkes, A. S. (1939). Observations on secondary sexual characters in monkeys. *Journal of Endocrinology, 1*, 430–439.

Zuckerman, S., van Wagenen, G., & Gardiner, R. H. (1938). The sexual skin of the rhesus monkey. *Proceedings of the Zoological Society of London, 108*, 385–401.

CHAPTER 7 | The New Human Science

Sound, New Evolutionary Theory Gives Us Ultimate Causal Understanding of Human Origins, Behavior, History, Politics, and Economics

JOANNE SOUZA AND PAUL M. BINGHAM

WE HAVE AN AMBITIOUS goal for this short chapter: To argue that the study of all the diverse aspects of humans, from our evolutionary origins to our present behaviors, has recently reached the point where we can generate and test coherent, falsifiable theories with the same rigor traditionally thought of as possible only with the "hard" natural sciences. We are entering an exciting new period.

Specifically, we now possess an apparently complete reductionist, Darwinian theory of human origins, behavior, and history, an approach that has recently come to be called "social coercion theory." This theory proposes that humans rapidly evolved to become a unique animal as the result of an evolutionary change in their social behavior, in turn, creating a social revolution that put strong selective force on the genetics of our ancestors. The essential novelty of the human social revolution was the capacity to cost-effectively coercively enforce extensive cooperation independent of close genetic kinship. This unprecedented coercive capacity resulted from the immediately preceding evolution of the ability to conjointly project coercive threat from a distance (many body diameters away) for the first time in the history of animal life on Earth.

This novel coercive capacity initially arose through the evolution of elite aimed throwing ~2 to 2.5 million years ago; social coercion continues to be essential to human social cooperation, currently in the form of the advanced projectile weapons routinely deployed in law enforcement. Indeed, over the course of our history, dramatic changes in the scale of this novel social adaptation have apparently been driven solely by inventions of new coercive technology. In turn, these increases in scale are apparently responsible for the adaptive revolutions so conspicuous throughout human history (e.g., the Neolithic revolutions and the rise of the modern state).

We should be very clear about what social coercion theory is saying and about what it is not saying. Many (including Darwin) came to the intuitive conclusion that the importance of throwing allowed group selection for cooperative hunting and/or warfare groups to form. However, this view is no longer considered viable (see Pinker, 2012, for more information regarding the deceptive allure of group selection). In contrast, social coercion theory argues that conjoint projected threat must be cost-effective to each self-interested individual, in turn, *allowing* cooperative groups to form beyond close kin.

There is a subtle, yet crucial distinction to be made here that any curricula in evolution could capitalize on as a point of discussion. What began as an individual and kin-selected adaptation (elite throwing), most likely allowing successful kin power-scavenging, then, "inadvertently," permitted cooperative groups to form beyond close kin. This power-scavenging adaptation for throwing was redeployed against conspecifics who would attempt to free-ride on the benefits of the cooperative efforts, in pursuit of individual self-interest. This permitted nonkin potential cheaters on any cooperative enterprise to be either ostracized or "kept in line" with the interests of all cooperators within the group, allowing kinship-independent social cooperation to emerge as a byproduct.

To lay the groundwork for understanding and testing claims of this or any theory, we should understand that good theory unifies, simplifies, and massively empowers a scientific discipline; think of quantum mechanics unifying chemistry or chemistry unifying fundamental biology (the recent "molecular revolution"), for example. Correspondingly, social coercion theory appears to have the analogous power to unify, simplify, and empower human biology, the social sciences, and the humanities, producing what we might call the "new human science." A mature science of ourselves presents special, severe challenges; only by the same self-conscious, rigorous application of the requirements of falsification and parsimony that we have learned in the simpler natural sciences can we hope to be successful (discussed in more detail in the following text).

We will strive to use this short format to briefly introduce social coercion theory to the large, diverse group of scholars (especially emerging scholars) and students who possess the power to implement the scientific revolution apparently enabled by social coercion theory over the next generation (see Bingham & Souza, 2009, for an extensive, detailed treatment).

It is difficult to objectively understand the evolution of human behavior from within the subjective human mind. That mind has evolved to *navigate* our social environment, with no evolutionary requirement to *understand* its ultimate causal origins. Thus, we must attempt to approach this task with as much dispassionate detachment as possible. Aspiring to achieve the perspective of an extraterrestrial anthropologist can sometimes make us sound slightly inhuman. However, this perspective also gives us the best chance to understand clearly what we must grasp if, together, we are to hope to build a humane future for all the world's peoples.

As our research community has sought to build a science of humans, we have also been confronted with a massive blizzard of seemingly complex and diverse empirical phenomena to be explained. To mention a few from among thousands: we communicate at a level other animals do not appear to; we invented the calculus, compose symphonies, and put footprints on the Moon, while no other Terran animal will ever do such things; we build social units (e.g., Paris, the United States, the "global economy") that are many orders of magnitude larger than any nonhuman social unit; we engage in vast economic/political enterprises capable of producing both great wealth and massive misery; we "believe" things, including that some people can have more "power" than others and use that power in particular ways. We will refer here to this vast set of apparently unprecedented empirical properties as the "uniquely human panoply."

We cannot hope to have a science of humans until we have a theory that explains all these things (and their cause/effect connections at a fundamental level) with the same simplicity that modern chemistry explains burning gasoline or the replication of DNA. We will review the argument here that social coercion theory apparently positions us to simply explain the human panoply.

We will progress through the chapter to a description of some of the ways in which both the research program and the curricula of our colleges and universities are expected to be transformed by this new level of insight. We will also attempt to provide a sense of the new capacities to enhance human welfare that we can begin to glimpse from this substantially different perspective. The institutions and each of the individuals within them that seize this opportunity will have the chance to contribute to shaping a newly humane future, we propose.

Before proceeding, we note that some investigators and colleagues may not agree with some of our assertions; however, such initial disagreement is inevitable at the emergence of potentially strong new theory. Indeed, active public disagreement over clearly defined perspectives is a central element of the social process of science. Moreover, the appropriate response by those who disagree with us is for them to produce alternative theory that can compete successfully with social coercion theory by the rigorous standards of the hard sciences (see the following discussion for more detail about the rules of such theory competition). Moreover, the study of these current disagreements (and their competitive empirical evaluation) is a powerful pedagogical opportunity for both students and young investigators.

Social Coercion Theory: The Fundamentals

The first simplifying element of social coercion theory is a now well-established observation from fundamental biology. The social relationships (cooperation or competition) between all animals are essentially always limited (or enabled) by conflicts (or confluences) of interest. Moreover, we understand clearly that the fundamental units of this interest are individual pieces of genetic design information, genes, for short (see Dawkins, 1976; Hamilton, 1996, Williams, 1966, 2008, for classic insights and reviews; also see Dugatkin, 1997, for a review of animal social behavior from this perspective, known as kin-selection theory). The implications of kin-selection theory are widely considered to be quite clear (and very well supported empirically) as follows. Animals evolve to behave socially as if they have confluent interests only with very close kin. Correspondingly, they are predicted (and observed) to behave toward nonkin conspecifics with patterns usually ranging from indifference to overt hostility (sometimes including violence) with only rare, special exceptions.

Even the rare cases of nonkin cooperation in nonhuman animals most commonly reflect unusual conditions in which even these nonkin individuals do not have conflicting interests (e.g., sexual mating in some cases), a pattern often referred to as "byproduct mutualism."

As a result of this large body of work, we can state that *nonkin* animals have pervasive conflicts of interest that preclude the evolution of cooperation between them under almost all circumstances. Technically, gene copies ("alleles") build animal minds to behave *as if* they have conflicts of interest with conspecifics with whom they do not share very recent common ancestry (close kin), with only a few, narrow exceptions, as previously noted. These same alleles also build animal minds to behave *as if* they have partially confluent interests with animals with whom they share very recent common ancestry. Specific alleles programming such behavior patterns leave more copies of themselves over time (through more successful replication) than alleles engendering alternative strategies. We use the widely deployed verbal simplification of saying animals evolve to have conflicts of interest with nonkin conspecifics. We will focus here on "public" or "kinship-independent" social cooperation and, thus, will refer to the key insight here as the "universal conflict of interest problem."[1]

The second simplifying element of social coercion theory is the proposal that all the rich diversity of uniquely human properties, the human panoply, emerges entirely from our unprecedented evolution of the capacity to control or manage this universal conflict of interest problem. Acquisition of this single novel capability, in turn, has created/allowed myriad unprecedented adaptations and consequences that ultimately produce the uniquely human panoply. As we will review in the following text, the capacity of this theory to predict the extensive empirical evidence is remarkable.

The third simplifying element of social coercion theory is the proposal that the unprecedented human solution to the conflict of interest problem is itself strikingly simple. Specifically, our last (most recent) *pre*human ancestor evolved the capacity to project credible coercive threat (to kill or credibly threaten death/serious injury) from a substantial distance, "remotely" (from many body diameters away). This capacity happened to result from the evolution of elite aimed throwing[2] (probably originally a component of an individually and kin-selected power-scavenging adaptation first arising

---

[1] Notice especially that cooperation between close kin conspecifics does *not* reflect management of *conflicts* of interest (our focus here). Rather, it reflects the *confluent* interests of genes they share by virtue of recent common ancestry (reviewed in Dawkins, 1976).

[2] First, humans have a phenomenal evolved capacity to throw with force and accuracy; this capacity becomes an actual elite skill any time we grow up in a cultural environment where elite throwing is a routine adaptive activity (see Isaac, 1987, and the following text).

Second, being a target of expertly thrown stones is a serious risk/cost to humans. We are vulnerable to immediate death from compression fractures of the cranium or rib cage produced by thrown stones. Moreover, disabling injuries to the jaw, shoulders, hips, and extremities would likely often have been substantially fitness-reducing (or even effectively fatal) in the highly peripatetic ancestral environment (Bramble & Lieberman, 2004; see following discussion), especially in the absence of professional "medical care."

Third, humans are very likely to be highly adapted to recognition of the implications (including threat) of a thrown projectile. While anecdotal evidence exists (e.g., the apparent uniqueness of the human capacity to track a high-speed projectile well enough to strike it firmly with a bat), it will be important to investigate this issue continuously and systematically.

> # BOX 7.1. THE GAME THEORY OF SOCIAL COERCION
>
> Fundamental statement of social coercion theory's casual description of the origin of kinship-independent cooperative behaviors (reviewed in detail in Bingham & Souza, 2009).
>
> *Fundamental inequality of kinship-independent social cooperation (where b is the individual benefit from a cooperative act), c_{coerc} is the sum of the individual costs of projection of the coercive threat necessary to render free-riding a maladaptive choice, and c_{coop} is the sum of all the noncoercive costs of cooperation, such as energy expended and opportunity costs:*
>
> $$b > c_{coop} + c_{coerc}$$
>
> *Essential features of the C_{coerc} term:*
> *First, costs of coercion can only become sufficiently low to allow this inequality to be fulfilled broadly, commonly (rather than rarely, idiosyncratically) when large sets of mutually self-interested individuals can synchronously project roughly the same average amount of individual coercion threat per unit time as can a free-riding target.*
>
> *Second, means of coercion must operate in a fashion such that contributing to conjoint social coercion (rather that free-riding on coercive acts of others; being a "second-order free rider") is the immediately, individually adaptive choice (see text for details).*
>
> *The only methods of coercion currently known that fulfill both these requirements involve the projection of decisive threat from a substantial distance (many body diameters away), which is to say, use of projectile weaponry.*

roughly 2 to 2.5 million years ago; see following discussion), inadvertently producing a social revolutionary (see following discussion). See Box 7.1 for a summary of fundamental social coercion theory.

More specifically, in an animal that can project threat from a distance (remotely), the costs of conjoint coercion to each individual in a set of cooperators drops by the square of the factor of their numerical advantage over a socially parasitic individual, a "free rider". (Thus, individual costs of coercion decrease as a function of the number of cooperators much more rapidly than do returns on their cooperation; Figure 7.1.) This effect arises because all the coercive cooperators can project threat simultaneously, rapidly incapacitating and eliminating return fire from an individual free rider. In contrast, nonhuman animals fight and kill "proximally" with tooth and claw, making the individual risk of being harmed or killed very high. (Specifically, costs of coercion decrease as a linear function of numerical advantage, the same factor by which individual

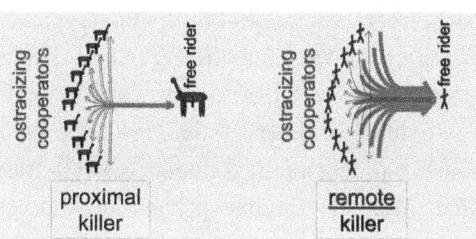

FIGURE 7.1. The Game Theory of Social Coercion. Illustrated is the fundamentally different coercive position of ostracizing cooperators, proximal and remote killers. *Proximal killers* can only engage with a free rider roughly one at a time. Thus, the amount of coercive threat a would-be free rider experiences per unit time, when there are 10 ostracizing cooperators, is roughly 10 times more than each cooperator experiences (on average). However, the free rider's share of a social asset (e.g., a cooperative hunting kill) is also 10 times greater than the return to each ostracizing cooperator if the would-be free rider is successfully repelled. Indeed, when the free rider is stronger than average (as here; as one half of animals will be) the costs to the individual cooperator of imposing ostracism on the free rider will exceed the returns from this ostracism to each cooperator. *Remote killers*, in contrast, can project far more coercive threat per unit time (10 times more in this case) on a would-be free rider. This shortens the time required to impose an unacceptable cost on the would-be free rider by this same 10-fold. Thus, each ostracizing cooperator obtains the same benefit as above, but with 1/10th the cost. With this reduced cost of ostracism, the original social act (e.g., a cooperative hunt) can often become individually adaptive, even when low genetic kinship eliminates inclusive fitness effects (see text and Box 7.1 for additional details).

benefit from ostracism of free riders simultaneously decreases in proximal killers.) Such a cost–benefit ratio generally produces costs too high to pay and is especially prohibitive if the target free rider animal is larger/stronger/more adept.

The effect of remote killing capacity on social coercion is *socially* revolutionary; it makes "law enforcement" (coercive management of the universal conflict of interest problem) a Darwinian adaptation for the first time in the history of large animals on Earth.[3] As we will see in the following discussion, we are the products of ~2 million years of evolution to exploit this revolutionary new *social* adaptation.

---

[3] Note that nonhuman animals can add very small amounts of individually self-interested social coercion to produce very slightly expanded scales (by human standards) of social cooperation, beyond what would be produced by kin-selection or byproduct mutualism, as expected on social coercion theory. See, for example, worker policing in social hymenoptera (reviewed in Wenseleers & Ratnieks, 2006). This nonhuman social scale remains small because it does *not* depend on remote projection of threat among adults under diverse circumstances, but, rather, inexpensive coercion of defenseless larvae by adults under one special circumstance.

Note that we still, today, engage in this species-typical behavior. Think of the gunpowder projectile weapons (killing from a distance) in the hands of sets of contemporary law enforcement professionals, for example. Moreover, there is extensive observational evidence that our contemporary economic (kinship-independent cooperative) systems rapidly collapse under massive free-riding if effective projectile-supported law enforcement is absent.

Because social coercion appears to be so fundamentally central to the uniquely human condition, we take the next five paragraphs to provide a deeper grasp of the salient quantitative elements of the theory through a simple thought example (see Bingham, 1999, for the first draft of social coercion theory; Okada & Bingham, 2008, for the definitive game theory; and Bingham & Souza, 2009, for an extensive review of theory, evidence and implications; more recently others have begun to converge on related views of the evolution of human cooperation; Boyd, Gintis, & Bowles, 2010).

Consider 10 ancestral cooperating individuals confronted by a single would-be free rider. This free rider aspires to capture a portion of the proceeds of the cooperators' efforts to which he or she did not contribute (e.g., portion of an antelope scavenged by chasing a cheetah from its kill). If the cooperators can ostracize this aspirational free rider by projection of credible threat, each gains an individual adaptive advantage (extra food); however, the costs of this ostracism to each of the cooperating individuals must not outweigh the benefits of the extra food. The advantage of the purloined food to the would-be free rider is high; he or she keeps all he or she can get, while each of the cooperators retains only ~1/10th of that amount if the free rider is successfully ostracized. The free rider is, thus, willing to incur significant individual cost to capture his or her booty—interests in conflict.

However, if all 10 ostracizing cooperators project threat *from a distance* at the would-be free rider at once, he or she experiences "10 units" of cost/risk/incoming fire per unit time. He or she returns "1 unit" of fire per unit of time, which is distributed (on average, over time) among the 10 ostracizers, who each experiences 0.1 unit of cost/risk per unit time (Figure 7.1). Thus, each cooperator can get ~1/10th as much benefit from coercive ostracism (e.g., extra food) as the would-be free rider would receive from successful parasitism, but with only 1/100th as much exposure to coercive cost (10/0.1 units of coercive risk per unit time).[4] Thus,

---

[4] It can be helpful to visualize the quantitative game-theoretic issues from a slightly different perspective. Before abandoning attempted parasitism, the free rider will be willing to absorb very slightly less coercive risk/cost than would correspond to his or her benefit from successful parasitism. Normalize this cost/risk to "1 unit." In the time required for the free rider to absorb this

coercive ostracism can be a very adaptive behavior *when the ostracizing animals in question can project threat remotely and, thus, synchronously (simultaneously)* (Box 7.1). (See following discussion for one other essential implication of remote threat capability.)

Contrast the previous paragraph with the situation for a nonhuman, *proximally* killing animal. In this case, roughly only one ostracizer can engage the target at any moment in time. Assuming that the 10 ostracizers can be relied on to share the burden of ostracism by randomly rotating into and out of attack position (in actuality, they cannot; see following discussion), the target experiences 1 unit of threat per unit time and each attacking ostracizer 0.1 unit of threat per unit time, on average. Thus, the relative costs of exposure to threat (0.1/1) and the benefit (0.1/1) for the individual ostracizers/free rider are the same (rather than 10-fold lower for the cooperating ostracizers, as for the previously mentioned remote killers). Thus, ostracism is a negligible adaptive advantage for these animals. *Indeed, any time the would-be free rider is a little bigger and stronger (as will be true one half the time), ostracism is actively maladaptive.* In contrast, the enormous relative advantage of numbers for the previously discussed remote killer generally swamps any differences in individual fighting ability.

There is one last detail essential to understanding why something as seemingly mundane as conjoint remote projection of threat (as with elite aimed throwing or modern handguns) actually has revolutionary social implications. Specifically, sharing costs of coercion with nonkin in an animal that can only project threat proximally (up close) presents its own intractable "second-order" free-rider problem; the optimal strategy for each actor would be to transfer these costs of ostracism to other nonkin individuals. As this attempted transfer of cost cannot be prevented (without incurring still more, untenable, new coercive costs), such proximal conjoint coercion is only very rarely, idiosyncratically adaptive in nonhuman animals.

In contrast, under many common circumstances, in a remote killing animal, refraining from contributing to individually adaptive conjoint

---

1 total unit of cost from the 10 ostracizing cooperators, he/she and each of the cooperators will have projected 0.1 unit of coercive costs on their various targets. While all the 0.1 units of cost projected by cooperators are focused on the free rider (totaling 1 unit), the 0.1 total unit of coercive cost projected by the free rider in this time is distributed among the 10 cooperators. Thus, each cooperator receives 0.1 unit of benefit (her or his share of what the free rider leaves behind when abandoning attempted parasitism) for the cost of exposure to only 0.01 unit of coercive cost (his or her share of the threat the free rider is willing to project).

coercive threat is costlier than contributing to ostracism. Specifically, failing to contribute to projection of threat extends the conflict (increasing exposure to return fire from the free rider) to any cooperator in the vicinity who refrains from throwing. Thus, in remote killing humans and proto-humans, mutually self-interested, conjoint social coercive management of nonkin conflicts of interest is not only cost-effective (see previous discussion), it is also the stable game theoretic strategy (Okada & Bingham, 2008). Such coercive management of conflicts of interest will evolve to become a broadly pervasive behavior, revolutionizing the remote killing animal's social adaptation and sweeping the population as a strongly selected adaptive behavior.

Focusing on individual cost–benefit, social coercion theory is an "individual selection" theory; no "group selection" is involved. (Group selection refers to models where individually "altruistic" traits that favor the fitness of the group can evolve because of their return effects on the altruistic actor. However, this approach ignores the best individual strategy of all, free-riding on the altruism of others in the group.) To our knowledge, there is no compelling theoretical or empirical reason to invoke group selection in accounting for the human panoply (including cultural group selection as recently invoked by Richerson et al., 2016).

Similarly, there remain some scholars looking to explain some elements of human social behavior as "altruistic" (Wilson, 2015). Social coercion theory explains the intuitive appearance of altruism as misleading. What looks like altruistic behavior is actually (almost always) immediately, individually self-interested in the context of coalitions insisting on such behavior from their members under intense coercive threat. Our proximate subjective experience of altruistic feelings is not evidence for actual altruism in the ultimate evolutionary logic of our behavior, of course.

In summary, the evolution of the capacity for synchronous projection of threat (e.g., elite aimed throwing in the specific case of the immediate ancestors of the first humans) will produce a revolutionary new kind of animal in a single specific way. This novel animal's social behavior will have the unprecedented feature of extensive, pervasive cooperation between nonkin (despite the universal conflict of interest problem). This will involve those diverse contexts in which coercive ostracism of would-be free riders is likely to be immediately, individually adaptive in a remote killing animal.

The intense coercive power of remotely threatening human coalitions means that we impose very strong selection on one another's social behavior. We are, thus, the products of over 2 million years of very strong "self-domestication." This self-domestication results in changing the logic of self-interest from purely selfish behavior to mutually self-interested cooperation with nonkin as the best available option. Thus, the empirical observation that human uniqueness evolved rapidly[5] is predicted by social coercion theory. More specifically, the extensive genetic redesign of our lineage (e.g., producing enlarged, ethical, speaking brains) is an *effect* of adaptation to this strongly selective new social environment and the adaptive opportunities is affords, not its ultimate cause. Thereby, social coercion theory clarifies the long intellectual struggle and deep confusion around the ultimate causes of human origins and behavior (see Scott-Philips, Dickins, & West, 2011, for review of this struggle).

Before reviewing some details of the extensive empirical support for social coercion theory, it is useful to briefly review the fundamental social process that apparently underlies all science and scientific progress.

The Fundamental Rules of All Successful Science

Historically, scientific knowledge has progressed by following specific operational rules (see Kuhn, 1962; Popper, 1959, for two influential perspectives on these long-standing practices). As the new human science is now maturing, these same rules can and must apply. Most fundamentally, these are two. First is theory "falsification" (not confirmation). This means that we are required to formulate our theories in ways that make many, diverse predictions of such precision and detail that they are readily testable ("falsifiable") on practically obtainable and diverse empirical evidence. A theory only begins to be accepted as it survives multiple, robust rounds of such potential falsification.

It is well known that if we fail to subject our theoretical work to this requirement of falsifiability, fundamentally wrong "stories" that "sound or feel right" can become accepted based on their intuitive attractiveness. This danger is particularly severe for the human sciences because, again, our minds have evolved to *navigate* our unique social environment, *not to understand* its origin or ultimate cause. When we are focused on human

---

[5] Recall that the chimp-human divergence occurred only ~6 million years ago. Nonetheless, chimps remain much more like the other rainforest apes, gorillas (~12 million years divergent), and orangutans (~16 million years divergent), than like the human lineage, rocketing off in its own novel direction.

social behavior, we must be particularly careful that our proximate social minds are not actively misleading us. Insistence on seeking falsification, rather than seeking confirmatory evidence (with its vast potential for selection bias) is vital.

The second element of the process of theory testing is "competitive parsimony." We may initially have multiple, mutually exclusive theories that, nonetheless, all survive falsification against some (generally narrow or incomplete) set of empirical data. To decide between such theories, we next ask them to explain ever broader bodies of evidence, in other words, to generalize beyond a very specific case they were originally designed to explain. They should continue to produce additional falsifiable hypotheses as they generalize in scope. The theory that both explains the broadest body of data most simply (i.e., most general and parsimonious) and that continues to survive falsification as its explanatory domain is enlarged is most likely to be correct. Notice that a theory that cannot be simply, economically generalized beyond its original narrow domain loses this competition, by default.

As we will summarize in the following text, social coercion theory is—apparently, by far—the most likely theory of human uniqueness we currently possess on these two criteria. With these two standards of theory evaluation in mind, we now proceed to briefly review a set of examples illustrating social coercion theory's causal picture as an apparently simple, unifying explanation of the entire uniquely human panoply. While we begin in our evolutionary past, we will continue with examples throughout human history to the present moment. This brief tour is designed to illustrate the broad range of falsifiable predictions generated by the theory, as well as its diverse implications for the future of the academic enterprise and for the welfare of the emerging global human coalition (see Bingham & Souza, 2009, for a detailed discussion of these theoretical and empirical issues).

Explaining the Fossil Record of Human Origins

Social coercion theory makes simple, clear, falsifiable predictions about human origins and the resulting fossil record. Specifically, at least one local population of prehuman bipedal australopiths should evolve some unprecedented capacity to project conjoint coercive threat synchronously, most likely remotely (see Box 7.1, Figure 7.1, and previous discussion). Given the fossil record and the ultimate capabilities we have inherited, evolution of elite aimed throwing is almost certain to have been this initial projectile technology. Moreover, to avoid the fallacy of evolutionary foresight, this capacity must have initially evolved under selection for inclusive fitness improvement (kin and individual selection).

It currently appears that the most likely path to this novelty is adaptation to a new local power-scavenging niche opened by climate fluctuation in Africa between 2 and 3 million years ago. These climate changes apparently stochastically created small, isolated savannah domains possessing australopiths but having limited scavenger competition (Dominguez-Rodrigo et al., 2010; Vrba, 1995). For example, one isolated region of savanna might have lost its dominant power-scavengers, like lions and hyenas, while retaining light-weight solitary hunters, like cheetahs and leopards. Under these conditions a bipedal ape (with neither carnivore teeth nor hunter's claws) might plausibly evolve elite throwing to threaten the solitary hunters and sharp flake and smashing stone technology to process scavenged prey. (For more detailed discussion of the empirical evidence for the power-scavenging hypothesis, see Bingham & Souza, 2009, Part II, Chapter 7.)

After the evolution of elite throwing in the context of kin selected power-scavenging, natural selection is predicted to rapidly "discover" kinship-independent social cooperation, as these animals now redeploy this adaption to prevent the stealing of the individual benefits of their cooperative efforts by other conspecifics. Of course, this novel social adaptation is driven by individually adaptive coercion (Box 7.1; also see previous discussion), an adaptive opportunity opened to an animal that can, thereby, cost-effectively manage the universal conflict of interest problem. This evolutionary process results in an unprecedented scale of social cooperation, well beyond what the processes of kin-selection, adaptation to byproduct mutualism, and other small-scale effects (Dugatkin, 1997) can produce in nonhuman animals.

Thus, the first humans are predicted to emerge rapidly from these proto-human elite throwers, and expanded social cooperation should be their first novel property. Paleoarchaeological and paleoanthropological evidence allows several usefully detailed tests of these predictions. The earliest humans are expected to begin evolving brain enlargement as an adaptation to expanded social cooperation (for more details, see the following text); brain expansion is a rapidly ensuing reporter for the enlarged, uniquely human social scale. As predicted, the earliest members of *Homo* to show substantial brain expansion were already elite throwers, based on their postcranial anatomy (Bingham & Souza, 2009, Chapter 7). More useful still are the transitional fossils, the very earliest members of *Homo* as follows.

As we originally noted (Bingham & Souza, 2009), the earliest known members of *Homo* to leave Africa display skeletal and lithic evidence of elite throwing, while some members still show australopith brain sizes, indicating that elite throwing evolved immediately before a new scale of social cooperation, as required by social coercion theory (the Dmanisi erectus individuals; Ferring, et al., 2011; Lordkipanidze et al., 2007, 2013). Moreover, one of the Dmanisi skulls was an individual who had lived for at least several years with only one tooth, a potentially suggestive detail. Such toothless individuals would likely require expanded social support (e.g., privileged access to soft, rich foods from nonkin).

More recently, the dramatic discovery of a large fossil population of *Homo naledi* in South Africa (Berger et al., 2015) is also potentially quite useful for theory testing. A large sample of naledi individuals show a mixture of ancestral australopith and derived (novel) *Homo* characteristics. The *Homo* characteristics include very human-like hands and feet. While hand changes might be explained equally well by other theories (elite tool use; see, e.g., Kivell et al., 2015; Napier & Tuttle, 1993), the coordinated evolution of the feet is not predicted by these competing theories.

In contrast, properly designed feet form the platform for elite throwing and properly designed hands are critical to violent, yet accurate projectile release; thus, their concordant redesign is readily accounted for by social coercion theory. *Homo naledi* individuals retain small, australopith-sized brains, again consistent with the essential, throwing-first prediction of social coercion theory. Finally, the large naledi fossil trove very strongly suggests deliberate burial of many individuals over time (deep in a cave; Berger et al., 2015). Such behavior is not seen in other primates and suggests human-like social behavior in these animals, in turn, consistent with the predictions of social coercion theory.

Remarkably, the naledi fossils are apparently less than 400,000 years old (Dirks, et al., 2017). The simplest interpretation of these data on social coercion theory is that the naledi lineage diverged from our lineage very early (at the point of early evolution of elite throwing) and was under selection to retain some australopith properties (shoulder anatomy and smaller brains). Such selection could result from commitment to an arboreal adaptation requiring light weight and elite tree climbing, for example. This interpretation predicts that the naledi lineage will prove to have diverged from our lineage around 2.2-5 million years ago, a prediction that might become testable by fossil DNA sequencing, for example.

In summary, social coercion theory predicts that elite aimed throwing must evolve first, driving subsequent enlargement of social cooperation (eventually scorable by brain size expansion in our lineage; see following discussion). The existence of at least three ostensibly very early human/proto-human lineages[6] sharing the foot and hand adaptations to elite throwing indicates that social coercion theory survives this very strong opportunity for falsification. Indeed, social coercion theory apparently accounts for the current fossil evidence of human origins better than any older theories. We invite our paleoanthropological and paleoarchaeological colleagues and emerging young student scholars to attempt to falsify this strong claim and/or to produce even better (more predictive) theory.[7]

A Coercive Manager of Nonkin Conflicts of Interest Will Inevitably Become a Talking, Cultural, Cognitively Gifted, Sexually Flexible Animal

Having established the fundamentals of humans as the first animal to be able to adaptively manage the universal conflict of interest problem, our next step is to understand how this animal will adapt to this profoundly new niche. We have an extensive grasp of how this approach predicts our biological properties (Bingham & Souza, 2009). We briefly review key insights from this work to illustrate, in parallel, the uniquely broad predictive power of social coercion theory (competitive parsimony) and its

---

[6] Despite other anatomical differences, Australopithecus sediba (Berger, et al, 2010), Homo naledi (Berger, et al, 2015) and the Dmanisi Homo individuals (Lordkipanidze et al., 2007) all share human-like, ostensibly elite-throwing hands and feet, in most cases without brain expansion. This strongly suggests an adaptive radiation rapidly following evolution of elite throwing, one or several of whose lineages probably contributed to modern humans. Such an adaptive radiation is fully consistent with the predictions of social coercion theory.

[7] Rick Potts and colleagues have proposed that the well-known climate instability in Africa 2 to 3 million years ago might be sufficient to *directly* select for human uniqueness (Anton, Potts, & Aiello, 2014). We note that this theory is quite unattractive on theoretical grounds; it provides no answer to how humans uniquely came to control the universal conflict of interest problem, for example. It is also unattractive empirically; it provides no answers as to why an australopith lineage was radically transformed, while other contemporary African lineages (savanna baboons, lions, and many others) were not. In contrast, social coercion theory posits that climate instability had an inadvertently lineage-specific effect by creating the power-scavenging niche to which only a bipedal primate (australopiths) could respond by the evolution of elite throwing (see previous discussion).

ability to account in detail for uniquely human properties (survival of robust potential falsification).

It is fruitful to begin with the question of cultural information. Humans are clearly genetically designed to be *cultural* animals. This feature is predicted quite directly and simply on social coercion theory. We expect transmission of potentially useful information to be limited by the universal conflict of interest problem. Most simply put, it is generally individually adaptive to actively mislead nonkin conspecifics. This theoretical prediction is remarkably well fulfilled when we look at nonhuman animal communication. Capacities to transmit useful cultural know-how (see Terkel, 1996, for a particularly clear, simple example) and valuable perishable intelligence (as in the waggle dance of bees, reviewed in Dyer, 2002) evolve as if they are precisely limited by the universal conflict of interest problem. These informational assets are almost exclusively exchanged between close kin nonhuman individuals in the wild, for example. Moreover, exchange is extended beyond kin-selected cases only to the tiny extent (by human standards) to which residual conflicts of interest can be coercively managed (in social hymenoptera; Wenseleers & Ratnieks, 2006) or when such exchange allows individually self-interested byproduct mutualism (Dugatkin, 1997).

Thus, the human capacity to manage the conflict of interest problem associated with cultural information exchange (also called the "hostile manipulation problem" in this context) on a vast new scale is predicted to correspondingly expand our adaptation to this enlarged contingent[8] information exchange. Our novel communicative capacities offer opportunities to falsify this prediction. As predicted, for example, we are not merely the cultural animal; we are also the *pedagogical* animal. The teaching of nonkin by most of the authors of this book (and tens of millions of "teachers" worldwide) is precisely what social coercion theory would predict in humans.

Of course, as exchange of reliable information (coercively policed) rapidly enlarged from our origins around 2 million years ago, we would predict a massive evolutionary refinement and expansion of the information exchange capabilities inherited from our nonhuman ancestors. Uniquely human communicative abilities[9] apparently conform to this expectation remarkably well. Moreover, the combination of our genetic adaptation to

---

[8] Contingent information is information that is useful, if true, but potentially dangerous, if false.

[9] These include speech, but also elite signing and virtuosic teaching/learning by demonstration, along with enlargement of communication-related brain regions, like Broca's area.

productive use of larger amounts of reliable cultural information and the massively expanded access to such information our large-scale, coercively policed cooperation gives us is apparently sufficient to account for our unique individual problem-solving abilities.

Finally, it has become clear that human sexual/reproductive behavior has unique features. On the one hand, we are the planet's first large-scale, kinship-independent social breeders ("It takes a village to raise a human child"). This human village includes both males and females that are unrelated compared to the kin-selected matrilocal or patrilocal nonhuman animal social units that evolve where nonkin conflicts of interests cannot be managed. This unique village of kinship-independent social breeders and its effects on our life history redesign clearly dates from early in our 2 million year history as follows. For example, elevated social support and ecological dominance are required to permit the massive redesign of human life history allowing, in turn, the evolution of our dramatically larger brains (reviewed in detail in Bingham & Souza, 2009, Chapter 6). Our 2.5-fold larger full-term fetus and its late-term mother require substantially more social support than the corresponding stages in the other apes. In addition, our "babies," continuing to grow their brains at the very rapid fetal rate, are behaviorally dependent on 24/7 care for the first year of life; again, expanded social support is required. Lastly, our "children" were weaned at around three years of age in the ancestral environment but continue to grow their brains for an additional three years (unlike any postweaning nonhuman animal). This childhood can only be adaptive with the ecologically dominant control of choice foods (to replace mother's milk) provided by the uniquely human social scale. This profound life history redesign culminates in adult brain sizes ~3.5 times larger than in chimps and our most recent prehuman ancestors.[10]

On the other hand, humans are remarkably contingent maters. Under conditions of low adult/parental mortality risks, most humans are born from "monogamous" pairings (genetic fathers are known with relatively high confidence). Contemporary democratized states are examples. In contrast, when adult mortality risks are high, humans mate promiscuously (paternity is highly uncertain; see Beckerman & Valentine, 2002, for diverse examples of this mating strategy in the ethnographic record). Through promiscuous mating both sexes purchase a kind of life insurance against

---

[10] Our metabolically expensive large brains are presumably an adaptation to storing, monitoring, and using our vastly expanded repertoire of culturally transmitted information, again, a product of coercive control of conflicts of interest (see previous discussion).

premature parental demise (discussed in detail in Bingham & Souza, 2009, Chapter 8). What is relevant here is that *both* these mating systems are vulnerable to free-riding in dense human social communities: rape, prostitution, and "cheating" in monogamous systems and mate guarding in promiscuous ones, for example. Only in an animal that coercively manages such conflicts of interest can this strategically contingent mating adaptation arise.

The rich complexity and ambivalence of our evolved sexual psychology is predicted to emerge.

A Coercive Manager of Nonkin Conflicts of Interest Will Inevitably Become the Scientific Animal, Exploiting Social Doubt with Nonkin Cooperators

Social coercion theory predicts that we should be the cultural and pedagogical animal, as noted. Equally predictable on social coercion theory is our status as the "scientific" animal. The scientific method is often presented somewhat artificially, as something executed by isolated expert individuals. In fact, and in contrast, scientific practice is a fundamentally public, social process. Indeed, it is well accounted for as merely one more manifestation of our ~2 million years old, coercively managed, kinship-independent cooperative adaptation.

Specifically, the universal conflict of interest problem requires that we subject any cultural information provided by nonkin to skeptical assessment (not necessarily consciously). Moreover, the most efficient way to do this is through applying "social doubt," where multiple individuals who are equally suspicious (again, not necessarily consciously) of any information offered then attempt to falsify that information. Their mutually coercively policed social doubting (including evaluation of empirical tests) of proffered insights is the only real tool we have for gaining objective new insight into the world. The contemporary formal scientific method we previously discussed merely reflects the ancient rules whereby we evolved to subject any and all *potential* information to such conjoint social doubt.

Specifically, the theory falsification and subjection to competitive parsimony that are so well established in contemporary professional practice are merely the operational details of how we engage in social doubt. As we "publish" in "peer-reviewed" and "peer-read" sources, we are feeding into and submitting to this process of social doubt.

As we will discuss, the detailed history of the recent Scientific Revolution is quite consistent with this view. The Scientific Revolution over the last 400 years merely reflects the application of this ancient process of public social doubt on the massive new cooperative scale of the coercively managed democratized state.

It is also interesting (and predictable) to note that our cognitive systems are designed to learn most deeply and effectively through subjecting what we are "taught" to our own skepticism and to social doubt (reviewed in Koenig & Echols, 2003; Lake, 2012; Mills, 2013; Souza & Bingham, 2018).[11]

A Coercive Manager of Nonkin Conflicts of Interest Will Undergo Predictable Adaptive Revolutions Produced by Expanding the Scale of Our Ancient Social Adaptation

To understand the relatively recent periods of social change in the human lineage that we think of as the "historical era," we must return to the central predictions of social coercion theory. First, the scale of human cooperation is predicted to be limited/determined by the scale on which we can *cost-effectively* (i.e., individually adaptively) manage the universal conflict of interest problem. This adaptive constraint means that human social units supported by ancient, ancestral elite throwing are expected to have a sharply limited maximal scale, the scale on which throwing was a cost-effective policing behavior.

We can estimate this scale with some precision by examining the effective range of contemporary human throwing. While healthy, skilled, young adult humans can chuck a stone as far as, perhaps, 100 meters, our capacity to put a thrown stone reliably on a small target is limited to shorter distances. The statistics of American baseball indicate that this limit is of the order of the 20 to 40 meters defining the "infield" of this game.[12] If we assume that an elite throwing adult human needs 1 to 2 meters of throwing room when encircling a free rider, this distance allows somewhere between ~30 and ~120 individuals to throw at a single target (or a small group of "organized criminals") simultaneously. Thus, we expect the ancestral human social unit to have had a core of adults

---

[11] There are important pedagogical insights here for contemporary teaching (Souza & Bingham, manuscripts in preparation).

[12] Note that ethnographic evidence indicates that humans who routinely use thrown stones in their daily social adaptations throw with skill approaching that of professional athletes (Isaac, 1987).

in this size range.[13] Ethnography of the smallest prestate societies is broadly consistent with this size range prediction (Scupin, 2012; Wenke & Olszewski, 2007).

Second, thus, social coercion theory predicts that humans should be able to substantially expand the scale of their social cooperation *if and only if* they acquire new technologies rendering policing of conflicts of interest adaptive on a new scale. Generally, this is predicted to involve projectile weapons of greater range (though other approaches are possible; see following discussion).

Third, many approaches, including social coercion theory, predict that human adaptive sophistication (ecological dominance, technological prowess) will be strictly limited by the scale of our social cooperation. This scale determines the amount of cultural information we can develop, store, and deploy with expert skill; that is, the extent and scope of individual specialization. We might imagine that the evolution of our genetically built individual cognitive capabilities is also important here; no doubt they were at some point in the very ancient past. However, our increasing understanding of our evolutionary and social history makes it quite clear that such genetic evolution of individual cognitive capacity has played very little, if any, role in the massive social changes of the last ~15,000 years of our history (and probably in the last 500,000 years thereof).

Specifically, human historical change beginning with the many global Neolithic revolutions from ~15,000 to 1,500 years ago (depending on location) occur very rapidly (over a few tens of generations) and encompass far too diverse populations (e.g., peoples of entire continents) to have been fundamentally genetic events (see Bingham & Souza, 2009, for a review; see Bingham, Souza, & Blitz, 2013, and other papers in the same issue for discussion of the particularly illuminating example of the prehistoric North American Neolithic). The Neolithic Revolutions and subsequent cases, like the rise of the first states or the modern economic miracle (reviewed in Bingham & Souza, 2009), are, likewise, too fast and broad to be genetic events. Again, these transitions were social, not genetic, processes. By extrapolation it is also quite likely that the behaviorally modern human revolution beginning roughly 50,000 years was also a social revolution (reviewed in Bingham & Souza, 2009; see Mellars, 2005,

---

[13] When social units become substantially larger than this, the logic of self-interested participation in social coercion deteriorates. Moreover, social coercion of now-larger free-riding organized criminal subunits approaches highly expensive "one-on-one" conflict (e.g., think of 500 individuals against 200 free riders). Thus, these oversized units are predicted to fission into internally policeable, externally co-hostile coalitions.

for the conundra confronting older theories here, apparently resolved by social coercion theory).

The archaeological and written historical evidence strongly supports the prediction that all the human adaptive revolutions correlate precisely in space and time with increases in the scale of human social cooperation. For example, the behaviorally modern human revolution, with its new adaptive sophistication, correlates with the emergence of the first large, cooperatively built "ceremonial" architecture, the Neolithic revolutions with the first large permanent settlements, and the rise of the first states with the knitting together of dispersed Neolithic settlements and the formation of the first large cities (reviewed in Trigger, 2003; Wenke & Olszewski, 2007).

Fourth, social coercion theory predicts that such formidable increases in social scale can only be driven by development/acquisition of coercive weapons of fundamentally new capabilities. There is robust evidence consistent with each instantiation of this robust prediction. This extensive evidence includes the coincidence of the invention of atlatl/spear-thrower with rise of behaviorally modern humans; of the local invention/acquisition of the bow and arrow with the many Eurasian and North American Neolithic revolutions[14]; of local development of advanced, professional body armor and shock weapons with the first states (see following discussion); of local acquisition of gunpowder artillery with stabilization of cycling archaic state to producing the early modern state; of local deployment of gunpowder handguns with the partial redemocratization of states (see following discussion); and of development of aircraft/missile technology of intercontinental range with the incipient consolidation of a pan-global economy (all these cases are reviewed in detail in Bingham & Souza, 2009, Chapters 11–17).

Notice especially that all the diverse events taken as *causes* in traditional theories of history (plant domestication, invention of written language, rise of religions, printing press invention, etc.) are predicted by social coercion theory to be, instead, merely *effects* of a single, universal cause of historical transitions, increased scale of cost-effective coercive threat, in turn, driving expansion of social scale.

---

[14] The various Neolithic revolutions consist of many independent chances to falsify social coercion theory. As predicted, each such revolution is preceded *immediately* by the local acquisition of the bow as this novel weapon (but not local domesticates, cultures, or religions) spreads from the Middle East across Eurasia and eventually into North America over ~12,000 years (reviewed in detail in Bingham and Souza, 2009, Chapter 12; also see Bingham, Souza, & Blitz, 2013 and other papers in the same issue).

Moreover, each of the competing older theories of history is hyperlocal; for example, the invention of the printing press *could*, in principle, have something to do with rise of the modern state but not with the rise of behaviorally modern humans or the Neolithic revolutions. In contrast, social coercion theory predicts every one of these historical transitions as a recurrence (at a new, larger scale) of precisely the same casual process (local invention of an improved coercive technology). Social coercion theory is superior to all these older theories on the criterion of competitive parsimony.

Thus, a clear, simple, highly predictive theory of history emerges from social coercion theory. No other theories of history of which we are aware have such broad predictive capacity. This feature of the theory emphasizes its powerful parsimony and further illustrates the very large number of cases of its survival of potential falsification. Again, we encourage our archaeologist and historian colleagues and emerging young student scholars to attempt to falsify these predictions in any case(s) that interests them.

Our Ancient History as the Democratic, Xenophobic Animal

The logic of the scale dependence of our management of the universal conflict of interest problem (see previous discussion) has strong implications for our evolved social psychology. We discuss two of these predictions of social coercion theory here that are particularly important to understanding our contemporary behavior and options for building a humane future.

First, until a recent, transient episode we will discuss, individual access to the tools of decisive coercive threat was broadly (more or less universally) distributed throughout the vast majority of our evolutionary history (thrown stones, atlatl, bow, and, most recently, gunpowder handguns). Under these conditions, the only stable game theoretic equilibrium is expected to be conjoint enforcement of mutual interests (Okada & Bingham, 2008; Souza & Bingham, 2009). Thus, we are highly adapted to democratized social cooperation; it is quite natural to us. We are predicted by social coercion theory to be the "democratic animal." Moreover, because humans are predicted on social coercion theory to always exploit their coercive assets in pursuit of individual self-interest, social cooperation that serves the humane interests of a majority can *only* be sustained by democratized control of decisive coercive power, *never* by "cooperation" policed by small elites with monopoly control of coercive assets.

As expected from the evidence for democratized access to coercive threat throughout most of our ~2 million year history, our evolved psychologies respond to increasing hierarchy as an actively toxic threat to our well-being (Case & Deaton, 2015; Rubin, 2000; Wilkinson & Pickett, 2010).

Given these considerations, it is predictable that democratized cooperation is observed to be much more effective in producing the assets supporting human welfare (more economically productive) than the alternative, wherever we are in the position to collect comparative economic data. Indeed, the massive acceleration of economic growth over the last several centuries (the "modern economic miracle"; see Deaton, 2013; McCloskey, 2016, for recent empirically powerful, if theoretically limited, reviews) correlates precisely in space and time to the partial redemocratization of control of coercive power (resulting from development of potentially widely available gunpowder handguns) within the massive social scale of the state (reviewed in detail in Bingham & Souza, 2009, Chapter 15), conforming very well to the predictions of social coercion theory.

Second, because of the limited scale on which thrown stones allow policing of nonkin cooperation (see previous discussion), the presumptively highly democratized ancestral units of humans would grow when successful, then fission, ultimately into separate, mutually co-hostile coalitions. Thus, social coercion theory predicts that we should also be the "xenophobic animal." We are predicted to be loyal to our own ingroup, while aggressively suspicious of outgroups, as is widely observed empirically (reviewed in Hewstone et al., 2002). This powerful ancient evolved psychology is predicted to render us particularly vulnerable to hostile political manipulation in the context of the adaptively novel scale of massive modern states (see following discussion).

"Power" and the Adaptive Novelty of Massive Recent Human Coalitions

Applying social coercion theory to the contemporary world represents a special challenge of extra difficulty for all of us. Each of us is embedded directly in this world as adapted actors with powerful, evolved proximate psychologies, originally designed to defend our individual interests in the ancestral social world. These interests are so evocative and compelling (consciously and unconsciously) that detached consideration of our social environment is extraordinarily effortful. Moreover, adaptively novel contemporary institutions, vast compared to the ancestral scale of our cooperation, exist largely to the extent that they manipulate these psychologies, often without our full understanding and against our optimal interests. Generations of additional scholarship need to be done in this area; however, we summarize here what we believe to be significant initial progress.

The word "power" is widely used in our everyday conversations; the word has rich intuitive meaning for us. However, our scientific understanding of power has been poor. As reviewed in Winter (2016), the seeming paradox and duality of power has been a subject of social science interest for decades. Social coercion theory allows us to make clearer predictions about this issue.

First, the only theoretically plausible source of what we think of intuitively as power is predicted by social coercion theory to be actual, credible coercive threat. Our politics ultimately consists of negotiation of the policies and practices that will be coercively supported or suppressed; this occurs currently among groups comprised of individuals with competing interests and differential control of coercive assets. Our ancient evolved psychology is predicted to be designed to defend our individual interests in these negotiations as effectively as our individual access to coercive power allows. As ancestral coercive power resided in conjoint coercive threat projected by consensual majorities, most of our evolved political behavior is predicted to consist of influencing the opinions of others in our individual interests and of assessing their coercive assets and interests. This process of influence and negotiation is predicted to capture most of our conscious awareness in the political arena, though we may be no more conscious of overt coercion than we are of genetic replication during sexual behavior.

In view of this evolved psychology, in turn, the only plausible ancestral definitions of the "right" (ethically) are predicted to have been the consensually arrived at beliefs of a democratized coercive majority, shaped to serve the mutual self-interests of its members. The processes of social doubt (see previous discussion) are expected to be the primary mechanism for arriving as such consensual beliefs.

It is vital to recognize that, throughout most of our evolutionary history, power ("might"; majority self-interested coercive threat) and "right" (majority consensual beliefs) are predicted to have had the same locus/cause; they would have been operationally synonymous. We are predicted to have evolved to see power and the right as synonymous, with diverse, profound implications for our contemporary political psychology.

Second, throughout most of our evolutionary history, majority consensus was dominant and served the interests of both sexes more or less equally, because the weapons of social coercion were inherently democratically available (thrown stones, atlatl bolts, arrows). Not only did we increase the size of our social units to include nonkin of both sexes, we also escaped the dominance hierarchies exhibited by nonhuman animal social units. Our proximate ethical feelings, so subjectively familiar to us, are those that social coercion theory predicts to evolve within this democratized social environment. These include concern with fairness, empathy, moral outrage, guilt, and a desire to contribute to the greater good (see Bingham & Souza, 2009, Chapter 10, for additional detail). However, beginning around 5,500 years ago, a fundamental adaptive novelty arose: the concentration of decisive coercive power in the hands of small subgroups of professional male warriors. This development is predicted to have been driven by the technical invention of advanced body armor based on metallurgy (bronze, then iron) in the Eastern Hemisphere (see, e.g., Roth, 2009) and, independently, on Kevlar-like quilted cotton body armor in Central and South America (see, e.g., Hassig, 1998).

This development, in turn, rendered elite armored warriors relatively impervious to contemporary projectile weapons (atlatls and bows), allowing the advanced shock weapons they also possessed (swords, spears, and maces) to be decisive in social coercion. These weapons are both expensive to manufacture and are associated with extremely high opportunity costs to deploy effectively (military training and drilling), limiting access to small groups of elite males. Thus, the effect of elite body armor was to take away decisive coercive power from a majority consensus, placing it in the hands of a small minority of males (professional soldiers and their de facto functionaries).

The historically and archaeologically documented consequence of this technical development is the rise of the first states, often referred to as "archaic states." Social coercion theory makes a quite specific prediction about the structure of these states. Again, humans are predicted to have evolved to project social threat in pursuit of individual self-interest. To be clear, kinship-independent social cooperation is not a product of selection for group performance; rather, it is a byproduct of the individual advantage of projecting threat in defense of personal interests mutually, conjointly held with others, in turn, sharing/creating decisive coercive power. Thus, we expect social units policed by elite male warriors to very selectively serve the interests of these elite armed males (and their de facto functionaries).[15]

---

[15] The use of shock weapons rather than projectile weapons here fundamentally changes the game theory of coercion. Policing of elite conflicts of interest among elite armored warriors will be limited and poor. Though it is beyond our scope here to discuss this issue in detail, the pervasive

Remarkably, archaic states of predicted structure arise, thusly, completely independently in the Eastern and Western Hemisphere (Claessen & van de Velde, 1991; Coe & Houston, 2015; Moseley, 1992; Parkinson & Gallaty, 2009; Yoffee, 2004). Moreover, the core features of such units across these massive areas are strikingly consistent, as predicted by social coercion theory. Archaic states are elite male dominated (females are never systematically enfranchised in these states[16]) and fiercely hierarchical (e.g., slavery and serfdom on a massive scale was endemic, including in the Athenian "democracy" and the Roman "republic"). Again, this is precisely what social coercion theory predicts. Human social units are predicted to serve the interests of those individuals who control decisive coercive threat. It follows from this requirement of social coercion theory that their political structure (democratized or authoritarian/hierarchical) will emerge *exclusively* from the distribution of control to this decisive threat, as is widely observed empirically. Political belief systems are merely proximate effects of this ultimate cause.

Again, we invite our historian and archaeologist colleagues and emerging young student scholars to attempt to falsify this causal description of the archaic state. Broad recognition of these insights is predicted to be vital to a scientific approach to the present and the future (see following discussion).

As Mao Zedong famously remarked, "All power flows from the barrel of a gun." Social coercion theory predicts that his empirical insight is fundamentally correct. However, our evolved psychology gives us the intuitive sense that power is something much more complex and subtle than this. This intuition is predicted by social coercion theory to be partially correct but also fundamentally misleading. In addition to the effects discussed at the beginning of this section, this partially misleading intuition is predicted to arise from the adaptively novel, massive scale of recent human social units as follows.

First, throughout most of our evolutionary history we were members of kinship-independent coalitions numbering in the hundreds (see previous discussion). Our contemporary coalitions have grown to hundreds of millions, explosively over a mere few thousand years.

---

instability of archaic states (see, e.g., Halsall, 2006; Aimers, 2007) is predicted to arise from this limitation (reviewed in detail in Bingham & Souza, 2009; Chapters 13–14).

[16] Note that, in contrast, throughout *most* of human evolutionary history *before* these states and with the development of the modern handgun *after* these early states, women have apparently had the strong political enfranchisement enforceable by their coequal access to coercive threat, precisely as social coercion theory predicts.

On the ancestral scale of hundreds of individuals, we are predicted to have had continuous, real-time access to the interests of all our social confreres. In contrast, on the massive, new scale of hundreds of millions (or even of tens of thousands), the coercive power we *potentially* possess as a majority is obviously limited by our capacity to exchange essential coercively relevant information in real time. The novel, immense scale of our social units compromises our ability to identify our actual confluent interests, on the one hand, and then to *actualize* (cost-effectively project) the potential majority coercive threat we hold, on the other.

Thus, we are predicted to be vulnerable to being deprived of actual coercive power by any interest groups with control over our exchange of information. As predicted, when coercive power becomes hierarchically monopolized, this coercive asymmetry is empirically observed to be immediately exploited to control commoner access to information exchange—further exaggerating elite control of coercive power. There are many well-known examples of this predicted phenomenon, from the enforcement of commoner illiteracy in archaic and early modern states to the contemporary "Great Firewall of China." An especially thoroughly documented case is the Nazi state (powerfully reviewed in Ullrich, 2016).

Second, a related effect of the adaptively novel scale of contemporary social units (states) is predicted to be their processes of arriving at social consensus. Rather than a single relatively homogenous interest group, modern states are amalgams of huge numbers of diverse interest groups. Thus, when we engage in public policy discussions, we commonly obscure some components of our narrow interests, using obliqueness and code wording (metaphorically sometimes called "dog whistling") to speak to our narrow confreres, while minimizing alienation of the other interest groups in the coalition of the whole. This complex, partially hidden negotiation process further confuses our understanding of where real power lies.

On social coercion theory, we argue that these effects of adaptively novel scale distort our intuition about the actual locus of power in the short term. However, we are highly adapted to ferret out actual power over the long term. We have evolved to challenge would-be wielders of power when we doubt their power and/or when they threaten our interests. Thus, the capacity to bluff about the possession of power (by blocking actualization of commoner coercive power) is predicted to be significant but ultimately limited. In the following section, we turn attention to another important confusion about power, the role of "money."

Money and Power in the Adaptively Novel Scale of the Contemporary State

We tend to believe intuitively that money and power are somehow connected in massive contemporary states. Understanding the actual mechanics and limitations of this connection will be vital to our ability to take majority consensual control of a humane future.

First, note that social coercion theory predicts that money is not power, sensu stricto. If one interest group possesses power, but no wealth, while a second possesses wealth, but no coercive power, the first group will inevitably seize the wealth of the second. For example, recall the seizure of the North American continent by coercively powerful Euroamerican states from relatively less powerful, smaller scale Native American Neolithic interest groups. Analogous is the aggressive displacement and cultural suppression of Tibetans, Uyghurs, and other relatively powerless ethnic groups by the majority Han in contemporary China.

The casual connection between wealth and power is tautological in crony/criminal capitalist economies like contemporary Russia and China, where the elite state interest groups have near monopoly control of advanced coercive technologies. Individuals are predicted to be able to accumulate wealth only to the extent that they feed the state military/police apparati and their functionaries (see, e.g., Ledeneva, 2013; Satter, 2016, for empirical support for this prediction in the relatively well documented case of contemporary Russia).

Second, the subtler question is the causal connection between wealth and power in partially democratized states like those of the North Atlantic basin, including the United States and United Kingdom. In economic systems policed by relatively large majorities, it is nonetheless possible for small interest groups to capture very large fractions of the wealth these systems generate (Alperovitz & Daly, 2008; Hacker & Pierson, 2010, 2016; Lazonick, 2014; Mayer, 2016; Piketty, 2014, 2015; Veblen, 1899/2007; Zucman, Fagan, & Piketty, 2015). It seems intuitively self-evident that such wealthy interest groups hold disproportionate power.

To see more deeply into this issue, consider the options of members of wealthy interest groups ("oligarchs" henceforth) within these large partially democratized social units. These concentrations of wealth are highly vulnerable to at least partial confiscation by potentially coercively powerful majority interests. Moreover, the capacity of oligarchs to directly bribe the majority is very limited; oligarchs are wealthy but cannot possibly be that wealthy.

Thus, the only option available to financial oligarchs in politically democratized countries is to bribe other small interest groups who substantially control the capacity of the majority electorate to exchange the information between its members necessary to arrive at consensus about mutual self-interest (in turn, allowing projection of the majority's decisive coercive threat in prosecution of those interests). Two of the most obvious such targets of oligarchic control are the press and the academy.

On the one hand, the original universities in the West were designed to train the scions of elite early state power (how to read among an illiterate commoner population, as well as behavioral lessons of elite power like "never apologize, never explain"). The hierarchical, caste-like tenure system of the contemporary academy continues to feed this role. Most academics consider themselves (consciously or otherwise) as gifted individuals deserving of special privileges funded by wealthy donors and the state. All of us, as academics, have the obligation to transcend this history and speak as clearly and honestly as we can to the global coalition of the whole.[17]

On the other hand, the press is likewise extraordinarily vulnerable to oligarchic bribery. A "news" organization might be owned outright by an oligarch and actively serve elite interests (Fox News in the United States, owned by Rupert Murdoch, is one obvious example). Alternatively, news organizations are highly dependent on advertising revenue, much of which is controlled by aspiring oligarchs, directly or indirectly (see, e.g., Mayer, 2016). Again, any news organization wishing to be a prosocial part of the human future has an obligation to transcend this vulnerability and speak as clearly and honestly as it can to the global coalition of the whole.

Attempts to take ownership of commoner communication channels by financial oligarchs have become extremely aggressive in the North Atlantic basin over the last several decades (see MacLean, 2017, for an outstanding history of this most recent attack in the United States). The complicity in and feeble opposition to the resulting deluge of "fake news" (Streitfeld, 2016) indicates considerable success in this effort to date.

Despite our being potentially tainted, as academics, we can help the members of the majority center reclaim their partially compromised control of the coercive power they actually possess. Being clear in our descriptions of how human institutions actually work is one way in which

---

[17] See Koch (2016) for a representative attempt to manipulate the academy by an aspiring American oligarch. The pose is "Libertarian," predicted to represent cover for antidemocratic governance positions.

we can contribute. For example, it is quite predictable that manipulative oligarchs will attempt to exploit our ancient evolved xenophobia (see previous discussion) to obscure their manipulation of us. Thus, any interest group that uses racism, sexism, homophobia, classism, religious intolerance, etc. is immediately identifiable as serving oligarchic interests.[18] Many members of the potentially powerful center of democratized states may be capable of recognizing this pattern and responding constructively.

The Continuing, Inevitable Struggle to Democratize the State: The Most Important Challenge to the Human Future

We can reasonably define evil as the inhumane treatment of people by other people. The study of evil and its place in the human social world presents seeming paradoxes that have long been a subject of social science research (see Zimbardo, 2008, for one example). The insights from social coercion theory previously discussed can be interpreted as giving us a crisp, clear theory of evil. Specifically, evil is predicted to arise from hierarchical control of coercive power, in turn, allowing powerful minority interest groups to exploit less powerful interest groups, up to and including their use as domesticated animals (slaves, de jure or de facto).

Of course, elites will not view themselves as evil, but believe they are in the "right" since in the more democratized setting in which these proximate mechanisms originally evolved, it was democratized might that made right. When coercive power shifts to a minority, the proximate mechanism of might and right stays intact, complete with self-justification of inhumane elite behavior.

This insight has been arrived at many times intuitively by political actors, especially beginning with the democratization of the modern state gaining momentum in the 18th century. As previously noted, newly invented gunpowder handguns wrenched decisive individual coercive threat from elite warriors, potentially returning it partially to majority commoner coalitions.[19]

---

[18] Note that contemporary "terrorists" are merely aspiring military oligarchs of a state they have not yet been able to subdue or consolidate. They are also convenient fall guys for xenophobic manipulation of democratized electorates.

[19] The democratized states of the North Atlantic basin can be interpreted on social coercion theory as direct descendants of this initial coercive redistribution. As handguns have become more sophisticated, their increasingly capital-intensive manufacture has allowed their re-monopolization

Such evil hierarchical distributions of coercive power arise from two practical sources in the contemporary world. First, is the existence of disproportionately powerful geographical entities (states). These powerful states can be internally democratized and relatively humane toward their citizens. However, even when democratized, these units are predicted to use their disproportionate power to exploit less powerful external entities (again, the might-makes-right proximate psychology here engenders ready self-justifications). Before the last half of the 20th century, this effect was a source of enormous suffering and deprivation (e.g., think of the Native North American genocide by the ostensibly democratized early United States). However, with the ongoing consolidation of coercive pan-global coalitions of states, such exploitative behavior between individual states has become progressively more difficult.

Second, in contrast, the hierarchical control of coercive power *within* nation states (especially in authoritarian polities) remains a major source of contemporary human suffering. Arguably, problems like epidemic disease, poverty, global warming, and others are not solved, ultimately, due to hierarchical distributions of power. This observed political effect is predicted to result from the fact that solutions to such problems benefit the many, but impose significant extra costs (e.g., contributing to public investment) on the elite few.

The global solution to this state-internal hierarchical distribution of control of coercive power is predicted to have two components on social coercion theory. On the one hand, technical penetration of elite control of commoner information access/exchange can allow majority coalitions to better actualize the significant coercive power they hold, despite authoritarian near monopoly control of advanced weapons. However, this effect, alone, is likely to be insufficient. Threatened hierarchies will exploit their weapons monopolies to escalate state intimidation when threatened (e.g., note such an escalation ongoing in China at this writing in winter 2018). However, mobilization of domestic commoner coercive power is predicted to be an essential component of a larger international effort.

On the other hand, transnational coercion by coalitions of states is predicted to be capable of contributing profoundly to the democratization of remaining authoritarian states, if executed with patience and in concert with majority coalitions within authoritarian states. The fundamentals are predicted to be simple. The overwhelming coercive power in the hands

---

in the modern authoritarian state. A humane future is predicted to require our collective coercive demilitarization of such hierarchical states (see following discussion).

of large coalitions of mutually self-interested nation-states can be used to impose economic sanctions on individual outlaw states. Such sanctions do not directly threaten state sovereignty and are, thus, unlikely to provoke full-blown warfare. This approach is currently being used on a modest scale with Russia following its invasion of Crimea and the Ukraine and was relatively effective recently against the nuclear weapons program in Iran. Coalitions of democratized states will need to gradually become more strategic and assertive in this way.

In practice, democratized states have an enormous advantage over authoritarian states in such an endeavor. The economic productivity of democratized states is systematically and substantially superior to that of authoritarian states. Humans are highly adapted to democratized coalitions and perform much better under such systems than under exploitative hierarchical systems, including the crony/criminal capitalist systems that are favored by the remaining large authoritarian states (especially Russia and China). For example, the per capita gross domestic products (GDPs) of the democratized United States, the United Kingdom, and Japan ($53,000, $39,000, and $39,000, respectively) are dramatically higher than China's ($6,800) or Russia's ($11,000). Moreover, the lower per capita GDPs of authoritarian regimes are disproportionately driven by selling to functional economies of democratized countries rather than from internal economic activity. For example, Russia's $11,000 per capita GDP is very dependent on selling gas and oil to the economically functional West. Likewise, China's limited GDP is substantially dependent on exporting relatively cheap manufactured goods to the West.

Thus, functional, internally self-sufficient democratized economies regulated by majority consensus can afford to impose economic sanctions on authoritarian regimes, while authoritarian regimes cannot survive such sanctions if they are strong and consistent. Modest, consistent trade sanctions can be imposed on authoritarian regimes with little risk of recession to the members of the imposing coalitions, but very high risk of serious economic consequences to the targets. Moreover, it is often possible to target the holdings of tiny financial elites in authoritarian systems, while minimizing the consequences to commoner citizens of those states.

Of course, implementing such a systematic coercive dismantling of the remaining authoritarian states will present myriad technical challenges; patience, wisdom, and statesmanship will be required. Moreover, we predict that a global commoner consensus will need to be built contemporaneously with this effort. This brings us back to the first element of continuing global democratization. The global community will need to improve its means for penetrating elite control of information (both to defend existing democracy from aspiring internal oligarchs and to build domestic support for democratization within authoritarian regimes). This effort will have a technical component (providing access channels difficult for states to obstruct) and a social component (demonstrating that these efforts are controlled by majority coalitions, not by aspiring oligarchs within democratized states). Indeed, preventing collusion between our own domestic would-be oligarchs and authoritarian oligarchs will be a continuing and essential facet of this portion of the project. The danger here is real and imminent, as especially well illustrated by the 2016 American presidential election (Flegenheimer & Shane, 2017; Krugman, 2016, 2017; Shear & Sangerjan, 2017).

In summary, the world war for redemocratization on the massive scale of the state has been ongoing since at least the 17th century. This effort has produced very dramatic improvements in the human condition. However, it remains to the next several generations of humans to complete and extend the process, opening the door to a vastly wiser, richer, more peaceful and humane panhuman future.

These examples and this discussion illustrate how an understanding of human evolution is essential to understanding our current social world at all levels, including our everyday social behavior; our educational, economic, and political systems; and how this understanding can give us insight, not only into the human past, but also into a more humane and productive future.

Implications for the Academy

On the research side, we need more senior human biologists, social scientists, and humanists to elevate their theoretical efforts. Historically, these disciplines have been empirically deep but, understandably, theoretically shallow, as expected of the natural history phase of their development. This natural history phase is over now. Designing the empirical/investigative work we do to directly serve evaluation of potentially strong theory by subjection to robust falsification and intense competitive parsimony is vital.

This new level of theoretical discipline and sophistication will enhance the efficiency and impact of what we do. Accepting this discipline will also win back the societal respect we have lost under comparison with the mature STEM disciplines, where aggressive respect for and use of robustly arrived at theory is well established.

On the curricular side, we need to teach aggressive theory falsification and competitive parsimony to the students who will shortly mature and take our places as investigators. Historically, acceptance of light, local theory in the original human sciences has been inescapable during their early development; however, again, that time is now past. It is vital to help young students understand that the vast majority of creative activity in a mature science consists of falsifying wrong or incomplete theories. Such "creative destruction" of older theory is how we "fail our way to success" in all mature sciences. Falsification of theory is an inevitable step to newer theories so strong they escape falsification, becoming the tools with which we understand and deal with the world. Young social scientists and human biologists, armed with this new level of theoretical sophistication (and the vast body of empirical evidence from the last two centuries), can make the looming revolution in the new human science.

Academic institutions and individual investigators that are first to rise to this new challenge have the opportunity to make unique contributions to the newly emerging pan-global, democratized human condition.[20]

Implications for the Human Future

We hope that the previous discussions have made the following argument clear. We are apparently engaged in the final stage of global struggle to reacquire ancestral, humane democratized social behavior from the recent aberration of the hierarchical state. Social coercion theory predicts there to be only two mutually exclusive, highly stable game theoretic outcomes. The world will become vastly wiser, wealthier, and more humane through universal ascendency of democratization, or, alternatively, the world can universally succumb to the hierarchical rule of oligarchs. Such a world would inevitably devolve into brutality and extreme poverty (North Korea is an excellent contemporary natural experiment, we argue).

The contemporary United States is an excellent illustration of the current status of this struggle. On the one hand, the United States has diverse,

---

[20] In the short term, as other institutions come up to speed, students can gain access to sophisticated instruction in social coercion theory through a globally available, interactive undergraduate summer online course we offer Faculty and graduate students may prefer the globally available, interactive online graduate version of this course, given in fall, spring, and summer. [See Courses button at www.deathfromadistance.com.]

robust institutions standing in defense of republican governance. On the other hand, American democracy is under attack from aspiring oligarchs attempting to take control of the vast resources and military strength originally produced by the democratized efforts of generations of citizens. The priming of the ancestral mechanisms of xenophobia, racism, and sexism are conspicuous in the oligarchic political message every day in the popular press. Moreover, oligarchs have been aggressive in their attempts at manipulation and control of information needed to mobilize the potential coercive threat of the majority (see, e.g., Hasan, 2016; Mayer, 2016; MacLean, 2017). These overt symptoms of an attempted hierarchical power grab are predicted by social coercion theory.

Which of the two alternative, permanent outcomes of this global struggle our descendants inherit will very likely be settled by the end of this century; that outcome is still very much in doubt. Deciding the human future falls to us, our children, and our grandchildren. We have urgently necessary science and teaching to do.

References

Aimers, J. J. (2007). What Maya collapse? Terminal classic variation in the Maya lowlands. *Journal of Archaeological Research, 15*, 329–377.

Alperovitz, G., & Daly, L. (2008). *Unjust deserts: How the rich are taking our common inheritance.* New York, NY: New Press.

Anton, S. C., Potts, R., & Aiello, L. C. (2014). Evolution of early *Homo*: An integrated biological perspective. *Science, 345*, 1236828.

Beckerman, S., & Valentine, P. (2002). *Cultures of multiple fathers: The theory and practice of partible paternity in Lowland South America.* Gainesville, FL: University Press of Florida.

Berger, L. R., de Ruiter, D. J., Churchill, S. E., Schmid, P., Carlson, K. J., Dirks, P., & Kibii, J. M. (2010). Australopithecus sediba: A new species of homo-like Australopith from South Africa. *Science, 328*, 195–204.

Berger, L. R., Hawks, J., de Ruiter, D. J., Churchill, S. E., Schmid, P., Delezene, L. K., ... Zipfel, B. (2015). Homo naledi, a new species of the genus *Homo* from the Dinaledi Chamber, South Africa. *eLife, 4*, e09560.

Bingham, P. M. (1999). Human uniqueness: A general theory. *Quarterly Review of Biology, 74*, 133–169.

Bingham, P. M., Souza, J., & Blitz, J. H. (2013). Social complexity and the bow in the prehistoric North American record. *Evolutionary Anthropology, 22*(3), 81–88.

Bingham, P. M., & Souza, J. (2009). *Death from a distance and the birth of a human universe.* Charleston, SC: Booksurge.

Boyd, R., Gintis, H., & Bowles, S. (2010). Coordinated punishment of defectors sustains cooperation and can proliferate when rare. *Science, 328*, 617–620.

Bramble, D. M., & Lieberman, D. E. (2004). Endurance running and the evolution of Homo. *Nature*, *432*, 345–352.

Case, A., & Deaton, A. (2015). Rising morbidity and mortality in midlife among white non-Hispanic Americans in the 21st century. *Proceedings of the National Academy of Sciences of the United States of America*, *112*, 15078–15083.

Claessen H. J. M., & van de Velde, P. (1991). *Early state economics*. London, England: Transaction.

Coe, M. D., & Houston, S. D. (2015). *The Maya* (9th ed.). New York, NY: Thames & Hudson.

Dawkins, R. (1976). *The selfish gene*. New York, NY: Oxford University Press.

Deaton, A. (2013). *The great escape: Health, wealth, and the origins of inequality*. Princeton, NJ: Princeton University Press.

Dirks, P. H. G. M., Roberts, E. M., Hilbert-Wolf, H., Kramers, J. D., Hawks, J., Dosseto, A., . . . Berger, L. R. (2017). The age of Homo naledi and associated sediments in the Rising Star Cave, South Africa. *eLife*, *6*, e24231. doi:10.7554/eLife.24231.001

Dominguez-Rodrigo, M., Mabulla, A. Z. P., Bunn, H. T., Diez-Martin, F., Baquedano, E., Barboni, D., . . . Yravedra, J. (2010). Disentangling hominin and carnivore activities near a spring at FLK North (Olduvai Gorge, Tanzania). *Quaternary Research*, *74*, 363–375.

Dugatkin, L. A. (1997). *Cooperation among animals: An evolutionary perspective*. New York, NY: Oxford University Press.

Dyer, F. C. (2002). The biology of the dance language. *Annual Review of Entomology*, *47*, 917–949.

Flegenheimer, M. A., & Scott, S. (2017). Intelligence chief criticizes "disparagement" of findings on Russian hacking. *New York Times*. Retrieved from https://www.nytimes.com/2017/01/05/us/politics/armed-services-committee-john-mccain-russia-hacking.html

Ferring, R., Oms, O., Agusti, J., Berna, F., Nioradze, M., Shelia, T., . . . Lordkipanidze, D. (2011). Earliest human occupations at Dmanisi (Georgian Caucasus) dated to 1.85–1.78 Ma. *Proceedings of the National Academy of Sciences of the United States of America*, *108*, 10432–10436.

Hacker, J. S., & Pierson, P. (2010). *Winner-take-all politics: How Washington made the Rich Richer and turned its back on the middle class*. New York, NY: Simon & Schuster.

Hacker, J. S., & Pierson, P. (2016). *American amnesia: How the war on government led us to forget what made America prosper*. New York, NY: Simon & Schuster.

Halsall, G. (2006). *Movers and shakers: The barbarians and the fall of Rome*. London, England: Routledge.

Hamilton, W. D. (1996). *Narrow roads of gene land: The collected papers of W. D. Hamilton*. Oxford, England: W. H. Freeman/Spektrum.

Hassig, R. (1988). *Aztec warfare: Imperial expansion and political control* (1st ed.). Norman, OK: University of Oklahoma Press.

Hasen, R. L. (2016). *Plutocrats united: Campaign money, the Supreme Court, and the distortion of American elections*. New Haven, CT: Yale University Press.

Hewstone, M., Rubin, M., & Willis, H. (2002). Intergroup bias. *Annual Review of Psychology*, *53*, 575–604.

Isaac, B. (1987). Throwing and human evolution. *African Archaelogical Review*, 5, 3–17.

Koch, C. (2016, July 21). The closing of the American mind [Opinion]. *Wall Street Journal*. Retrieved from http://www.wsj.com/articles/the-closing-of-the-american-mind-1469141624

Kivell, T. L., Deane, A. S., Tocheri, M. W., Orr, C. M., Schmid, P., Hawks, J., Berger, L. R., & Churchill, S. E. (2015). The hand of Homo naledi. *Nature Communications*, 6, 8431. 1–9. Retrieved from https://www.nature.com/articles/ncomms9431

Koenig, M. A., & Echols, C. H. (2003). Infants' understanding of false labeling events: the referential roles of words and the speakers who use them. *Cognition*, 87(3), 179–208.

Krugman, P. (2016, July 22). Donald Trump, the Siberian candidate. *New York Times*. Retrieved from http://www.nytimes.com/2016/07/22/opinion/donald-trump-the-siberian-candidate.html

Krugman, P. (2017, January 2). American becomes a Stan. *New York Times*. Retrieved from https://www.nytimes.com/2017/01/02/opinion/america-becomes-a-stan.html

Kuhn, T. S. (1962). *The structure of scientific revolutions*. Chicago, IL: University of Chicago Press.

Lake, R. (2012). *Vygotsky on education*. New York, NY: Peter Lang.

Lazonick, W. (2014). Profits without prosperity. *Harvard Business Review*, 92(9), 46–55.

Ledeneva, A. V. (2013). *Can Russia modernise? Sistema, power networks and informal governance*. Cambridge, MA: Cambridge University Press.

Lordkipanidze, D., de Leon, M. S. P., Margvelashvili, A., Rak, Y., Rightmire, G. P., Vekua, A., & Zollikofer, C. P. E. (2013). A complete skull from Dmanisi, Georgia, and the evolutionary biology of early *Homo*. *Science*, 342, 326–331.

Lordkipanidze, D., Jashashvili, T., Vekua, A., de Leon, M. S. P., Zollikofer, C. P. E., Rightmire, G. P., . . . Rook, L. (2007). Postcranial evidence from early *Homo* from Dmanisi, Georgia. *Nature*, 449, 305–310.

MacLean, N. (2017). *Democracy in Chains*. Penguin Random House. New York.

Mayer, J. (2016). *Dark money: The hidden history of the billionaires behind the rise of the radical right* (1st ed.). New York, NY: Doubleday.

McCloskey, D. N. (2016). *Bourgeois equality: How ideas, not capital or institutions, enriched the world*. Chicago, IL: University of Chicago Press.

Mellars, P. (2005). The impossible coincidence: A single-species model for the origins of modern human behavior in Europe. *Evolutionary Anthropology*, 14, 12–27.

Mills, C. M. (2013). Knowing when to doubt: Developing a critical stance when learning from others. *Developmental Psychology*, 49, 404–418.

Moseley, M. E. (1992). *The Incas and their ancestors: The archaeology of Peru*. New York, NY: Thames and Hudson.

Napier, J. R., & Tuttle, R. H. (1993). *Hands*. Princeton University Press.

Okada, D., & Bingham, P. M. (2008). Human uniqueness-self-interest and social cooperation. *Journal of Theoretical Biology*, 253, 261–270.

Parkinson, W. A., & Galaty, M. L. (2009). *Archaic state interaction: The Eastern Mediterranean in the Bronze Age* (1st ed.). Santa Fe, NM: School for Advanced Research Press.

Piketty, T. (2015). *The economics of inequality*. Cambridge, MA: Belknap Press of Harvard University Press.

Piketty, T., & Goldhammer, A. (2014). *Capital in the twenty-first century*. Cambridge, MA: Belknap Press of Harvard University Press.

Pinker, S. (2012). The false allure of group selection. *The handbook of evolutionary psychology*. Wiley.

Popper, K. R. (1959). *The logic of scientific discovery*. New York: Basic Books.

Richerson, P., Baldini, R., Bell, A. V., Demps, K., Frost, K., Hillis, V., . . . Zefferman, M. (2016). Cultural group selection follows Darwin's classic syllogism for the operation of selection. *Behavioral and Brain Sciences, 39*, e58.

Roth, J. P. (2009). *Roman warfare*. Cambridge, England: Cambridge University Press.

Rubin, P. H. (2000). Hierarchy. *Human Nature, 11*, 259–279.

Satter, D. (2016). *The less you know, the better you sleep*. New Haven, CT: Yale University Press.

Scott-Phillips, T. C., Dickins, T. E., & West, S. A. (2011). Evolutionary theory and the ultimate-proximate distinction in the human behavioral sciences. *Perspectives on Psychological Science, 6*, 38–47.

Scupin, R. (2012). *Cultural anthropology: A global perspective* (8th ed.). Boston, MA: Pearson.

Shear, M. D., & Sanger, D. E. (2017, January 6). Putin led a complex cyberattack scheme to aid Trump, report finds. *New York Times*. Retrieved from https://www.nytimes.com/2017/01/06/us/politics/donald-trump-wall-hack-russia.html

Streitfeld, D. (2016, December 25). For fact-checking website Snopes, a bigger role brings more attacks. *New York Times*. Retrieved from https://www.nytimes.com/2016/12/25/technology/for-fact-checking-website-snopes-a-bigger-role-brings-more-attacks.html

Souza, J., & Bingham, P. M. (2018). The rachet effect in student learning. Unpublished manuscript.

Terkel, J. (1996). Cultural transmission of feeding behavior in the black rat (Rattus rattus). In C. M. Heyes & B. G. Galef, Jr. (Ed.), *Social learning in animals: The roots of culture* (pp. 17–47). New York, NY: Academic Press.

Trigger, B. G. (2003). *Understanding early civilizations*. McGill University Press, Montreal.

Ullrich, V. (2016). *Hitler: Ascent, 1889–1939* (1st American ed.). New York, NY: Knopf.

Veblen, T. (2007). *The theory of the leisure class*. Oxford, England: Oxford University Press. (Original work published 1899)

Vrba, E. S. (1995). *Paleoclimate and evolution, with emphasis on human origins*. New Haven, CT: Yale University Press.

Wenke, R. J., & Olszewski, D. (2007). *Patterns in prehistory: Humankind's first three million years* (5th ed.). New York, NY: Oxford University Press.

Wenseleers, T., & Ratnieks, F. L. W. (2006). Comparative analysis of worker reproduction and policing in eusocial hymenoptera supports relatedness theory. *American Naturalist, 168*, E163–E179.

Wilkinson, R. G., & Pickett, K. (2010). *The spirit level: Why Greater equality makes societies stronger*. New York, NY: Bloomsbury.

Williams, G. C. (1966). *Adaptation and natural selection: A critique of some current evolutionary thought*. Princeton, NJ: Princeton University Press.

Williams, G. C. (2008). *Group selection*. New Brunswick, NJ: Aldine Transaction.

Wilson, D. S. (2015). *Does altruism exist? Culture, genes, and the welfare of others*. New Haven, CT: Yale University Press.

Winter, D. G. (2016). Taming power: Generative historical consciousness. *American Psychologist, 71*, 160–174.

Yoffee, N. (2004). *Myths of the archaic state: Evolution of the earliest cities, states and civilizations*. New York, NY: Cambridge University Press.

Zimbardo, P. G. (2008). *The Lucifer effect: Understanding how good people turn evil*. New York, NY: Random House.

Zucman, G., Fagan, T. L., & Piketty, T. (2015). *The hidden wealth of nations: The scourge of tax havens*. Chicago, IL: University of Chicago Press.

CHAPTER 8 | Controversies Surrounding Evolutionary Psychology

GLENN GEHER AND VANIA ROLÓN

EVOLUTIONARY PSYCHOLOGY (GEHER, 2014) IS an approach to understanding behavior that is rooted in evolutionary principles such as natural selection. In short, evolutionary psychologists address questions of mind and behavior by first asking how said behavior or mental process could have arisen via evolutionary forces. For instance, when Ekman and Friesen (1968) conducted their landmark research on the universal nature of human facial expressions, they hypothesized that emotional expressions should be consistent across human groups as emotional expressivity is strongly connected to our biology and as it has clear adaptive functions, such as, for instance, connecting with other members of a group and warning others of threats in the vicinity. An evolutionary psychologist essentially steps back and asks "ultimate" questions in regard to mind and behavior—and, as such, is able to not only understand proximate causes of behaviors (such as how the amygdala plays a role in fear responding) but is also able to understand ultimate causes of behaviors (such as why fear responses evolved in the first place and have come to characterize emotion systems across all human groups).

The field of evolutionary psychology has grown considerably over the past several decades (Garcia et al., 2011). This growth has led to new insights into many domains of human functioning, such as aggression, mating, social relationships, emotions, and more (Geher, 2014). In a now-classic paper on the powerful nature of the evolutionary perspective in psychology, Ketelaar and Ellis (2000) argued that a good scientific framework is one that (a) leads to new questions and, as a result, (b) leads to new understanding. Evolutionary psychologists (Geher, 2014) study behavior—seeking to apply evolutionary principles to help us, in fact, best understand why different behavioral phenomena exist. The field of evolutionary psychology is full of new findings about human behavior that we simply would not have without the mountain of research that this field has cultivated. Such findings include the tendency for young males to be more likely to die of various causes compared with young females (Kruger & Nesse, 2006), the fact that step-parents are much more likely to commit filicide than are biological parents (Daly & Wilson, 2005), and the fact that, across the globe, men show stronger preferences for variety in partners relative to women's preferences (Schmitt et al., 2003). And more.

A Powerful Framework for Understanding Who We Are

A scientific framework is only useful if it clearly helps provide new insights into the phenomena being studied (Ketelaar & Ellis, 2000). Evolutionary psychology is such a scientific framework. If you think about the findings described herein, evolutionary psychology sheds light on major questions of humanity, including the nature of family violence, causes of death in young adults, and the nature of sex differences in mating and relationship patterns. In fact, this brief list of phenomena that have been elucidated by the field of evolutionary psychology represents simply the tip of the iceberg (Carmen et al., 2013; Geher, 2014). As we see it, evolutionary psychology is a powerful scientific framework that sheds enormous light on all facets of who we are.

With its dual focus on evolution and on human behavior, evolutionary psychology is importantly positioned within the broader area of evolutionary studies. Given the powerful nature of evolutionary psychology, one might think that this approach to understanding behavior would be fully embraced by the broader academic community. After all, Darwin's ideas are nearly universally accepted among academics across the world, and the field of psychology (or the behavioral sciences) is a basic area of intellectual inquiry in academic institutions across the globe. Thus, the marriage of evolution and psychology (i.e., *evolutionary psychology*), in many ways, just seems like a natural area of intellectual synergism in the modern landscape of academia.

In addition to being a powerful approach to understanding behavior in a conceptual sense, the field of evolutionary psychology has demonstrated itself to be highly advanced in terms of the use of multiple methodologies (Schmitt & Pilcher, 2004). In making claims about the structure of human nature, evolutionists need to use a variety of methods so as to substantiate the questions and claims that they study. In an analytical treatise on this topic, Schmitt and Pilcher carefully demonstrate how evolutionarily informed research includes such varied methods as self-report, cross-cultural surveys, markers of autonomic arousal, behavior genetics, comparative psychology, phylogenetic analyses, and more. Further, the work by evolutionary psychologists has demonstrated itself to be more likely to actually collect cross-cultural data compared with other areas of psychology, allowing scholars in this field to home in on truly universal features of human behavior.

This said, as it turns out, the field of evolutionary psychology has been, since its origins, fraught with controversy. The remainder of this chapter delineates the various controversies that surround this area of inquiry with an eye toward developing ways to better integrate evolutionary psychological approaches into other academic areas of inquiry moving forward.

The Evolutionary Psychology Controversies

"Evolutionary psychology (is) . . . subject to a level of implacable hostility, which seems far out of proportion to anything even sober reason or common politeness might sanction" (Dawkins, 2005, p. 975).

As a perspective—rather than a content area of psychology—evolutionary psychology is differently positioned than other areas of psychology. We make big claims—and not everyone likes that! Evolutionary psychologists make claims about the ultimate origins of all psychological processes—and we make claims about all subareas of the behavioral sciences—and beyond. While evolutionary psychologists see this approach as integrative, others seem to find it off-putting.

There are many controversial aspects of evolutionary psychology—and understanding these controversies and voices on different sides of these issues can be very helpful in allowing students to take content away from a course in evolutionary psychology while concurrently having healthy skepticism and an open mind when it comes to the controversies surrounding this field.

As authors of this chapter, we believe that we are actually uniquely positioned to comment on this particular topic. Our home institution, SUNY New Paltz, has heard the expression of many voices regarding evolutionary psychology—including several critical perspectives (see Chapter 4). In fact, in one highly publicized instance, a guest speaker related to evolutionary psychology, Rutgers' Lionel Tiger, was met with resistance and protest due to the nature of his views and his research (for a detailed summary of this situation, see Geher & Gambacorta, 2010). This feature of our academic environment has been extremely useful for us in understanding the criticisms of this field and helping to forge intellectual bridges with other academic areas. In short, being an evolutionary psychologist at SUNY New Paltz has provided us with a unique window into the controversies that surround this area—and this chapter is our attempt to share what we have seen through this window with readers of this book.

Major Controversies Surrounding Evolutionary Psychology

As you'll see, our research has shown that one of the most contentious controversies in evolutionary psychology relates to the idea of evolved behavioral sex differences—the idea that men and women show some natural differences as a result of evolutionary processes. Including this particular issue, controversies that will be addressed in this chapter include the following:

- The *evolved behavioral sex differences* controversy;
- The *religion* controversy;
- The *genetic determinism* controversy;
- The *bad science* controversy; and
- The *eugenics* controversy.

The Evolved Behavioral Sex Differences Controversy

While several controversies regarding evolutionary psychology exist, the largest controversy (to our mind) seems to revolve around the idea of evolved behavioral sex differences. Such sex differences have been documented by evolutionary psychologists in such domains as cognition (Geary, 2015), human mating (Buss, 2016), life history strategies (Figuredo et al., 2013), and more. The evolutionary perspective suggests that males and females, largely due to differentiated required parental investment (Trivers, 1972), show marked behavioral sex differences across a battery of dimensions. This said, this topic has been fraught with controversy, as explicated in the research described in the following discussion.

In one study, to empirically examine this question regarding how prominent this particular controversy regarding evolutionary psychology is in the landscape of higher education, Gambacorta and Geher (2010) created a survey with several kinds of questions, and they disseminated this survey to hundreds of adults. About half the participants were professors at varying institutions, and others were employed in other professions. Authors also asked participants what academic area they were in (if they were professors) and if they had children.

Dependent variables represented whether participants believed that several attributes are primarily the result of biological evolution versus socialization. Variables addressed attitudes about (a) sex differences in adults, (b) sex differences in children, (c) sex differences in chickens, (d) human universals, and (e) differences between dogs and cats. Using a Likert-scale, participants were asked to rate the degree to which they believed items were due to "nature" versus "nurture."

Sample items from these different subscales are:

1. Attitudes about whether human behavioral sex differences in adults are shaped by biological evolution (nature) versus socialization (nurture).
   - Sample item: *Women are more responsive than men to the cries of infants*.
2. Attitudes about whether human behavioral sex differences in children are shaped by biological evolution (nature) versus socialization (nurture).
   - Sample item: *Girls develop language skills earlier than boys*.
3. Attitudes about whether behavioral sex differences in chickens (between hens and roosters) are shaped by biological evolution (nature) versus socialization (nurture).
   - Sample item: *Roosters seem to prefer copulating with more than one hen while hens don't seem to mind copulating with a single rooster*.
4. Attitudes about whether human universals that are not related to sex differences are shaped by biological evolution (nature) versus socialization (nurture).
   - Sample item: *Feces and vomit are found to be universally disgusting among humans*.
5. Attitudes about whether behavioral differences between dogs and cats are shaped by biological evolution (nature) versus socialization (nurture).
   - Sample item: *Dogs are more pack-oriented than cats*.

In composite, these measures allowed for the examination of the degree to which participants believed that these different kinds of phenomena are differentially due to biological or socialization-based causes.

An important caveat to this research is that participants were not asked to make fine and nuanced distinctions regarding the interactions between "nature" and "nurture"—this is an important point. This particular methodology was chosen to simplify the presentation of the materials for the participants—so the methodology was designed to strike something of a balance on this issue.

The primary independent variables included parental status and academic employment status. We studied parental status thinking that the experience of raising children might influence beliefs about innate causes of behavior. Also, given that scholars in the fields of women's studies and sociology tend to be particularly likely to endorse a social constructionist perspective, academic participants were divided into categories of either "sociology or women's studies" or "other."

Academic employment status was independently predictive of the belief that sex differences are the result of "nurture." This effect was exacerbated for academics who came from sociology or women's studies backgrounds. The effect of academic employment status also corresponded to seeing behavioral differences between roosters and hens as caused by "nurture." Further, parents were more likely than nonparents to endorse "nature" for the sex-difference variables. Beliefs about differences between cats and dogs and beliefs about causes of human universals (that are not tied to sex differences) were not related to these independent variables, suggesting that the political resistance seems largely localized to the idea of evolved behavioral sex differences.

Interestingly, the dependent variable regarding human universals was not related to parental or academic status. In short, this implies that the areas of evolutionary psychology that are aside from issues of male–female differences (e.g., research on the evolutionary psychology of emotions, universal fear responses in humans, etc.) are different in terms of political volatility than are the areas of evolutionary psychology that do relate to male–female differences.

This all provides further evidence that while evolutionary psychology often finds itself surrounded by controversy, the issue of evolved sex differences is, at least within the halls of academia, near the top of this list.

The Religion Controversy

Interestingly, the primary controversy that people often think of when they hear about evolutionary psychology relates to religion. This is likely because evolution and religion have a long history of disagreements—going back to Darwin's own internal conflicts on this issue. However, as it turns out, conflicts regarding the origins of life, which sit at the heart of most religious concerns regarding evolution, don't tend to emerge in discussions regarding evolutionary psychology. Some fundamentalist Christians will vocally dismiss the field outright, given its connection with evolution writ large, but generally, evolutionary psychologists don't find themselves butting heads with religious fundamentalists.

When it comes to resistance to evolution, people talk about *resistance from the right* and *resistance from the left*—with the former pertaining to conservative, right-wing resistance (often hand in hand with religious fundamentalist resistance) and the latter pertaining to resistance from political liberals. To the surprise of many, the resistance from the left tends to be much more palpable in the realm of evolutionary psychology. And for this reason, this section is actually quite brief. This is not because people on the left don't "believe in evolution"; rather, social liberalism often is connected with a negative reaction to the idea of innate behavioral qualities—and evolutionary psychology often gets lumped into a group of perspectives that are painted as "genetically deterministic." It is to this issue that we turn next.

The Genetic Determinism Controversy

One reason for the resistance to evolutionary psychology pertains to *genetic determinism*. This is essentially the idea that people do things because of their genes. Taken to an extreme, this idea implies that people actually have no control over their behavior and that only genes matter in determining behavior. This is, without question, an inaccurate and dangerous view of human behavior. It's also not consistent with the basic premises of evolutionary psychology, which focus extensively on the impact of situational and contextual factors in shaping the nature of behavior.

One of the beliefs that many people tend to hold about evolutionary psychology is that it is a non-situationist doctrine, suggesting that organisms have just a few immutable, invariant ways of responding, which are under the direct control of genes. This portrait of evolutionary psychology is simply inaccurate (Kurzban & Haselton, 2005). Evolutionary psychology posits that species-typical psychological design features with some heritable component have been shaped by natural and sexual selection. Often, many (but not all) evolutionary psychologists will conceive of such design features as *adaptations*. In any case, such adaptations are rarely understood by evolutionary psychologists as being context-independent.

Evolutionary psychologists and biologists make an important distinction between *nonconditional* and *conditional strategies* that describe the phenotypes of different organisms. A classic example of a nonconditional, fully genetically determined (and immutable) strategy is found in male sunfish (Gross, 1982), which come in discrete varieties (that vary as a function of the lifespan). The first variety includes large males who have the ability to acquire sufficient territories in intrasexual competition. The second variety includes smaller, *sneaker* males, who are nearly indiscernible from females and who do not elicit aggressive responses from territory-holding males. While territory-holding males reproduce by honestly attracting females, sneaker males use a somewhat dishonest strategy: They blast their gametes after a female has released her eggs in a large male's territory, thereby using deception as a tool for reproduction. It turns out that the differences between these kinds of males is attributable to genetic differences. As such, the strategies employed are nonconditional.

The notion of *conditional strategies*, on the other hand, corresponds to situations in which an organism modifies its strategy vis a vis variability in situational factors. For instance, male tree frogs (Perrill, Gerhardt, & Daniel, 1978) use strategies similar to the male sunfish when it comes to mating. Sometimes, a male will carve out a territory and croak loudly. At other times, a male will hide near a territory-holding male and try to mate with females that are attracted to the croaking, territory-holding male. Importantly, in this species, males have been documented to show *strategic pluralism* (Gangestad & Simpson, 2000); they modify their choice of strategy depending on the nature of such situational factors as the number of male territory-holders at a given time.

The use of a variety of strategies by male wood frogs does not suggest that their repertoire of mating behaviors is somehow outside the bounds of natural law or that these strategies are not designed for "purpose" of reproduction. Clearly, these mating strategies are related to optimal reproduction, a fact that speaks to their selection by evolutionary processes. As such, evolutionary geneticists (e.g., Maynard Smith, 2002) and evolutionary psychologists (e.g., Gangestad & Simpson, 2000) have come to apply evolutionary reasoning to our understanding of mixed behavioral strategies that are highly context-sensitive.

In fact, modern-day evolutionary psychology is an extraordinarily situationist perspective. Consider, for instance, evolutionarily informed research on homicide and familial violence. All of the most highly cited work in this area focuses on situational factors that underlie family violence. For instance, Daly and Wilson's (1988) often-cited work on violence toward children is all about contextual factors that covary with this atrocious act. Simply, the presence of a step-parent in a household has been shown to be the primary contextual factor that predicts fatal violence toward children. Another contextual factor that Daly and Wilson document as having a significant relationship with such violence is the age of a given child (another contextual factor). In fact, their research, which is, in this regard, very prototypical of much work in evolutionary psychology overall, is all about contextual factors that underlie behaviors.

Consider, as another example, research on factors that predict promiscuous behavior on the part of women. Evolutionary psychologists have uncovered such important contextual factors as localized sex ratios, ovulation cycles, a woman's age, and the presence of children from prior mateships (Buss, 2016)—each such contextual factor serving as an important statistical predictor of female promiscuity. In short, evolutionary psychology is, in fact, a highly situationist perspective, generally conceiving of human behavioral strategies as being extremely flexible and as falling within the realm of this general idea of strategic pluralism.

Evolutionary psychology does not conceptualize humans as genetically guided automatons whose conscious decision-making processes are irrelevant or non-existent. Rather, this perspective sees humans as capable of extraordinary conscious decision-making. Further, with its roots in strategic pluralism, evolutionary psychology is situationist at its core. Importantly, evolutionary psychology has lessons to provide regarding the nature of situationism as an epistemological doctrine. While situationism in the social sciences is often framed as conceiving of human behavior as highly under the influence of situational influences (both small and large; see Ross & Nisbett, 1991), this generic brand of situationism has generally been framed in a manner that is devoid of any insights into how important psychological design features have been ultimately shaped by evolutionary forces for the purpose of reproduction.

The kind of situationism that characterizes modern-day evolutionary psychology may be thought of as a sort of *evolutionary situationism*. This particular brand of situationism suggests that while human behavior is largely under the control of situational influences, the particular situational factors that should matter most in affecting behavior are ones that bear directly on factors associated with survival and reproductive success. As such, Daly and Wilson (1988) did not document just any factors that underlie familial violence—they specifically uncovered the role of step-parenting, a situational factor with clear and theoretically predictable relevance to issues tied to genetic fitness (from a strictly genetic-fitness perspective, a stepchild shares no genes with a step-parent, and is, thus, costly).

Given the tremendous potential for evolutionary psychology to inform the search for contextual factors that underlie human psychological outcomes, this idea of evolutionary situationism has the potential to create extraordinary bridges between traditional social psychology and evolutionary psychology.

A general kind of criticism that underlies the genetic determinism controversy relates to conceptions of evolutionary explanations of behavior as justifications for said behaviors. Thus, for instance, if researchers discuss an evolutionary explanation for warfare (Smith, 2008), someone who hears this might interpret this research as saying that warfare is "in our genes" and that warmongers should not be held responsible as a result. This same kind of reasoning can apply to sexism, racism, sexual harassment, murder, and a large battery of social ills that have been studied from an evolutionary perspective. Evolutionary psychologists argue (a) that such "evolved" behaviors actually have an enormous amount of flexibility (per evolution-based situationism) and (b) that such accusations are, logically, representative of the naturalistic fallacy—which is essentially hearing that something is framed as natural—as how something *is*—and then concluding that said phenomenon was being framed as how something *should be*. This naturalistic fallacy, in fact, may well be responsible for a large proportion of the controversy surrounding evolutionary psychology.

The Bad Science Controversy

In a comment on the state of evolutionary psychology within academia, Richard Dawkins (2005), an evolutionary biologist with some sympathies toward evolutionary psychology, suggests that the bar for the quality of science in evolutionary psychology may actually be set too high. From his perspective, he agrees that extraordinary claims should require extraordinary evidence, but he believes that the basic premises of evolutionary psychology do not constitute extraordinary claims. That is, the idea that human behavior is ultimately the result of evolution is not, from his angle, a controversial idea that should require extraordinary proof. This said, many scholars have raised issues with the science element of evolutionary psychology. Perhaps most notably, philosopher of science David Buller (2005) presented a critique of evolutionary psychology that addresses both the theoretical underpinnings and several methodological points pertaining to research in this area.

A major element of Buller's (2005) critique pertains to what has been called a "just so story"—or the idea that evolutionary psychologists will often take any finding and mold it into an evolutionary explanation in an after-the-fact (post hoc) manner. This accusation is important and is something that we need to strongly consider.

Other critiques (summarized well by Grossi, Kelly, Nash, & Parameswaran, 2014) consider evolutionary psychology "dangerous" and see it as a tool to keep the status quo in society, supporting a male-dominated and patriarchal approach to society. This largely feminist-based critique has permeated responses to evolutionary psychology—particularly in light of research related to evolved sex differences, which is a major area of inquiry in the field. Acknowledging the concerns that such scholars raise, along with addressing them in a scientific rather than political manner, is critical to helping the field advance.

The idea of "bad science" can be taken in various ways. Here, the specific points of criticism often revolve around (a) non-cross-cultural samples while making universalist claims, (b) small sample sizes, and (c) an overreliance on nonexperimental methodologies. Further, issues of philosophy of science, such as falsifiability of the main premises of the evolutionary paradigm have been broached as well.

As evolutionary psychologists, we are unable to conduct experimental research going back into deep time and changing the course of human evolution to test the effects of certain variables. In fact, to make inferences regarding the evolutionary underpinnings of behavior, we must rely on our understanding of the principles of evolution and existing research to help us explain what we are studying. And, to be honest, sometimes evolutionary psychologists may well overstep the boundaries, finding something simple that fits with an evolutionary explanation and weaving together a just-so story. Importantly, this is likely not intentional—and critics such as Buller are helpful in keeping us honest.

How can we address the just-so story critique? For one, if we are making inferences about human universals (as evolutionary psychologists often are), then we should collect data in a way that best captures this point. As Schmitt and Pilcher (2004) indicate, the better we can collect data on a single topic, the better positioned we are to make claims about such evolutionary concepts as adaptations. For Schmitt and Pilcher, collecting data from samples from multiple cultures and multiple modes (e.g., self-reported data along with physiological data) goes a long way toward helping us make evolution-based inferences.

Critics such as Buller (2005) will often also critique our theoretical approaches—sometimes calling evolutionary psychology, in the terms of the great philosopher Karl Popper (1984), as *unfalsifiable*. That is, some point out that whatever the outcome of a study is, evolutionary psychologists may have a tendency to essentially say, "Yup, that makes sense from an evolutionary perspective!" This criticism essentially suggests that we overapply evolutionary principles and explanations and that we don't have any safeguards against such overapplication.

Perhaps the most powerful response to these kinds of claims, for us, anyway, comes from Ketelaar and Ellis (2000), who argue that an area of scientific inquiry shouldn't be judged exclusively by whether it is falsifiable. Falsifiability is important in terms of specific elements of a large theory—you need to be able to falsify basic ideas in a theory that comes from a paradigm. This said, Ketelaar and Ellis suggest that we go beyond falsifiability in assessing the usefulness of a scientific approach. From their reasoned perspective, a large-scale paradigm (as evolutionary psychology is) should be judged primarily by how *progressive* it is and how able that field is to *digest anomalies*. In this context, progressivity refers to the ability to generate novel research questions and, accordingly, the ability to provide new answers and new information about the world that would not be known otherwise. In fact, evolutionary psychology is famous for this kind of work. Research topics such as the influence of step-parenting on family violence (Daly & Wilson, 1988), the evolutionary function of infantile crying and other attachment behaviors (Bowlby, 1969), and the universal nature of human emotion expression (Ekman & Friesen, 1968) all are topics that have been enormously illuminated by an evolutionary approach to behavior—and, in fact, without an evolutionary framework to guide this research, these topics would remain poorly understood to this day. This said, ideas that are grounded in evolutionary principles need to be testable and, thus, falsifiable so as to remain true to the empirical nature of this field.

Evolutionary psychology, in a similar vein, has shown the ability to guide research that digests anomalies. Before Daly and Wilson (1988) applied an evolutionary framework to understanding the role of being a step-parent in family violence, the effects of being a step-parent on violence was simply poorly understood and didn't fit in with then-existing theories of violence. This phenomenon was an anomaly. It was the evolutionarily informed work that Daly and Wilson did that digested this anomaly—and more.

So, while the issues of falsifiability and just-so stories are valid, raising these matters to the consciousness of evolutionary psychologists has important potential to help strengthen our science. So our comment to the critics is this: Keep it up—and thank you!

The Eugenics Controversy

An atypical, ardently negative criticism of evolutionary psychology that we have become aware of (from several of students) suggests that evolutionary psychology is, in fact, a form of eugenics. As we argue in this section, evolutionary psychology is absolutely not synonymous with eugenics. Period. Eugenics is all about how human societies *should* selectively breed people so that only relatively fit individuals are the ones to reproduce. The goal of eugenicists is to create an optimal species. What a disturbing idea this eugenics is! Further, how far from evolutionary psychology it is! Consider, for instance, male sexual jealousy (Daly, Wilson, & Weghorst, 1982)—the tendency, documented across cultures, for males to be particularly upset by thoughts of their female romantic partners engaging in sexual infidelity coupled with a proclivity toward committing relatively aggressive acts tied to sexual infidelity in violent ways (relative to females). Evolutionary psychology is interested in how this phenomenon may be species-typical and how it may have been shaped by natural selection. Further, evolutionary psychologists are interested in understanding the detrimental impact of this phenomenon on society and are interested, further, in using knowledge gleaned from evolutionarily guided research to help solve social problems associated with this phenomenon.

On the other hand, someone adopting a eugenics perspective would be focusing on improving the species in terms of optimizing the gene pool—thus, a eugenicist would see such jealousy as bad insofar as it may work to preclude the most fit among us from having more mates than others!

Generally speaking, an evolutionary psychologist is focusing on human behavior as shaped to optimize individuals' own chances of reproduction. Note that while important advances in evolutionary theory on this topic have been made, which allow for various kinds of group-selection processes to be conceptualized as evolvable (Wilson, 2007), the kind of "species-level" selection that eugenicists would espouse is very different from the kind of multilevel selection processes that modern evolutionists study. For most intents and purposes, modern evolutionary psychology is a nonspecies-selectionist approach to understanding behavior. It very much focuses on behavior as largely serving the purpose of getting one's own genes into the future—with essentially no regard for *saving the species*. A eugenicist, on the other hand, believes that we *should* use our understanding of the effects of genes on behavior and bodies to consciously choose who should reproduce and who should not for the good of the species. This perspective suggests that we should optimize the gene pool of the species via selective breeding—that is the goal of eugenics. That is not at all the goal of evolutionary psychology.

From the perspective of eugenics, we should all work to have people like Arnold Schwarzenegger and Beyoncé do all the mating for our species. From the perspective of evolutionary psychology, people were shaped by natural selection to endorse nothing of the kind; rather, from this perspective, we were shaped to work to reproduce our own particular genes, regardless, in fact, of whether we believe ours may actually be the best in the pool! As is delineated in Table 8.1, evolutionary psychology and eugenics differ in:

(a) *the level of selection* (for evolutionary psychology, selection happens at the level of the individual, whereas eugenics is a naïve-species-selectionist idea).
(b) *the selector* (for evolutionary psychology, the selector of heritable qualities is blind natural selection; for eugenics, the selector is a group of humans with conscious intent).
(c) *their basic goals* (the goal of an evolutionary psychologist is to use insights gleaned from evolutionary theory to understand human behavior; the basic goal of eugenics is to improve the human gene pool for the purposes of some small, powerful group).

In thinking about eugenics, a reader might be thinking about who modern-day eugenicists are. While we are clearly arguing that anyone looking to evolutionary psychology for hints of eugenics is barking up the wrong tree (so to speak), there are clearly eugenicist implications found in many modern social movements. Given the historical atrocities associated with eugenics and the potential misuse of modern technologies, we think it

TABLE 8.1. Distinguishing Evolutionary Psychology from Eugenics

	EVOLUTIONARY PSYCHOLOGY	EUGENICS
Level of selection	Natural selection happens at the level of the individual organism. Psychological qualities viewed as "adaptations" are qualities that confer survival and/or reproductive benefits to the organisms possessing the particular qualities. The entity that is presumably "benefiting" here is the individual.	In large part, eugenics is a naïve-species-selectionist doctrine. It suggests that people should work together in selectively breeding humans to make it so that the species will benefit in the future. The entity that is presumably "benefiting" here is the species.
The selector	The process of natural selection (and, perhaps, other evolutionary processes such as sexual selection). Natural selection is a blind process with no intention and no plan. The selector here is a natural process fully devoid of human intentions and political agendas.	Individuals or groups of individuals with particular intentional plans/objectives and, often, particular political agendas. The selector here is a fully human entity, replete with intentions and political agendas.
Basic goal	Evolutionary psychology represents a basic scientific endeavor. The goal is to use our understanding of evolutionary principles so as to optimize our ability to understand human behavior and psychological processes. This basic scientific paradigm does not have a specific political agenda; increasing understanding of human psychology is the agenda.	The goal of eugenics is quite applied in nature. The point of this perspective is to apply our understanding of genes to a program of selective breeding of humans. This applied perspective has a very specific agenda.
Consciousness	Many psychological processes that are studied by Evolutionary Psychologists are unconscious in nature. For instance, Cosmides and Tooby (1992) argue that we differentially apply rules of logic, unknowingly, when we are faced with highly evolutionarily relevant versus relatively evolutionarily nonrelevant judgments. Such unconscious processes were shaped by natural selection to serve the purposes of individual reproduction.	The basic idea of eugenics is a highly conscious one. There is not a focus on unconscious psychological processes. Rather, from this perspective, there is a clear and highly conscious plan. The plan is for members of society to selectively breed in a way that would lead to an optimized gene pool for the society at large in the future.

(*continued*)

TABLE 8.1. Continued

	EVOLUTIONARY PSYCHOLOGY	EUGENICS
Thoughts on Arnold Schwarzenegger	From the perspective of evolutionary psychology, this man has been endowed with highly adaptive genes. Good for him. Evolutionary psychologists do not want (consciously or not) him to out-reproduce them. Heterosexual male evolutionary psychologists involved in monogamous relationships would not prefer that their female partners would mate with Arnold rather than with themselves.	A eugenicist might see Schwarzenegger as a horse breeder would see a blue-ribbon stallion: He should be used as a stud and should be encouraged, from this perspective, to mate with as many (relatively fit) females as possible in hopes of improving the species.

is very much worth our time to consider current technological, social, and intellectual trends that may ultimately provide a basis for future eugenicist endeavors.

One strikingly large such social movement concerns observations in sperm-donation trends. In sperm donation, women are able to choose qualities of their offspring based on phenotypical features of genetic fathers who have donated sperm. Consider an article published in the *New York Times* (Egan, 2006) dealing with the prevalence of women choosing to have children via sperm donation with no paternal care to assist in the parenting process. According to this article, "the California Cryobank, the largest sperm bank in the country, owed a third of its business to single women in 2005, shipping them 9,600 vials of sperm, each good for one insemination."

In addition to the relatively large economic niche that sperm donation is filling in industrialized societies, this *New York Times* article addresses the nature of the donors who are selected as fathers. The results presented in this article are eye-opening. For instance, Egan (2006) writes, "Short donors don't exist; because most women seek out tall ones, most banks don't accept men under 5-foot-9." Further, the article goes on to describe a woman who chose sperm from a tall German rugby player (whom the mother in question describes as "Aryan"). One could argue that the mothers who are choosing sperm in this way are engaging in eugenicist practices. In fact, the parallels between sperm choice and eugenics are made quite explicitly by this allusion to the "Aryan" sperm donor.

This line of thought, interestingly, extends to all nonrandom mate-choice processes in any sexually reproducing species (Miller, 2000). Once individuals within a species are using criteria to selectively choose to mate with individuals based on the presence of certain phenotypic qualities, parallels regarding eugenics may become apparent. In writing on this topic, Miller (2000) writes, "Finding mates with good genes is one of the major functions of mate choice (across all sexually reproducing species)" (p. 431). He further writes (p. 431): "We could outlaw genetic screening for heritable traits, but I imagine that our jails would have difficulty housing all of the sexually reproducing animals in the world that exercise mate choice—the female humpback whales alone would require prohibitively costly, high-security aquariums."

Our point in describing the parallels between the sperm-donation industry, mate-choice in general, and eugenics is not to sound alarm bells (although this analysis does raise concerns that should be addressed!). Rather, our point here is that there are existing practices in all societies which potentially do have some eugenicist overtones. Further, importantly, work within the domain of evolutionary psychology that is conducted by scholars who are interested in helping us understand human nature, simply, has no conceptual and/or empirical overlap with eugenics whatsoever.

Evolutionary Psychology's Future

Evolutionary psychology is a powerful approach to understanding all aspects of human behavior (Geher, 2014). Further, this area of inquiry is a basic element of the broader field of evolutionary studies (Garcia et al., 2011). This said, controversies have surrounded the field of evolutionary psychology for decades, making progress in this field difficult in a variety of ways (Glass, Wilson, & Geher, 2012). In fact, several scholars in our field have written strong responses to criticisms—including responses by Hagen and Symons (2007), Confer et al. (2010), and Richard Dawkins (2005), among others. This suite of responses to criticisms provides various insights into the issues that have been raised along with research and logic-based responses that help provide a framework for moving our field forward.

Shrouded in controversy, evolutionary psychology often has run into resistance from angles across the political spectrum. From the political right, religious fundamentalists have expressed concerns about this perspective based on its ideas regarding the nature of human origins. And from the far political left, evolutionary psychology has encountered resistance regarding claims that this perspective espouses a genetically deterministic view of human nature.

Perhaps the largest controversy in the field pertains to the idea of evolved behavioral sex differences in humans—an issue that sparks large-scale discussions regarding the idea of men and women being naturally different from one another in certain ways. An additional controversy pertains to the idea of genetic determinism—or what it means for genes to affect behavior. Further, concerns have been raised regarding the quality of the science employed by evolutionary psychologists as well as the socially concerning notion that evolutionary psychology is a modern form of eugenics. While evolutionary psychologists have made several sharp and coherent statements speaking to these controversial issues, it's clear that having critics raise concerns about this field has the capacity to improve the mission, coherence, and work that characterizes the future of evolutionary psychology.

Given evolutionary psychology's central placement in the broader field of evolutionary studies (Garcia et al., 2011), we believe that future work in this area needs to continue to underscore connections with disciplines outside of psychology. Further, consistent with Wilson's (2007) ideas on the importance of underscoring the applicability of evolutionary constructs to pave the way for their acceptance, future work in the field of evolutionary psychology that emphasizes beneficial applications of this field for humanity (e.g., Geher & Wedberg, in press) should help bring this work closer to the main stream.

Copyright Notes

1. Parts of this chapter were adapted from Chapter 9 of Glenn Geher's (2014) book *Evolutionary Psychology 101* with the expressed written consent of Springer Publishing.
2. Parts of this chapter were adapted from Glenn Geher's article from 2006 published in *Entelechy* titled "An Evolutionary Basis to Behavioral Differences between Cats and Dogs" with the expressed written consent of the editor.
3. Parts of Chapter this chapter were adapted from Glenn Geher's article "Evolutionary Psychology is Not Evil!" published in *Psychological Topics*—with the expressed written consent of the editor.

4. Parts of this chapter were adapted from Glenn Geher's article "There are No Evolved Behavioral Sex Differences in Humans Because I Want it That Way!" published in *EvoS Journal* in 2010—with the expressed written consent of the editor.
5. Parts of this chapter are adapted from Glenn Geher's (2015) *Psychology Today* blog post titled "The Power of Evolutionary Psychology."

References

Bowlby, J. (1969). *Attachment and loss. Vol. 1: Attachment*. New York, NY: Basic Books.

Buller, D. J. (2005). *Adapting Minds: Evolutionary Psychology and the Persistent Quest for Human Nature*. Cambridge, MA: MIT Press/Bradford Books, 2005.

Buss, D. M. (2016). *The evolution of desire: Strategies of human mating*. New York, NY: Basic Books.

Carmen, R. A., Geher, G., Glass, D. J., Guitar, A. E., Grandis, T. L., Johnsen, L., . . . Tauber, B. R. (2013). Evolution integrated across all islands of the human behavioral archipelago: All psychology as evolutionary psychology. *EvoS Journal, 5*(1), 108–126.

Confer, J. C., Easton, J. A., Fleischman, D. S., Goetz, C., Lewis, D. L., Perilloux, C., & Buss, D. M. (2010). Evolutionary psychology: Questions, prospects, and limitations. *American Psychologist, 65*, 110–126.

Cosmides, L., & Tooby, J. (1997). The modular nature of human intelligence. In A. B. Scheibel & J. W. Schopf (Eds.), *The origin and evolution of intelligence*. Sudbury, MA: Jones and Bartlett.

Daly, M., & Wilson, M. (1988). *Homicide*. New York, NY: Aldine de Gruyter.

Daly, M., Wilson, M., & Weghorst, S. J. (1982). Male sexual jealousy. *Ethology and Sociobiology, 3*, 11–27.

Daly, M., & Wilson, M. (2005). The "Cinderella effect" is no fairy tale. *Trends in Cognitive Sciences, 9*, 507–508.

Dawkins, R. (2005). Afterword. In D. M. Buss (Ed.), *The handbook of evolutionary psychology*. New York, NY: Wiley.

Egan, J. (2006, March 19). Wanted: A few good sperm. *New York Times Magazine*. Retrieved from https://www.nytimes.com/2006/03/19/magazine/wanted-a-few-good-sperm.html

Ekman, P. & Friesen, W. V. (1968). Nonverbal behavior in psychotherapy research. In J. Shlien (Ed.), *Research in psychotherapy* (Vol. 3, pp. 179–216). Washington, DC: American Psychological Association.

Figueredo, A. J., Cabeza de Baca, T., & Woodley, M. A. (2013). The measurement of human life history strategy. *Personality and Individual Differences, 55*, 251–255.

Gangestad, S. W., & Simpson, J. A. (2000). The evolution of human mating: Trade-offs and strategic pluralism. *Behavioral and Brain Sciences, 23*, 573–587.

Garcia, J. R., Geher, G., Crosier, B., Saad, G., Gambacorta, D., Johnsen, L., & Pranckitas, E. (2011). The interdisciplinary context of evolutionary approaches to human behavior: A key to survival in the ivory archipelago. *Futures, 43,* 749–761.

Geary, D. C. (2015). The classification and cognitive characteristics of mathematical disabilities in children. In R. C. Kadosh & A. Dowker (Eds.), *Oxford Handbook of Numerical Cognition* (pp. 767–786). Oxford, UK: Oxford University Press.

Geher, G. (2006a). An evolutionary basis to behavioral differences between cats and dogs? An almost-serious scholarly debate. *Entelechy: Mind and Culture, 7.* http://www.entelechyjournal.com/glenngeher.html

Geher, G. (2006b). Evolutionary psychology is not evil . . . and here's why . . . *Psihologijske Teme, 2,* 181–202.

Geher, G. (2014). *Evolutionary psychology 101.* New York, NY: Springer.

Geher, G., & Gambacorta, D. (2010). Evolution is not relevant to sex differences in humans because I want it that way! Evidence for the politicization of human evolutionary psychology. *EvoS Journal, 2,* 32–47.

Geher, G., & Wedberg, N. A. (in press). *Positive evolutionary psychology: Darwin's guide to living a richer life.* New York, NY: Oxford University Press.

Glass, D. J., Wilson, D. S., & Geher, G. (2012). Evolutionary training in relation to human affairs is sorely lacking in higher education. *EvoS Journal, 4,* 16–22.

Gross, M. R. (1982). Sneakers, satellites, and parentals: Polymorphic mating strategies in North American sunfishes. *Zeitschrift fur Tierpsychologie, 60,* 1–26.

Grossi, G., Kelly, S., Nash, A., & Parameswaran, G. (2014). Challenging dangerous ideas: A multi-disciplinary critique of evolutionary psychology. *Dialectical Anthropology, 38,* 281–285.

Hagen, E. H., & Symons, D. (2007). Natural psychology: The environment of evolutionary adaptiveness and the structure of cognition. In S. Gangestad & J. Simpson (Eds.), *The evolution of mind: Fundamental questions and controversies* (pp. 38–44). New York, NY: Guilford.

Ketelaar, T., & Ellis, B.J. (2000). Are evolutionary explanations unfalsifiable? Evolutionary psychology and the Lakatosian philosophy of science. *Psychological Inquiry, 11,* 1–21.

Kruger, D. J., & Nesse, R. M. (2006). An evolutionary life-history framework for understanding sex differences in human mortality rates. *Human Nature, 17,* 74–97.

Kurzban, R., & Haselton, M. G. (2005). Making hay out of straw: Real and imagined debates in evolutionary psychology. In J. Barkow (Ed.), *Missing the revolution: Missing the revolution: Darwinism for social scientists* (pp. 149–162). New York, NY: Oxford University Press.

Maynard Smith, J. (2002). *Evolutionary genetics.* New York, NY: Oxford University Press.

Miller, G. F. (2000). *The mating mind: How sexual choice shaped the evolution of human nature.* New York, NY: Doubleday.

Perrill, S. A., Gerhardt, H. C., & Daniel, R. (1978). Sexual parasitism in the green tree frog (*Hyla cinerea*). *Science, 200,* 1179–1180.

Popper, K. R. (1984). Zwei Bedeutungen von falsifizierbarkeit [Two meanings of falsifiability]. In H. Seiffert & G. Radnitzky (Eds.), *Handlexikon der Wissenschaftstheorie.* Munich, Germany: Deutscher Taschenbuch Verlag.

Ross, L., & Nisbett, R.E. (1991). *The person and the situation: Perspectives of social psychology*. New York, NY: McGraw Hill.

Schmitt, D. P. (2014). Evaluating evidence of mate preference adaptations: how do we really know what Homo sapiens really want? In V. A. Weekes-Shackelford & T. K. Shackelford (Eds.), *Evolutionary perspectives on human sexual psychology and behavior* (pp. 3–39). New York, NY: Springer.

Schmitt, D. P., Alcalay, L., Allik, J., Ault, L., Austers, I., Bennett, K. L., . . . Zupanèiè, A. (2003). Universal sex differences in the desire for sexual variety: Tests from 52 nations, 6 continents, and 13 islands. *Journal of Personality and Social Psychology, 85*, 85–104.

Schmitt, D. P., & Pilcher, J. J. (2004). Evaluating evidence of psychological adaptation: How do we know one when we see one? *Psychological Science, 15*, 643–649.

Smith, D. L. (2008). *The most dangerous animal*. New York, NY: St. Martin's Griffin.

Trivers, R. (1972). Parental investment and sexual selection. In B. Campbell (Ed.), *Sexual selection and the descent of man 1871–1971*. Chicago, IL: Aldine.

Wilson, D. S. (2007). *Evolution for everyone: How Darwin's theory can change the way we think about our lives*. New York, NY: Delacorte.

CHAPTER 9 | Evolution, Religion, and Other Meaning Systems

DAVID SLOAN WILSON

RELIGIONS HAVE PUZZLED THE scientific and rational imagination since long before Darwin. Two aspects of religion demand an explanation: First, why do people believe in the existence of supernatural agents, despite the absence of empirical evidence? Second, why do religious beliefs cause people to behave in ways that are so detrimental to their welfare? Why would Abraham obey a command from his God to kill his own son Isaac, for example?

Attempts to answer these questions themselves fall into two broad categories. First, religious beliefs and practices might be just as irrational and costly as they seem, in which case they need to be explained as some sort of byproduct of more rational and utilitarian beliefs and practices. For example, the 19th-century religious scholars E. B. Tylor (1832–1917) and J. G. Frazer (1854–1941) theorized that religions are naïve scientific theories, attempts by simple people to explain the world around them that just happen to be wrong. If this interpretation is correct, then religions can be expected to wane in influence as more rational and scientifically supported explanations become available.

Alternatively, despite appearances, religious beliefs and practices might have a hidden secular utility after all. Emile Durkheim (1858–1917) was a champion of this view, which is embodied in his famous definition of a religion as "a unified system of beliefs and practices relative to sacred things . . . which unite into one single moral community called a Church, all those who adhere to them" (Durkheim, 1912/1995, p. 44). Not only did Durkheim postulate a functional basis for religion, but he specifically identified *the creation of a moral community* as its primary function.

If we regard these three figures as the founding fathers of religious scholarship, then the history of that subject has roughly the same vintage as the history of evolutionary thought. The two histories are entwined, since scholars of religion drew upon the ideas of evolutionists such as Charles Darwin and Herbert Spencer to a degree. However, the two histories are also largely separate, in the same way that all of the subjects associated with the human social sciences and humanities became detached from evolutionary theory during most of the 20th century (Wilson & Paul 2016; Wilson & Schutt, 2016). Hence, social scientists and humanist scholars of religion had over a century to reach a consensus on whether religious beliefs and practices are wasteful byproducts or useful adaptations without employing an explicitly evolutionary perspective. For the most part, they failed, in part because the study of religion was spread among many academic disciplines and lacked any kind of unity. The tradition of functionalism initiated by Durkheim peaked in the mid-20th century and was not associated with evolution even at its height (e.g., Evans-Pritchard, 1965; Wilson, 2002, Chapter 2). One of the most influential sociological theories of the late 20th century imagines religion as a wasteful byproduct of the economic mind, bargaining with imaginary agents for goods that can't be had (Stark & Bainbridge, 1987).

Against this background, the field of Evolutionary Religious Studies (ERS) was part of a more general rethinking of all human-related subjects that gave rise to terms such as evolutionary psychology, evolutionary anthropology, evolutionary economics, literary Darwinism, and so on. In this chapter, I will provide a progress report on ERS, stressing three major points.

1. During its short 20-year history, ERS has made more progress on the "adaptation versus byproduct" question than over a century of traditional religious scholarship.
2. ERS can be seamlessly integrated with the study of nonreligious cultural meaning systems.
3. Evolutionary theory can itself provide the foundation for a meaning system that strongly motivates action while remaining fully scientific.

Progress on the Adaptation versus Byproduct Question

Three books marked the start of ERS during the dawn of the 21st century: Pascal Boyer's (2001) *Religion Explained*, Scott Atran's (2002) *In Gods We Trust*, and my own *Darwin's Cathedral* (Wilson, 2002). For the most part, Boyer and Atran explained the elements of religions as byproducts of adaptations that evolved for other purposes, such as a psychological tendency to attribute agency to events, which is adaptive on balance even if it results in the false attribution of some events to supernatural agents. I advanced the Durkheimian position that most enduring religions are primarily "for the good of the group," although my argument was based on group-level selection as the process whereby the functional elements of religion evolved.

Thus, starting out, evolutionists were as divided on the "adaptation versus byproduct" question as other modern religious scholars and the founding fathers of religious scholarship. If we assess the current crop of books on religion from an evolutionary perspective, however, such as Robert Bellah's (2011), *Religion in Human Evolution,* Ara Norenzayan's (2013) *Big Gods*, and Dominic Johnson's *God Is Watching You* (2015), along with the rapidly expanding academic journal literature, we can see that evolutionists have reached a consensus that Durkheim was on the right track. Most enduring religions are impressively designed to foster cooperation among the believers. In reaching this consensus, evolutionists have made more progress in 20 years than the rest of the religious scholarship community during the previous century.

Why should this be? One reason is that evolutionists expect to provide a unifying theoretical framework for a topic such as religion, along with other topics such as psychology, anthropology, and economics. The very idea of a unifying theoretical framework has been abandoned by many other branches of scholarship in the human-related sciences and humanities. Another reason is that evolutionary theory is designed to distinguish between adaptation versus byproduct explanations. In one of my articles co-authored with the comparative religious scholar William Scott Green titled "Evolutionary Religious Studies (ERS): A Beginner's Guide" (Wilson & Green, 2011), we list six major hypotheses that need to be addressed for all products of genetic and cultural evolution, as shown in Table 9.1, which therefore serve as a blueprint for the study of religion.

The most important question is whether a given trait counts as an adaptation that evolved by contributing to survival and reproduction (the left hand column of Table 9.1). If so, what was the unit of selection? Did the

TABLE 9.1. Six Major Evolutionary Hypotheses about Religion

RELIGION AS AN ADAPTATION	RELIGION AS NONADAPTIVE
Group-level adaptation (benefits groups, compared to other groups)	Adaptive in small groups of related individuals but not in modern social environments
Individual-level adaptation (benefits individuals, compared to other individuals within the same group)	Byproduct of traits that are adaptive in nonreligious contexts
Cultural parasite (benefits cultural traits at the expense of human individuals or groups)	Neutral traits (drift)

NOTE: The table provides three adaptive explanations (first column) and three are nonadaptive explanations (second column).

trait evolve by virtue of increasing the fitness of individuals, relative to other individuals within the same group (*within-group selection*) or by increasing the fitness of groups, relative to other groups in a multigroup population (*between-group selection*)? A third possibility is that the trait evolved for its own benefit at the expense of both individuals and groups (*a parasitic gene or meme*). Each of these is a plausible hypothesis that must be empirically evaluated on a trait-by-trait basis.

If the trait is not an adaptation, then why does it persist in the population (the right hand column of Table 9.1)? Perhaps it is a *byproduct* of an adaptation, similar to the triangular space called a spandrel that forms whenever arches are placed next to each other in the construction of a building. Arches have a function but spandrels do not, although they can acquire a secondary function such as a decorative space. Gould and Lewontin (1979) famously used this example to distinguish between adaptations and byproducts in biological evolution. Another possibility is an *evolutionary mismatch*, whereby a trait that was adaptive in a previous environment has become maladaptive in the current environment. A third possibility is *drift*. Whenever alternative traits have the same effect on fitness, what evolves is a matter of chance. All three of these nonadaptation hypotheses are also plausible and must be evaluated empirically, along with the three adaptation hypotheses, for any phenotypic trait associated with religion.

Testing among six major hypotheses is not always easy. Moreover, questions about the function (or lack thereof) of a trait are only part of a fully rounded evolutionary approach. We also need to understand the mechanisms, development, and phylogeny of the same traits (Tinbergen,

1963). To make matters even more complex, we need to employ this theoretical framework for products of cultural evolution in addition to products of genetic evolution, which operate on different time scales and co-evolve with each other. Thus, studying religion from an evolutionary perspective is not a simple matter, but the theoretical framework provides much more guidance than most (I would argue any) other perspective.

One of my articles titled "Testing Major Evolutionary Hypotheses about Religion with a Random Sample" (Wilson, 2005) shows how I employed the theoretical framework during the early days of ERS. With the help of a class of students, we selected 35 religions at random from a 16-volume encyclopedia of religion and evaluated the six hypotheses for each one using information available in the academic literature. The literature was almost entirely qualitative (descriptive accounts of a given religion), but our study had the virtue of avoiding cherry-picking examples to support one's favored view of religion. Whatever conclusions can be drawn from the sample hold for the population of religions from which the sample was drawn. It was clear from the ethnographic descriptions of the religions in our sample that most of them were patently designed to foster cooperation within the communities of believers. This was true even for religious movements that ended up failing or that superficially appeared highly costly and dysfunctional for their members (e.g., Jainism).

It is important to stress that ERS does not lead to a monolithic conclusion about the nature of religion. Religions are an inherently fuzzy set of cultural traits with genetic underpinnings. Each trait associated with religion must be evaluated on a case-by-case basis. All six hypotheses are plausible and examples of all can be found, just as for the study of genetically encoded traits in nonhuman species. There are religious traits that have evolved by within-group selection or that parasitically spread at the expense of both individuals and groups. Most adaptations result in functionless byproducts, evolutionary mismatches invariably occur in changing environments, and some measurable religious traits can probably be explained by drift (one candidate for drift is the positive versus the negative form of the golden rule). Also, a trait associated with religion can qualify as a byproduct as far as its genetic underpinnings are concerned (such as agency detection) and an adaptation as far as cultural evolution is concerned (conceptions of Gods that foster cooperation, as opposed to many other possible conceptions of Gods that could have evolved). It is only on balance that we can make a statement such as, "Most enduring religions are impressively designed to foster cooperation among the community of believers."

My two most recent contributions to ERS illustrate the current state of the discipline. The first is a review article titled "The Nature of Religious Diversity: A Cultural Ecosystem Approach" (Wilson, Hartberg, MacDonald, Lanman, & Whitehouse, 2016), which is published with commentaries in the journal *Religion, Brain, and Behavior*. The commentaries provide a snapshot of the discipline as a whole. The second is titled "Sacred Text as Cultural Genome: An Inheritance Mechanism and Method for Studying Cultural Evolution" (Hartberg & Wilson, 2017), which begins to go beyond questions about function to address questions about proximate mechanisms.

In sum, I regard ERS as an outstanding success story demonstrating the "added value" of approaching a major topic area such as religion from an evolutionary perspective.

Integrating ERS with the Study of Nonreligious Meaning Systems

When religion is studied from an evolutionary perspective, it leads to the larger and more inclusive study of *meaning systems*, which I define as "sets of beliefs and practices that receive environmental information as input and results in action as output" (discussed under the label of "Unifying Systems" in Wilson, 2002, Chapter 7). We are such a cultural species that all of us have a meaning system, regardless of whether or not it counts as religious. Note that my definition of a meaning system bears an intriguing resemblance to the definition of a nervous system, which also receives environmental information as input and results in action as output. Thus, in functional terms, a meaning system can be regarded as the nervous system of a group of people sharing a given culture.

Thinking about meaning systems in general terms can help put one of the main enigmas about religion—counterfactual beliefs—into perspective. All beliefs can be evaluated according to two criteria: (a) how well they correspond to factual reality in a rationalistic and scientific sense and (b) how they cause the believers to act (Wilson, 1990). These can be termed factual realism and practical realism, respectively, and it is interesting that the word *realistic* is used in both senses. We describe a portrait as realistic when it corresponds closely to the person that was painted. We describe a business improvement plan as "realistic" if it is likely to work. We toggle back and forth between the two meanings of the word, depending upon the context, without consciously needing to think about it.

Evolution is sensitive only to how people act—practical realism—and is sensitive to factual realism only insofar as it leads to practical realism. Our ability to apprehend factual reality therefore depends critically on the relationship between factual and practical realism. The most fundamental question that can be asked about epistemology (a theory of knowledge) is therefore: "When does practical realism *require* factual realism and when does it require *departing* from factual realism?" It is easy to imagine examples of both positive and negative tradeoffs, depending upon the context. A hunter needs to know the exact location of its prey to hit it (a positive trade-off between practical and factual realism). In a territorial dispute, it might be more motivating to regard one's enemy as an inhuman monster than as someone much like oneself competing for the same square of ground (a negative trade-off between practical and factual realism). Just as we effortlessly toggle back and forth between the two meanings of the word *realistic,* we should also expect human meaning systems to toggle back and forth between close adherence to factual realism and wanton departures from factual realism, depending upon the context.

Importantly, *this prediction should apply to nearly all meaning systems, not just religious meaning systems.* Nonreligious meaning systems might not invoke supernatural agents (which is why we classify them as nonreligious), but most of them still wantonly depart from factual reality in other respects. Consider the beliefs that people form and defend about their own past. They are shot through with falsehoods, as any comparison between patriotic histories and scholarly histories attest, along with disputes that take place among the scholars (Hobsbawm & Ranger, 1983). Another example is modern economic theory, which for all its secular trappings functions as a religion and wantonly departs from factual knowledge about human psychology and social systems (Cox, 1999, 2016).

It follows that what puzzles the scientific and rational imagination about religion—counterfactual beliefs and the actions they motivate—should be extended to all human meaning systems. And the same evolutionary toolkit that has made such progress for the study of religion can make equal progress for other meaning systems. Wilson et al.'s (2016) article cited earlier takes a step in this direction. Although it is titled "The Nature of Religious Diversity," it could equally be titled "The Nature of Cultural Diversity": It showcases a distinction between "tight" and "loose" cultures that can be applied to traditional societies and modern nations in addition to religions (Gelfand, Nishii, & Raver, 2006, 2011).

Toward a Meaning System Informed by Evolutionary Science

All meaning systems recognize the need for factual realism in at least some contexts, even while flagrantly departing from factual realism in other contexts. In modern times, a firm grasp of factual reality is needed more than ever to solve problems such as global climate change and sustainable economies. Is it possible for a meaning system to fully respect factual realism—avoiding adaptive fictions altogether—and still function as an effective "nervous system" for a group of people?

This question is not new. It has been the goal of some philosophical systems since antiquity, along with 19th century thinkers such as Herbert Spencer and August Comte, who strove to create a secular "Religion of Humanity" (Comte, 1851) and gave rise to the modern Humanist movement. Consider the American Humanist Association's (n.d.) current manifesto:

> Humanism is a progressive philosophy of life that, without supernaturalism, affirms our ability and responsibility to lead ethical lives of personal fulfillment that aspire to the greater good of humanity.
>
> The lifestance of Humanism—guided by reason, inspired by compassion, and informed by experience encourages us to live life well and fully. It evolved through the ages and continues to develop through the efforts of thoughtful people who recognize that values and ideals, however carefully wrought, are subject to change as our knowledge and understandings advance.

By its own account, humanism strives to be a meaning system—or "lifestance"—that motivates action. I would replace the phrase "without supernaturalism" to "without departing from factual realism in any way," but the manifesto as a whole is in this spirit, showcasing rational analysis and science as the best methods for deriving knowledge of the world and for solving problems.

Unfortunately, the modern humanism movement and other attempts to create a fully scientific meaning system have yet to succeed for two major reasons. First, while avoiding supernaturalism, humanists do not necessarily avoid other departures from factual realism. The most glaring example is the so-called New Atheist Movement, which not only makes assertions about the lack of evidence for supernatural agents but also about the nature of religion as a human construction that ignore developments in ERS described earlier in this chapter (Wilson, 2016). In particular, many New Atheists find it very difficult to accept the neo-Durkheimian view of religion that is supported by the evidence. As another example, Ayn Rand, who was the "new atheist" of her day, delighted in calling her meaning system of objectivism a "stylized universe," which can be shown to have the same rhetorical structure as a fundamentalist religion (Wilson, 1995). In general, it is unsurprising that humanists are more vigilant about avoiding supernatural claims than many other claims about the world that are equally counterfactual.

Second, humanism does not function well as a meaning system except for a very narrow segment of the human population. You can convince yourself of this fact by performing the following experiment. Type "humanist" into Google Image and count the number of people in all images that have one or more people. Type "evangelical" into Google Image and repeat the exercise. As shown in Figure 9.1, the vast majority of images of humanists feature a single individual, while the vast majority of images of evangelicals feature a crowd. In addition, most of the humanists in the images are white and quite old, while the evangelicals are all ages and an ethnic rainbow.

Figure 9.1 strongly suggests that humanism is a "turn on" for only a narrow segment of the human population and even then does not motivate action that takes place in groups. The slang phrase "turn on" is appropriate because that is exactly what a meaning system is supposed to do—strongly

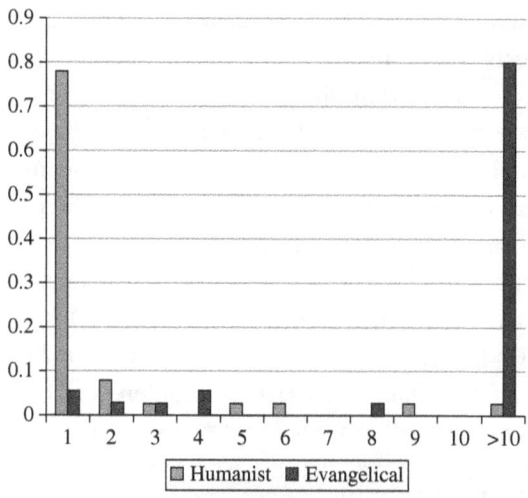

FIGURE 9.1. A frequency distribution of the number of people in images containing people obtained by typing either "humanist" or "evangelical" into Google Images.

motivate a suite of actions based on the perception and processing of information provided by the meaning system. It is humbling for scientists and rationalists such as myself to realize that we have a long way to go before a meaning system that scrupulously adheres to factual realism can rival those that capitalize on adaptive fictions in all their forms.

Nevertheless, when it comes to cultural evolution, the past does not predict the future. It is possible that modern evolutionary theory can do for humanism what it has already done for the academic study of religion—make rapid progress where decades and even centuries of previous efforts have failed. Before describing some encouraging developments, it is important to stress that the basic objective of using evolutionary theory to create a meaning system goes beyond the academic study of religion or meaning systems. Scientific knowledge by itself cannot inform action; to suggest otherwise is to commit the naturalistic fallacy (Curry, 2006; Wilson, Dietrich, & Clark, 2003). Scientific knowledge must be combined with values to inform action. In addition, a meaning system informed by science must be psychologically motivating, easily taught to children and other neophytes, communicated through stories and other artistic media, perhaps reinforced by rituals, and so on. All of this must result in people working effectively in groups to accomplish the goals motivated by the meaning system, which is presumably some version of global sustainability.

The enterprise of creating a meaning system takes some scientists beyond their comfort zone. While it is fine for some people to confine themselves to the generation of scientific knowledge, it would be wrong to criticize the enterprise of creating a meaning system as inappropriate for a scientist or somehow inherently unscientific. Evolutionary science can inform all aspects of the creation of a meaning system. With this in mind, I will end this chapter by describing two encouraging developments.

The first is a remarkable book titled *Beyond Religion: Ethics for a Whole World* by His Holiness the Dalai Lama (2011). As most readers will know, the Dalai Lama is the equivalent of the Catholic Pope for Tibetan Buddhism and has lived in exile since the Chinese invaded Tibet in 1950. During this period, the Dalai Lama has become a revered spiritual leader internationally, receiving the Nobel Peace Prize in 1989, among many other honors. He has also taken an exceptional interest in science and has educated himself by organizing a series of conferences on diverse topics, from neuroscience to evolutionary biology, over a period of decades. In *Beyond Religion*, he observes that the world is too culturally diverse for everyone to share the same religion. Furthermore, all religions (including his own) rely upon ethical principles that can be understood in purely secular (i.e., scientific) terms. A scientific understanding of concepts such as love, compassion, suffering, and interdependence can provide an ethics for the whole world, which can enlighten rather than threaten the teaching and practices of any particular religion or other meaning system. I highly recommend this book for its simplicity, clarity of thought, and unique perspective of its author.

The second development is a movement called interspirituality (Johnson & Ord, 2013; Wilson & Johnson, 2015), which notes that all religious and spiritual traditions converge upon a common awareness of rich interconnectedness. Scientific disciplines such as physics, environmental biology, and complex systems theory converge upon the same awareness. No matter how one arrives at a deeply systemic view, certain ethical conclusions follow. Namely, it becomes difficult to defend one part of the system against another part of the system. Almost all problems require solutions that take the whole system into account.

A deeply systemic view provides common ground for dialogue across all religious, spiritual, and scientific traditions. This is sometimes called "second-tier consciousness," in contrast to "first-tier consciousness," which is confined to a particular tradition. In keeping with the Dalai Lama's book, the two tiers of consciousness can exist in a positive relationship rather than one threatening the other. It is notable that the interspiritual movement is especially active among people and groups associated with the United Nations, where the need for dialogue across world views is at a premium (Johnson & Ord, 2013).

As someone who has been involved in the academic study of religion from an evolutionary perspective (ERS) since its modern inception, I regard the construction of a strong meaning system informed by evolutionary theory as by far the more important enterprise and the primary object of my own current efforts (e.g., http://www.evolution-institute.org/article/tvol1000 and http://www.prosocial.world). If comparable progress can be made, then 20 years from now the world will be a better place than it is at the moment.

References

American Humanist Association. (n.d.). *Humanism and its aspirations: Humanist Manifesto III, a successor to the Humanist Manifesto of 1933*. Retrieved from https://americanhumanist.org/what-is-humanism/manifesto3/

Atran, S. (2002). *In gods we trust: The evolutionary landscape of religion*. Oxford, England: Oxford University Press.

Bellah, R. (2011). *Religion in human evolution: From the Paleolithic to the Axial age.* Cambridge, MA: Belknap.

Boyer, P. (2001). *Religion explained.* New York, NY: Basic Books.

Comte, A. (1851). *Systeme de politique positive.* Paris.

Cox, H. (1999, March). The market as god. *Atlantic Monthly.* Retrieved from https://www.theatlantic.com/magazine/archive/1999/03/the-market-as-god/306397/

Cox, H. (2016). *The market as god.* Cambridge, MA: Harvard University Press.

Curry, O. (2006). Who's afraid of the naturalistic fallacy? *Evolutionary Psychology, 4,* 234–247.

Dali Lama, His Holiness. (2011). *Beyond religion: Ethics for a whole world.* New York, NY: Houghton Mifflin Harcourt.

Durkheim, E., & Fields, K. E. (1912). *The elementary forms of religious life.* New York, NY: Free Press.

Evans-Pritchard, E. E. (1965). *Theories of primitive religion.* Oxford, England: Oxford University Press.

Gelfand, M. J., Nishii, L. H., & Raver, J. L. (2006). On the nature and influence of cultural tightness-looseness. *Journal of Applied Psychology, 91,* 1225–1244.

Gelfand, M. J., Raver, J. L., Nishii, L., Leslie, L. M., Lun, J., Lim, B. C., . . . Yamaguchi, S. (2011). Differences between tight and loose cultures: A 33-nation study. *Science, 332,* 1100–1104.

Gould, S. J., & Lewontin, R. C. (1979). The spandrels of San Marco and the panglossian paradigm: A critique of the adaptationist program. *Proceedings of the Royal Society of London, B205,* 581–598.

Hartberg, Y. M., & Wilson, D. S. (2017). Sacred text as cultural genome: an inheritance mechanism and method for studying cultural evolution. *Religion, Brain & Behavior, 7,* 178–190.

Hobsbawn, E., & Ranger, T. O. (1983). *The invention of tradition.* Cambridge, England: Cambridge University Press.

Johnson, D. (2015). *God is watching you: How the fear of God makes us human.* Oxford, England: Oxford University Press.

Johnson, K., & Ord, D. (2013). *The coming interspiritual age.* Vancouver, BC: Namaste.

Norenzayan, A. (2013). *Big gods: How religion transformed cooperation and conflict.* Princeton, NJ: Princeton University Press.

Stark, R., & Bainbridge, W. S. (1987). *A theory of religion.* New Brunswick, NJ: Rutgers University Press.

Tinbergen, N. (1963). On aims and methods of ethology. *Zeitschrift Für Tierpsychologie, 20,* 410–433.

Wilson, D. S. (1990). Species of thought: a comment on evolutionary epistemology. *Biology and Philosophy, 5,* 37–62.

Wilson, D. S. (1995). Language as a community of interacting belief systems: A case study involving conduct toward self and others. *Biology and Philosophy, 10,* 77–97.

Wilson, D. S. (2002). *Darwin's cathedral: Evolution, religion and the nature of society.* Chicago, IL: University of Chicago Press.

Wilson, D. S. (2005). Testing major evolutionary hypotheses about religion with a random sample. *Human Nature, 16,* 382–409.

Wilson, D. S. (2016). Atheism as a stealth religion: Fiver years later. *This View of Life*. Retrieved from https://evolution-institute.org/article/the-new-atheism-as-a-stealth-religion-five-years-later/

Wilson, D. S., Dietrich, E., & Clark, A. B. (2003). On the inappropriate use of the naturalistic fallacy in evolutionary psychology. *Biology and Philosophy, 18*, 669–682.

Wilson, D. S., & Green, W. S. (2011). Evolutionary religious studies (ERS): A beginner's guide. In E. Slingerland & M. Collard (Eds.), *Creating consilience: Integrating science and the humanities: Interdisciplinary approaches* (pp. 225–242). Oxford, England: Oxford University Press.

Wilson, D. S., Hartberg, Y., MacDonald, I., Lanman, J. A., & Whitehouse, H. (2016). The nature of religious diversity: A cultural ecosystem approach. *Religion, Brain & Behavior, 7*, 134–153.

Wilson, D. S., & Johnson, K. (2015). The coming interspiritual age: A conversation with Kurt Johnson. *This View of Life*. https://evolution-institute.org/article/evolution-and-the-coming-interspiritual-age-a-conversation-with-kurt-johnson/

Wilson, D. S., & Paul, R. (2016). Cultural anthropology and cultural evolution: Tear down this wall! A conversation with Robert Paul. *This View of Life*. https://evolution-institute.org/article/cultural-anthropology-and-cultural-evolution-tear-down-this-wall-a-conversation-with-robert-paul/

Wilson, D. S., & Schutt, R. (2016). Why did sociology declare independence from biology (and can they be reunited)? An interview with Russell Schutt. *This View of Life*. https://evolution-institute.org/article/why-did-sociology-declare-independence-from-biology-and-can-they-be-reunited-an-interview-with-russell-schutt/

CHAPTER 10 | From Physical Culture to the Primal Life

Evolutionary Health Movements in Historical Context

HAMILTON M. STAPELL

IN 1973 THE BIOLOGIST Theodosius Dobzhansky famously said, "Nothing in biology makes sense except in the light of evolution." Human health—including nutrition, exercise, lifestyle, and disease—is no different. Evolutionary Studies can attempt to tackle the problem of human health from exactly this perspective. The ultimate goal of this interdisciplinary approach is to optimize human health and well-being, not just identify the baseline minimum requirements necessary for survival.

These evolutionary health theories are primarily based on the *mismatch hypothesis* (Lieberman, 2013). The theory is that we—as humans—are now poorly adapted to our modern environment. Predictably, the result is disease and ill health (Lindeberg, 2010). Basic evolutionary principles of natural selection, sexual selection, and cultural evolution offer us a lens through which to identify the ultimate source of these problems and to recommend possible solutions. The potential result is improved health outcomes for both individuals and society as a whole.

An evolutionary studies approach to human health begins with the fact that from the inception of the human genus, *Homo*, approximately 2.4 million years ago, our ancestors lived as hunter-gatherers. This equals approximately 84,000 generations. Survival within all types of hunter-gatherer environments required a large amount of daily energy expenditure, in activities such as food and water procurement, social interaction, escape from predators, and maintenance of shelter and clothing.

However, dramatic improvements in technology that spawned the agricultural revolution (350 generations ago), the industrial revolution (7 generations ago), and the digital age (2 generations ago) have produced large systematic reductions in the amount of physical work required by humans. These changes have also fundamentally altered the composition of our diet, the kinds of social interactions we experience, and the types of stressors we face.

Homo sapiens, of course, are still evolving. Yet, at the same time, the human genome has remained largely unchanged since the agricultural revolution of 10,000 years ago, with lactase persistence being one notable exception (Stock, 2008). Thus, our innate exercise and nutritional needs that evolved via natural selection over thousands of millennia remain largely the same today. The mismatch hypothesis suggests that marked deviation from ancestral exercise, nutritional, and lifestyle patterns predictably results in physical disability and the disease that we see today. It also suggests that greater alignment with ancestral patterns of living will improved diseased states and lead to greater health and well-being. In other words, not only do evolutionary principles help us identify the root problem, but they also provide a possible template from which to consider modern lifestyle choices.

It should be emphasized that this approach is not simply another fad diet or fleeting fitness trend. Rather, an evolutionary approach to health and disease offers a true "unifying field theory" for human well-being. In other words, it is a unified framework for guiding research, scientific inquiry, and experimentation. It also links together a broad spectrum of disciplines and specialties, ranging from anthropology and psychology to microbiology and epidemiology. Evolution provides the framework for organizing all of these fields of knowledge. Likewise, health, fitness, chronic disease, reproduction, and aging are all interconnected through the processes of evolution. The history of the twentieth century has also demonstrated how difficult it is to tackle each of those issues separately. Despite many advances, chronic diseases still plague our society, and healthcare costs are ever increasing. An evolutionary approach offers the opportunity to transcend the divisions inherent in the "Ivory Archipelago" of academia and medical research and to offer new insights into some of the most pressing health problems facing both individuals and our society (Platek et al., 2011). It provides explanations for our vulnerability to diseases. And it has the potential to offer practical medical applications from fields such as evolutionary genetics, molecular phylogenetics, and evolutionary epidemiology. In fact, this approach has already proven helpful in understanding the evolution of antibiotic resistance, virulence, autoimmune diseases, the microbiome, aging, and some forms of cancer (Nesse & Stearns, 2008; Stearns, Nesse, Govindaraju, & Ellison, 2010; Stearns, 2012).

An evolutionary studies approach to human health has pedagogical advantages as well. Namely, it provides a framework for organizing and teaching medical knowledge. Evolutionary perspectives on health and disease can also serve as a unifying framework within undergraduate health education. Right now, students too often learn about human health in a variety of separate academic disciplines. Instead, the fundamental ideas of natural selection, sexual selection, and cultural evolution can be applied broadly across the disciplines not only to better understand how the body works but also to investigate the ultimate causes of—and potential solutions to—many of our health problems today. For example, students should be asking why human biological systems are the way they are, and why those systems are potentially vulnerable to failure. In addition, the necessity and value of a basic evolutionary understanding can be coupled with the capacity for students to ask basic scientific questions in regard to human health. Put another way, evolutionary principles can be used to help teach basic science, and vice versa. Finally, the factors that lead to human wellness, or to disease, are complex. The evolutionary studies' methodology helps organize and unify all of the various strands. The result is a systems approach that is both more easily taught and understood.

By recognizing the wide variety of factors that lead to the success or failure of an organism in a given environment, this evolutionary lens offers a broad, or "holistic," approach to human health and well-being. So, in addition to focusing on nutrition and exercise, it also includes such issues as social connections, stress, sleep, sun exposure, circadian rhythm entrainment, and the gut microbiome. Again, in each of these cases, the idea is to better align our current lifestyle choices with the ancestral environment(s) in which our genome evolved. Yet, this is not a "one-size-fits-all" approach. Current evolutionary health theories acknowledge that not all individuals or groups evolved in the same environment. In fact, many researchers today stress the significant differences between ancient environmental niches (Kuipers, 2012). The result is "personalized" disease risk and lifestyle choices based on one's own personal ethnic and genetic makeup.

Both academic and popular interest in this evolutionary approach to health has grown dramatically over the past decade, and especially within the past five years (Schwartz & Stapell, 2013). Google searches for the term "Paleo diet" skyrocketed between 2010 and 2014,[1] with Google Trends even announcing it "The Diet of the Year" in its annual Zeitgeist list of 2013.[2] The *Paleo Solution* by Robb Wolf, *Practical Paleo* by Diane Sanfilippo, and *Paleo Cooking from Elana's Pantry* by Elana Amsterdam each hit the New York Times Bestseller Lists in 2011, 2012, and 2013, respectively. There is even a *Living Paleo for Dummies* (Joulwan & Petrucci, 2012) and *The Complete Idiot's Guide to the Eating Paleo* (Quinn & Glaspey, 2012). Hundreds of Paleo websites and podcasts now cover every aspect of the ancestral health movement, from sleep and nutrition to exercise and fertility. Advocates for living an ancestral lifestyle even appear in mainstream media outlets, such as *The Dr. Oz Show*.[3] The online Paleo Physicians Network (http://www.paleophysiciansnetwork.com) lists hundreds of medical professionals who embrace an evolutionary approach to medicine in the United States and abroad. The academic-based Ancestral Health Society (http://www.ancestralhealth.org), founded in 2010, has thousands of active members, publishes a peer-reviewed journal (*Journal of Evolution and Health*), and sponsors an annual conference. In addition, Randolph Nesse, one of the original founders of "Darwinian Medicine" in the 1990s, recently created the Center for Evolution and Medicine at Arizona State University, and Barbara Natterson-Horowitz, the author of the 2012 book, *Zoobiquity: The Astonishing Connection between Human and Animal Health*, currently co-runs the UCLA Evolutionary Medicine Program. Her interdisciplinary work has significantly increased the dialogue between doctors, veterinarians, psychiatrists, and evolutionary biologists.

Today, this large ancestral health movement (sometimes referred to as the *Paleo* or *primal* diet) is also quite heterogeneous. In general, it is populated by four main groups. First, academic researchers are developing both the theories and mechanisms of action to build and test evolutionary-based models of human health and disease. Second, a group of clinicians—including doctors, dentists, physical therapists, chiropractors, and mental health providers—are already using various evolutionary approaches in their practices. Third, there is a group of well-known "popularizers," who

---

[1] https://www.google.com/trends/explore#q=paleo%20diet
[2] http://www.google.com/trends/topcharts?zg=full
[3] On April 22, 2013, Dr. Loren Cordain, the author of several popular books, including *The Paleo Diet* (2010), and Nell Stephenson, the author of *Paleoista* (2012), appeared on *The Dr. Oz Show*; see Chris Kresser, the popular blogger and author of *The Paleo Code* (2013), also appeared on *The Dr. Oz Show* on February 17, 2014; see http://www.doctoroz.com/episode/your-personal-paleo-code-diet-lose-weight-and-get-healthy-life

have written best-selling books or run popular websites and podcasts. Some of the best examples of this include Mark Sisson, the creator of the website Mark's Daily Apple and the author of numerous primal books; Robb Wolf, who created one of the most popular podcasts and is a best-selling author; and Dallas and Melissa Hartwig, who created the Whole 30 Program and regularly appear on a number of mainstream media outlets. Fourth, educated laypersons make up the bulk of the movement. They buy the books, listen to the podcasts, attend annual conferences, and then spread the message through social media. Recent survey data suggest that these participants tend to be White, female, middle-aged (mean: 38 years old), in a committed relationship, highly educated, relatively affluent, and motivated by weight loss and health concerns. As for size, it has been estimated that the current ancestral health movement is between 1 million and 3 million persons (Schwartz & Stapell, 2013). Of course, these four main categories are not rigid, as some individuals play multiple roles. For instance, the academic researcher Loren Cordain has also published several best-selling books, and the clinician Chris Kresser is also a very popular author and podcast host.

There is a diversity of opinion within the movement as well. In other words, there is currently no single—or simple—evolutionary prescription for good health. For example, some proponents advocate the avoidance of all dairy products, while others see butter, yogurt, and other fermented daily products as healthy additions to the diet. There is also disagreement over the level of fruit and carbohydrate consumption, wine and alcohol, and legumes. There is another, and perhaps more significant, area of disagreement as well. In general, there is a divide between those, such as Randolph Nesse, who want to wait for further evidence from clinical trials before making any practical recommendation based on evolutionary principles, and those who want to use evolutionary theories to begin guiding lifestyle choices today.

It is also true that the degree of scientific rigor is not uniform throughout the movement. While most of these evolutionary approaches are well grounded in science, there is, of course, a degree of quackery out there as well. As is typical in any large and diverse movement, some proponents are better informed than others, while still others are more interested in creating a profitable business than in sticking to the scientific facts. And, inevitably, some well-intentioned beliefs and practices get out ahead of the science. The ancestral health movement also has not been shy about questioning conventional medical opinions, especially when it comes to such things as the benefits of whole grains, the dangers of saturated fats, and the appropriate amounts of exercise. All of this has led to criticism and to a number of stereotypes about the movement (see the following section Differences). Some of its followers have even been labeled as "kooks" by the mainstream media and the medical establishment (Schwartz & Stapell, 2013).

But where and when did all of this begin? And what kinds of societal changes have helped propel an interest in evolutionary approaches to human health and disease? S. Boyd Eaton's and Melvin Konner's 1985 paper, "Paleolithic Nutrition. A Consideration of Its Nature and Current Implications," in the *New England Journal of Medicine* is typically cited as the start of the current ancestral health movement, and, in many ways, justifiably so.[4] The article did attract significant new attention to the idea of an evolutionary approach to diet and lifestyle. However, this chapter argues that its historical antecedents stretch back more than 100 years to what is known as the *physical culture movement* of the late 19th and early 20th century. The interrelated and disruptive forces of industrialization, urbanization, and modernization produced this earlier movement and, in turn, helped to establish many of our modern notions of health, illness, and fitness.[5] Specifically, this period of rapid social, economic, and technological change led to a great deal of anxiety and uncertainty in the West about the future development of both individuals and nations as a whole, especially during a time of increasing competition and conflict before World War I (Budd, 1997).

In this context, the physical culture movement can be understood as a conscious attempt by certain segments of society, particularly the middle classes, to overcome this anxiety and disorientation. It was a highly diverse and amorphous movement but can be defined as the effort to build health, strength, and beauty in the human body through proper diet, exercise, and "a return to nature." And, although few remember it today, the physical culture movement was widely popular both in Europe and the United States, much more popular, in fact, than the ancestral health movement of today.

---

[4] While Eaton's 1985 paper, "Paleolithic Nutrition," is often considered the beginning of the current ancestral health movement, it was not the first work on this subject. Two important earlier books include *The Stone Age Diet: Based on In-depth Studies of Human Ecology and the Diet of Man* (Voegtlin, 1975) and *Primitive Man and His Food* (Vries, 1952).

[5] A large body of literature on the physical culture movement exists in several languages. One good example of the recent work on this topic includes Ina Zweiniger-Bargielowska's (2011) *Managing the Body: Beauty, Health, and Fitness in Britain 1880–1939*.

In Europe, the physical culture movement included everything from the creation of the YMCA (1844), which was founded on the principles of Muscular Christianity, and the Boy Scouts (1910) in England to the birth of the modern sport of gymnastics in the region of Germany. King George V of England even had his own personal instructor of physical culture—the popular author, performer, and strongman, Eugene Sandow (Chapman, 2006). In the United States, the physical culture movement included a wide variety of groups and individuals: Sylvester Graham and his cracker-eating Grahamites in the Northeast, John Harvey Kellogg and his sanitarium in the Midwest, and the athletic Turner Clubs, which imported gymnastic routines from Germany and could be found across the country (Pfister, 2009). Other notable physical culture advocates from this period include Bernarr Macfadden, Charles Post, Catharine Beecher, Dudley Allen Sargent, and Charles Atlas.

Fearing the deleterious effects of mass urbanization and industrialization associated with the transition from traditional, agrarian economies to modern, industrial economies, proponents of the physical culture movement offered many familiar Paleo recommendations of today: the consumption of foods that are subject to minimal refinement or processing, periodic fasting, outdoor exercise, weight training, drugless medical treatments, and exposure to sunlight. Also similar to today, some of these early physical culturalists were often branded as "kooks" by the medical establishment and mainstream society because they bucked the conventional wisdom of the period. In addition, these two movements share another a key commonality. The physical culture movement of 100 years ago and the ancestral health movement of today are similar in that they are both fundamentally responses to rapid social, economic, and technological change, namely, to the Industrial Revolution and the Digital Revolution, respectively. Seen from this broader historical perspective, the ancestral health movement should not be understood as an entirely new phenomenon of the late 20th and early 21st century. Rather, it should be seen as part of the ongoing response to the processes of modernization, which constantly shape and reshape the way individuals live, work, eat, move, and communicate. The processes of modernization include both the transition from premodern (agrarian) to industrial societies and from industrialized nation-states to postindustrial, globalized communities.

To fully demonstrate the relationship between these two movements, the remainder of this chapter is divided into three sections. The first part discusses the many notable similarities between the two movements. The second section highlights some of the differences between the physical culture and ancestral health movements. The third part concludes by offering a discussion of the fundamental origins and appeal of both movements within the context of the ongoing process of modernization and social change. The third section also outlines the necessary limitations of this chapter.

Similarities between the Physical Culture and Ancestral Health Movements

As for similarities between the two movements, first and foremost, there is a desire to create what is perceived to be a more "natural" way of life. Of course, defining exactly what is natural is highly problematic. It also can lead to the common naturalistic fallacy: Just because something existed in the past—or because it can be found in nature—does not mean that it is good generally, or good for human health specifically. Nevertheless, both movements often look to nature and to our ancestral past as a guide to making better lifestyle choices in the present. Or, to use a phrase that was common 100 years ago, it was believed that "the natural way is the best way."

Such feelings within the physical culture movement are exemplified by this quote from Theodore Knauff's 1894 book, *Athletics for Physical Culture*: "It is reasonably certain that man was originally made to live and exercise in the open air, bathe in rivers, [and] expose his body to the healthful action of the sun without even the protection of clothing" (p. 1). In other words, men and women were "designed," or had evolved, to live in accordance with nature. However, due to rapid industrialization and urbanization around the turn of the 20th century, an increasingly large percentage of the population no longer lived such a lifestyle. Millions of people across the United States and Europe moved from small, rural communities to crowded, fast-growing cities, such as Chicago, London, and Paris. Just like today, this led to the idea of a "mismatch." Specifically, a mismatch between the way humans had historically lived over hundreds and thousands of years and the way men and women were living in a rapidly modernizing world. Thus, just like in the ancestral health movement of today, "modern civilization" was understood as the real culprit, contributing to diseases, deformities, and mental illness.

One hundred years ago, the complaints often focused on tobacco and alcohol use, the overconsumption of food and drink, "patent" drugs (which were the forerunners to prescription drugs), adulterated foodstuffs, unsanitary living conditions, and ill-fitting clothing, especially corsets. It was believed that these problems were leading to the decline or, to use the language of the time, the "degeneration" of individuals, nations, and even the species as a whole (Zweiniger-Bargielowska, 2011). Charles Darwin's new theory of evolution played the key role here as well. With the publication of *The Descent of Man* in 1871, humans were no longer seen as something static, created by God and unchanging over time. Instead, individuals, races, and entire nations were now understood as capable of change, either improving or declining (Conlin, 2014). With this came a great deal of anxiety about the future health and well-being of European societies especially. Out of this very real anxiety arose the desire to alleviate the ills of civilization by returning to a more "natural" way of life, even though any such return was—and continues to be—impossible.

Also similar to the ancestral health movement of today, supporters of the physical culture movement looked to contemporary hunter-gatherer populations outside of the United States and Europe for clues to improving health and well-being. Today, the focus is often on the Kitavans of Papua New Guinea and the Hadza of Tanzania. In the period leading up to World War I, explorers, scientists, and colonial administrators traveled the world to locate, examine, and often "civilize" various indigenous populations.[6] The data, images, physical evidence, and even indigenous people they brought back home often become the physical standards by which physical culturalists judged themselves and those around them (Hau, 2003). Moreover, the traditional lifestyles of these contemporary hunter-gatherers appeared to hold to the secret to beauty, health, and vitality in what was seen as an increasingly unhealthy modern world.

As for the specific recommendations, they were remarkably similar to what is currently found in the ancestral health movement. For example, physical culturalists advocated the consumption of fresh foods and the avoidance of all processed foods, including white sugar and white flour. The drinking of raw milk and supplementing with fish oil, particularly cod liver oil, were also popular. And intestinal health was even recognized as being important 100 years ago, just as there is a growing focus on the gut microbiome today. In fact, many health advocates of that time recommended the use of probiotics, such as yogurt and other fermented food, to promote the growth of "good" gut bacteria. One good example of

---

[6] The turn of the 20th century marked the peak of Western colonialism and what historians call "New Imperialism." Parts of Latin America, a large portion of Asia, and almost all of Africa were colonized by European countries and the United States between 1870 and 1914.

such advice can be found in James Empringham's 1936 book, *Pandora's Box or Our Invisible Foes and How to Conquer Them and What to Eat and Why*.

In general, greater scrutiny and critical analysis was applied to all food choices, with the more the "natural" choice being seen as the best for promoting health. The key point here is that this was the beginning of a truly scientific approach to nutrition. While the standards of what constitutes "science" are clearly different today, it was nevertheless a clear attempt to explore the connections between human health and disease and nutrition in a new and systematic way. This stood in especially sharp contrast to the previous practice of simply eating what was dictated by tradition, or religion, or culture. And, exactly like the ancestral health movement of today, those in the physical culture movement believed in the power of good food choices to influence health outcomes. They believed that eating a more natural diet could optimize health and help cure a variety of diseases, from heart disease to cancer and depression. For example, the popular American health entrepreneur, Bernarr Macfadden, founded a series of health resorts and a very popular magazine, *Physical Culture*, based on this message. Although largely forgotten today, Macfadden attained both fame and fortune by preaching the power of eating "clean" (Adams, 2010).

There was also a great deal of interest in fasting. Both 24-hour and multiday fasts were popular, along with compressed eating schedules. Even handy pocket watches were sold at the beginning of the 20th century to instruct the wearer when to exercise—once in the morning and once in the afternoon—and when to eat—the first meal at 11 AM and the second and final meal at 5 PM. This kind of punctuated eating, often known today as intermittent fasting, restricted eating to a relatively small window each day, typically ranging from six to seven hours, with the idea that it would improve health and vitality. Today, researchers such as Valter Longo at the University of Southern California and Satchidananda Panda at the Salk Institute promote exactly this kind of eating approach (Longo & Panda, 2016).

The exercise prescriptions of the physical culture movement were also surprisingly similar to today. Namely, there was an emphasis on strength training and on full-body exercises, such as back squats and deadlifts. Running and jumping, preferably done out of doors, along with functional body-weight exercises, such as push-ups and pull-ups, were also popular. To compliment these more intense exercises, physical culturalists called for as much walking as possible, because this was understood to be the best and most "natural" form of exercise. A poem from 1859 clearly makes this point:

> Horseback riding is good
> But *walking* is better.
> Carriage riding is not bad
> But *walking* is better.
> Sailing is good
> But *walking* is better.
> Rowing is good
> But *walking* is better.
> Swinging dumb-bells is good
> But *walking* is better. ("How To Live," 1859, p. 11)

Bernarr Macfadden, one of the most outspoken proponents of walking, even promoted barefoot walking and the use of minimalist footwear when going shoeless was not practical.[7] There were also fears about new, "unnatural" forms of exercise, like cycling, which was increasingly popular during the first two decades of 20th century. Riding a bike was thought to promote an improper curvature of the spine and skeletal deformities of the lower extremities. So, as far as physical culturalists were concerned, the more an exercise was perceived to be "natural," the better it was deemed to be. Within the current ancestral health community, the evolutionary appropriateness of cycling has also been questioned. For example, at the annual conference for the Ancestral Health Society, held at Harvard University in August 2012, Jamie Scott's presentation, "High Performance Evolutionary Fitness—Using EvoBio to Optimize Training for Endurance Sports," compared such evolutionary novel activities as cycling with more "natural" movements like running.

Supporters of the physical culture movement also emphasized outdoor exercise and sun exposure. It was believed that sunlight and fresh air were necessary for good health and that nature was the best arena for physical training. For example, the book *Pandora's Box*, published in 1936, offers the following recommendation:

---

[7] Bernarr Macfadden encouraged his employees to walk as far and as often as possible. He himself even led multiday company walks from the main office of his publishing empire in New York City to Boston, more than 215 miles away.

> To expose the entire body to the direct rays of the sun with all of the clothing removed is very conducive to Health. This is the best method of obtaining an adequate ration of vitamin D. . . . Begin with a short exposure of ten or fifteen minutes and prolong the duration of the sunbath an additional few minutes each day. (Empringham, 1936, p. 91)

This kind of advice about the importance of moderate sun exposure is now commonplace and could have been taken right out of one of today's Paleo bestsellers, such as the *Primal Blueprint* (Sisson, 2012) or the *Paleo Solution* (Wolf & Cordain, 2010). On the other end of the thermal spectrum, it was also widely believed that ice baths and cold exposure promoted both recovery from exercise and increased longevity.[8] Although the evolutionary adaptiveness of cold exposure may not be clear today, these very same beliefs are commonplace within the current ancestral health movement (Greenfield, 2014; Kruse, 2013).

The second half of the 19th century also saw the creation of the first modern, urban gymnasiums across Europe and the United States. These new indoor training facilities, which sprang up from Sweden to Michigan, looked remarkably similar to many of the CrossFit "boxes" of today. The open-floor plans typically featured free weights, climbing ropes, pull-up bars, and gymnastics apparatus. Similar to many gyms today, the focus was not just on physical fitness. Members were also given instruction on proper diet, stress management, posture, and breathing. In many ways, these new gyms became important community centers and social hubs in a context of growing secularization and the collapse of traditional (agrarian) ways of life, including religious affiliations.

Formal competitions are another striking similarity. In 1905, the first Physical Culture Exhibition was held in Madison Square Garden, which might be called very first "CrossFit Games."[9] The goal was to "crown the fittest man and woman alive." Seventeen events, including the high jump, 50-yard dash, 5-mile run, weighted carry, deadlift, and rope climb, were held over multiple days.[10] All of these events represented a range of exercises that, together, mimicked the natural workout humans

---

[8] In fact, it is widely believed that Barnarr Macfadden founded the first Polar Bear Club in Coney Island in 1903, which is still in existence today.

[9] The CrossFit Games began in 2007 with goal of selecting the world's fittest athletes. CrossFit now enjoys wide-spread popularity, with more than 10,000 affiliates across the globe.

[10] The Physical Culture Exhibition, created by Bernarr Macfadden, was announced in the magazine, *Physical Culture*, in April 1905. Postevent coverage can be found in the October 1905 edition of the same magazine.

would have gotten in a premodern environment. To use the language of the fitness world today, all of these exercises were highly "functional," rather than isolating specific muscle groups or body parts. Hundreds of participants entered the contest, which took place before a full house. This type of competition—both now and in the past—serves to dual purpose of increasing the enthusiasm among current participants and spreading the movement to new audiences.

As just noted, the Physical Culture Exhibition was open to both men and women. In fact, and just like the ancestral health movement of today, women were very much a part of the physical culture movement of 100 years ago.[11] Specifically, there was an emphasis on strength training, on the importance of natural child birth and breastfeeding, and on dress reform (Stewart, 2000). Women were encouraged to walk, do body-weight exercises, participate in gymnastics, and even lift weights. Some women, such as Minnie Wells, even became known for their incredible feats of physical strength.

Also similar to the ancestral health movement of today, there was significant criticism of conventional medicine. In general, doctors were seen as "pill pushers," who only treated symptoms and failed to address underlying causes of diseases. In 1915, John Henry Tilden complained: "Regular medicine was forcing its fallacies down the necks of a gullible public, and there was a need of someone who could tell the truth about medical sophistry, and keep telling it" (Tilden, 1915, p. 2). As an alternative, supporters of the physical culture movement called for more preventative medicine and an end to patent drugs. These forerunners to modern prescription drugs came under heavy fire because they were usually ineffective and heavily marketed to the rapidly growing working classes. At the same time, there were also many concerns about the new use of vaccines. Echoes of some of these same concerns can be heard within the current ancestral health movement, even though there is no scientific basis for rejecting the use of vaccines today (DeStefano, Price, & Weintraub, 2013). In fact, it can be argued that vaccinations are a highly evolutionarily consonant invention, taking advantage of our naturally selected immune systems' ability to quickly learn to defend against novel pathogens. Likewise, it also demonstrates how our uniquely evolved species-typical adaptation of high intelligence and unprecedented cultural transmission can better improve our health and resist disease.

---

[11] Based on a recent survey of the ancestral health community, more women (56%) than men (44%) report practicing a paleo lifestyle (Schwartz & Stapell, 2013).

Physical culturalists also expressed concern over correct posture and breathing, and air quality. They argued that too much time was spent indoors and at sedentary occupations. This led to bad habits and ill health. In 1922 Edwin Checkley warned: "We are forgetting how to stand, and, above all—fatal error!—we are forgetting how to breathe" (Checkley, 1922, p. 15). Numerous books and articles demonstrated proper posture and breathing for men, women, and children alike. Similar work is being published today, with Kelly Starrett's 2016 book, *Deskbound: Standing Up to a Sitting World*, representing the best recent example. At the same time, there was also growing concern about both outdoor and indoor air quality, especially in rapidly industrializing areas of northern Europe and the northeastern United States. Goggles and nose filters were even recommended by some for city living, and especially train travel, which could be extremely dirty due to the coal-powered locomotives (Macfadden, 1907). Today, we see analogous concerns about the negative effects of environmental pollutants, including endocrine disruptors, volatile organic compounds, and heavy metals (Khetan, 2014).

Promoters of the physical culture movement increasingly made their criticisms—and their own alternative programs for health and wellness—known through the latest technologies and methods of communication of the day. In other words, and similar to the ancestral health movement of today, they used and were empowered by "new media." But, of course, the new media of 100 years ago were different than today. They consisted of photography (developed in the 1880s), motion pictures (developed in the 1890s), and mass-produced magazines and newspaper. In fact, the period of late 19th and early 20th century was the heyday of print publishing, when grand fortunes were made by the likes of William Randolph Hearst and Joseph Pulitzer.[12]

Out of this publishing frenzy came hundreds of popular physical culture books and magazines. Some of the most important books from the period include *The Gospel of Strength* (1902) by Eugene Sandow, *The Way to Live* (1911) by George Hackenschmidt, *Vitality Supreme* (1915) by Bernarr Macfadden, *Promotion and Conservation of Health, Strength and Mental Energy* (1920) by Lionel Strongfort, and *Super Strength* (1924) by Alan Calvert. In addition, some magazines, such as *Physical Culture*, published by Bernarr Macfadden, reached monthly circulation numbers of more than 300,000 by the 1930s (Adams, 2010, p. 182). Moreover, much

---

[12] Perhaps today we are living through a similar golden age of the digital media publishing, with millions of blogs and personal websites promoting and exploring every imaginable topic.

of the content of these books and magazines looked astonishingly similar to what can be found in popular Paleo books of today. For example, in *Natural Foods: The Safe Way to Health*, published in 1925, the table of contents outlines the following chapters:

Table of Contents
1. Sunlight: the Ultimate source of vital energy:
   Its beneficial influence on all the functions of the body
2. Fresh Air, Exercise and Rest:
   Their importance for maintaining health and vitality
3. Water—Nature's Universal Solvent:
   "Flushing the System" a mistaken idea
4. The Importance of Natural Foods for Life and Health:
   Why denatured and synthetic foods are destructive
5. Why the Calorie Theory is Misleading:
   Calories do not indicate real food values (Carque, 1925)

Each of these five chapters represents a key theme within the current ancestral health movement, yet the book was written over 90 years ago. Another similarity found in these books, and especially in the magazines, is the hundreds of "success stories," demonstrating everything from weight loss to recovery from tuberculosis. Of course, these testimonials are all anecdotal and do not represent any kind of empirical evidence supporting efficacy. Yet, they serve a very important function in both movements. Just like the hundreds of Paleo success stories on the Internet today, these printed physical culture success stories benefited both the submitter and the publisher by reinforcing individual success and providing a model to be emulated by others. The result was more book and magazine sales—or more "clicks" and page views—and greater public interest in the respective movements.

Differences between the Physical Culture and Ancestral Health Movements

While there are many important similarities between the ancestral health and physical culture movements, several differences should be noted as well. For example, unlike today, there was an emphasis on nudism, or *naturism*, 100 years ago. Wearing clothing was often seen as a weakness and unnatural. Supporters of naturism frequently looked to both classical Greek sculpture and contemporary hunter-gatherers, who appeared to enjoy excellent health, for inspiration and for justification for not wearing clothes. The first nudist clubs sprung up in Europe, particularly in Germany, and later in the United States (Ross, 2005). However, very few within the ancestral health movement of today openly embrace the idea of nudism.

Another difference was the concern over what was seen as poor "circulation." Many physical culturalists believed that "brain work"—work done sitting behind a desk—led to poor fluid circulation within the body. To combat this perceived infliction, they recommended a variety of cures, including massage, vigorously toweling off after a shower or bath, and *hydropathy*, which was the application of different temperatures and forces of water to the body (Weiss, 1967). The most famous practitioner of hydrotherapeutic techniques was Vincent Priessnitz from Austria (Claridge, 2007). Over the course of his career, he treated thousands of patients from around Europe, including the Austrian royalty. While sitting too much is often condemned today, none of these other techniques are similar to recommendations within the current ancestral health movement.

There were some food differences as well. Many physical culturalists were vegetarians, or at least advocated vegetarianism for others, and some promoted the consumption of raw foods.[13] While cereal grains are typically avoided on a Paleo diet today, unadulterated whole-wheat bread was often seen as healthful choice 100 years ago. This is mostly like due to the fact that the link between the protein gluten, found in wheat, and Celiac's disease was not made until the 1940s by the Dutch pediatrician, Dr. Willem Karel Dicke.

There was also much more talk about mastication, the chewing of food, and elimination. For example, Horace Fletcher published a number of popular books that urged people to chew their food to the point where swallowing was not necessary (Christen & Christen, 1997). If chewed sufficiently, he claimed, food simply slips down the back of the throat. Fletcher was also known for the famous quip: "Nature will castigate those who don't masticate." This message was so popular, in fact, that he was able to become quite wealthy and purchase a palazzo in Venice, Italy, with the proceeds of his books. On the other end of the digestive process, more attention was paid to excreta, which refers to human waste, such as sweat,

---

[13] While the paleo diet is typically associated with red meat consumption, nonmeat versions of the paleo diet have also emerged recently within the ancestral health community. See, for example, *Paleo Vegan: Plant-Based Primal Recipes* (Jones & Roettinger, 2014) and *Plant-Based Paleo: Protein-rich Vegan Recipes for Well-being and Vitality* (Zoe, 2015).

urine, and feces. Feces, in particular, received a lot of attention, including its size, consistency, and smell. Food, of course, played the key role in this process. Some physical culturalists believed that a proper natural diet would lead to proper human excrement: "Healthy human excreta are no more offensive than moist clay and have no more odor than a hot biscuit" (Fletcher, 1903, p. 11). Not only is this belief not present within current ancestral health movement, but this particular maxim is very unlikely to be true from the perspective of modern evolutionary theory. We now better understand that humans specifically evolved to find fecal matter malodorous, disgusting, and repulsive, since it is one of the most omnipresent sources of potential contamination in our ancestral environments (Curtis & Biran, 2001; Oaten, Stevenson, & Case, 2009).

Another important difference is that the ancestral health movement of today has far more scientific support. While the physical culture movement attempted to tackle the questions related to human health and disease through a scientific lens, the fact remains that scientific practices of 100 years ago were far different than today. Our standards for methodologies, evidence, and hypothesis testing were only first developing in the late 19th and early 20th century, and they often differed from country to country. Moreover, there were no randomized controlled trials, our "gold standard" for medical evidence today. As a result, the physical culture movement had to rely mainly on well-intentioned theories and anecdotes.

In contrast, the current ancestral health movement is informed by—and supported by—documented empirical evidence. Of course, much more work needs to be done, but there is now substantial body of literature supporting evolutionary approaches to human health and disease. For example, there have been at least 12 clinical trials on the effectiveness of a Paleo diet or lifestyle. The studies examined such things as glucose tolerance and heart disease, weight loss and blood pressure, cardiovascular risk in type 2 diabetes, metabolic markers and lipid levels, fat deposition and insulin activity, hypercholesterolemia, liver fat concentrations, fat mass and glycemic control in type 2 diabetes, obesity in postmenopausal women, fasting leptin levels in type 2 diabetes, and Paleolithic nutrition versus the recommended Australian diet (Fontes-Villalba et al., 2016; Frassetto et al., 2009; Genoni, Lyons-Wall, Lo, & Devine, 2016; Jönsson et al., 2009; Lindeberg et al., 2007; Masharani et al., 2015; Mellberg et al., 2014; Osterdahl, Kocturk, Koochek, & Wändell, 2008; Otten et al., 2016a, 2016b; Pastore, Brooks, & Carbone, 2015; Ryberg et al., 2013). In addition to these clinical trials, a recent meta-analysis has examined the effectiveness of using Paleolithic nutrition to treat metabolic syndrome (Manheimer, van Zuuren, Fedorowicz, & Pijl, 2015). All these studies have shown favorable results. Also, there are currently three academic, peer-reviewed journals dedicated to the field of evolutionary health: *Evolution, Medicine & Public Health Journal*, published by the International Society for Evolution, Medicine, and Public Health; *Journal of Evolution and Health*; and *Journal of Evolutionary Medicine*. Taken together, these three journals have published hundreds of peer-reviewed articles demonstrating the many varied connections between evolution and human health and disease.

It should also be noted that just like with any scientific theory, the mismatch hypothesis—and the ancestral health movement more generally—has come under criticism. For example, the evolutionary biologist Marlene Zuk has questioned both the underlining evolutionary principles of the movement and the application of those principles by clinicians and laypersons (Zuk, 2013). Zuk is particularly critical of what she sees as misunderstandings about how evolution works and the many assumptions, unfounded in her view, about how Paleolithic humans actually lived. The molecular anthropologist Christina Warinner has also taken the ancestral health movement to task in a popular TED talk entitled, "Debunking the Paleo Diet."[14] Specifically, she is critical of what she believes to be the movement's overemphasis on meat and its avoidance of grains. She also questions whether it is even possible today to identify any foods that resemble what our ancestors ate more 10,000 years ago, including even our current fruits and vegetables. However, despite such criticisms, neither Zuk nor Warinner deny the fundamental connection between evolution and human health. Rather, they question some of the specific theories and practical applications of the current evolutionary approaches.

Finally, in terms of differences, unlike the widely dispersed online virtual world of the Paleo blogosphere, the physical culture movement of 100 years ago had some important physical locations. Of course, this difference largely has to do with technology. Today, the Internet allows for unhindered communication and the dissemination of information throughout much of the world. Likewise, the effects brought on by the Digital Revolution are not limited to any particular region or continent. This has allowed the ancestral health movement to both easily spread and resonate with people around the global. For example, there is currently a

---

[14] This Christina Warinner's TED talk has been viewed more than 1.9 million times on YouTube: https://www.youtube.com/watch?v=BMOjVYgYaG8

large branch of the Ancestral Health Society in New Zealand. In contrast, the physical culture movement of 100 years ago was typically centered on rapidly urbanizing metropolitan areas, where the effects the industrialization were felt most acutely. In Europe, these included many of the major capitals, such as London, Paris, and Berlin. In the United States, New York City; Chicago; Battle Creek, Michigan; Dansville, New York; and Boston, Massachusetts were some of the main centers for physical culture activity.

It should be mentioned that Harvard College in Boston played a particularly important role here. For instance, Dr. George Barker Windship, a Harvard Medical School Graduate in 1857, lectured widely, demonstrated his impressive strength, and generally promoted the physical culture movement to combat the perceived ills of the modern world.[15] Also from Harvard, Dudley Allen Sargent is credited with the creation of the first modern physical education programs for primary and secondary schools, which was seen as now necessary as young people were increasingly sitting behind school desks rather than working on family farms. Sargent also created the Hemenway Gynmasium at Harvard College, which is still in use today and serves as a reminder of his efforts to improve the health and fitness of young adults. The importance of physical culture within this rapidly changing world was even recognized at the time by Harvard College's longest-serving president, Charles William Eliot (1869–1909). Eliot believed that, above all, college graduates needed:

> a wholesome diet, plenty of fresh air, and regular exercise. . . . A busy lawyer, editor, minister, physician, or teacher has need of greater physical endurance than a farmer, trader, manufacturer, or mechanic. All professional biography teaches that to win lasting distinction in sedentary, in-door occupations, which task the brain and the nervous system, extraordinary toughness of body must accompany extraordinary mental powers. (Blaikie, 1879, p. 90)

In other words, Eliot believe that the modern world of sedentary work and stressful urban living placed terrific demands on the human body, which could be met in part by embracing a more natural way of living. In this way, both the physical culture movement and the ancestral health movement represent responses to very similar kinds of problems, and they offer a set of similar solutions that connect the two movements together.

---

[15] Despite his diminutive size, standing only five feet tall, Dr. George Barker Windship was known for his remarkable strength, including his ability to do one-finger chin-ups.

Conclusions and Limitations

Even though the physical culture and the ancestral health movements developed more than 100 years apart, the links between the two are striking. Not only do they share many of the same solutions to parallel problems, but the fundamental origins and appeal of both movements appear to be similar as well.

Both movements are best understood as largely middle-class reactions to rapid social, economic, and technological change. First, 100 years ago, the physical culture movement was a response to increasing urbanization, industrialization, and the disruption of traditional agrarian lifestyles. Now, today, the ancestral health movement is a reaction to the Digital Revolution and everything that comes along with it: virtual friends, physical and social isolation, the automation of many careers and activities, and the increased disconnection from nature.

The key point here is that both movements seek to "return to nature" in a stressful and disorienting world. Even though any such return would be impossible, they offer what appears to be a "natural" alternative to a very "unnatural" world. They also may give some people a greater sense of autonomy and control. More specifically, both movements provide the satisfaction of exercising one's own body, feeling engaged with others, and comparing and noting improvement. These can be powerful daily practices that "ground" an individual, giving him or her a sense of place and purpose. To put it another way, and to borrow a phrase from Joseph Campbell, the physical culture and the ancestral health movements both offer a kind of "scared place" of one's own (Campbell & Moyers, 2011). They represent a new set of nonreligious "spiritual" practices in an uncertain environment where many traditional forms of faith and affiliation no longer hold sway. And, perhaps most significantly, they hold out the promise—true or not—of a kind of "inner awakening" in an apparently chaotic industrial or postindustrial world.

It is also important to be clear about what these evolutionary approaches to health do not attempt to do. First, they are not attempts at historical reconstruction. In other words, they do not advocate retuning to our Paleolithic past: exchanging blue jeans for loincloths, living in caves, or eating only raw meat (or any meat at all for that matter). Rather, the point is to develop a sharp and clear evolutionary lens through which to view contemporary lifestyle choices and to better optimize our health. Or, to put it another way, it is about considering choices that are more in line with the environments in which we evolved and are best adapted to. Again, the theory is that these choices will then, in turn, lead to better health and longevity outcomes. In short, the ultimate goal of this evolutionary approach is to better understand, prevent, and treat disease.

Second, an evolutionary approach to health does not represent just another diet fad. It is not an alternative to the Atkins Diet; the Zone Diet; a high-carb diet, a low-carb diet; Dr. Dean Ornish's diet; or the South Beach Diet. Rather, as suggested at the outset of this chapter, this evolutionary approach represents a true "unifying field theory" for human well-being. In other words, it is a paradigm for guiding research, inquiry, and experimentation—just like the big bang theory in the field of cosmology or continental drift and plate tectonics in the discipline of geology. It can also provide a useful framework for structuring and teaching undergraduate health education. Human health and evolution are inexorably linked. We cannot fully understand—or adequately teach—the former without the latter. The basic principles of evolution give us the tools both to analyze the ultimate causes of our diseases and to offer better solutions to those aliments. The power and potential of these tools have inspired researchers and laypersons alike for more than 100 years now.

Finally, this chapter has some clear limitations. First and foremost, it was written by a historian, and not by a medical researcher of even a social scientist. As such, the main purpose of the chapter has been to compare these two social movements across time, and to place their evolutionary approaches to health and disease within a broader historical context. Thus, my primary intention was not to determine whether or not these all of these different evolutionary approaches to human health are firmly based in science. Nor was the point to judge the effectiveness of the various theories and interventions. There are others better qualified to achieve those important objectives. Rather, my goal was to place those ideas and practices within the context of rapid social, economic, and technological change in the 19th, 20th, and 21st centuries.

So for the purposes of this chapter, the validity of each specific scientific claim is somewhat less germane. The point is that many people within these two movements *believed* in the veracity of these facts, theories, and interventions and thus did not rely simply on tradition, custom, or the conventional wisdom to guide their lifestyle choices. To put it another way, they have tried to actively use the principles related to the theory of evolution to solve the many problems related to human health and disease that come along with modernity. My chapter is thus more a story about how people have used the tools of evolutionary studies to make sense of—and attempt to adapt to—a rapidly changing world. Once again, it was not my objective to try to justify all of their various claims. Rigorous empirical testing must of course be applied to all evolutionary approaches and practices. I leave it to others to accomplish that important task.

References

Adams, M. (2009). *Mr. America: How muscular millionaire Bernarr Macfadden transformed the nation through sex, salad, and the ultimate starvation diet.* New York, NY: Harpers.

Amsterdam, E. (2013). *Paleo cooking from Elana's pantry: Gluten-free, grain-free, dairy-free recipes.* Berkeley, CA: Ten Speed.

Blaikie, W. (1879). *How to get strong and how to stay so.* New York, NY: Harper.

Budd, M. A. (1997). *The sculpture machine: Physical culture and body politics in the age of empire.* New York, NY: NYU Press.

Calvert, A. (1924). *Super strength.* Philadelphia, PA: Milo.

Campbell, J., & Moyers, B. (2011). *The power of myth.* New York, NY: Knopf Doubleday.

Carque, O. (1925). *Natural foods: The safe way to health.* Los Angeles, CA: Carque Pure Food Co.

Chapman, D. L. (2006). *Sandow the magnificent: Eugen Sandow and the beginnings of bodybuilding.* Urbandale, IL: University of Illinois Press.

Checkley, E. (1922). *Checkley's natural method of physical training.* Philadelphia, PA: Checkley Bureau.

Christen, A. G., & Christen, J. A. (1997). Horace Fletcher (1849–1919): "The great masticator." *Journal of the History of Dentistry, 45*(3), 95–100.

Claridge, R. T. (2007). *Hydropathy: Or the cold water cure as practiced by Vincent Priessnitz at Graefenberg, Austria.* Halifax, England: Nicholson & Wilson.

Conlin, J. (2014). *Evolution and the Victorians: Science, culture and politics in Darwin's Britain.* London, England: Bloomsbury Academic.

Cordain, L. (2010). *The Paleo diet: Lose weight and get healthy by eating the foods you were designed to eat* (Rev. ed.). Hoboken, NJ: Wiley.

Curtis, V., & Biran, A. (2001). Dirt, disgust, and disease: Is hygiene in our genes? *Perspectives in Biology and Medicine, 44,* 17–31.

Darwin, C. (1871). *The descent of man.* New York, NY: D. Appleton.

DeStefano, F., Price, C. S., & Weintraub, E. S. (2013). Increasing exposure to antibody-stimulating proteins and polysaccharides in vaccines is not associated with risk of autism. *Journal of Pediatrics, 163,* 561–567.

Eaton, S. B., & Konner, M. (1985). Paleolithic nutrition: A consideration of its nature and current implications. *New England Journal of Medicine, 312,* 283–289.

Empringham, J. (1936). *Pandora's box or our invisible foes and how to conquer them and what to eat and why.* Los Angeles, CA: Health Education Society.

Fletcher, H. (1903). *The A. B. Z. of our own nutrition.* New York: F. A. Stokes.

Fontes-Villalba, M., Lindeberg, S., Granfeldt, Y., Knop, F. K., Memon, A. A., Carrera-Bastos, P., . . . Jönsson, T. (2016). Palaeolithic diet decreases fasting plasma leptin concentrations more than a diabetes diet in patients with type 2 diabetes: A randomised cross-over trial. *Cardiovascular Diabetology, 15,* 80.

Frassetto, L. A., Schloetter, M., Mietus-Synder, M., Morris, R. C., & Sebastian, A. (2009). Metabolic and physiologic improvements from consuming a Paleolithic, hunter-gatherer type diet. *European Journal of Clinical Nutrition*, *63*, 947–955.

Genoni, A., Lyons-Wall, P., Lo, J., & Devine, A. (2016). Cardiovascular, metabolic effects and dietary composition of ad-libitum Paleolithic vs. Australian guide to healthy eating diets: A 4-week randomised trial. *Nutrients*, *8*, 314.

Greenfield, B. (2014). *Beyond training: Mastering endurance, health and life*. Las Vegas, NV: Victory Belt.

Hackenschmidt, G. (1911). *The way to live: Health and physical fitness*. London, England: Health & Strength.

Hau, M. (2003). *The cult of health and beauty in Germany: A social history, 1890–1930*. Chicago, IL: University of Chicago Press.

How To Live. (1859). The letter box. *A monthly health journal for the people*, *2*(2), 11.

Jones, E. J., & Roettinger, A. (2014). *Paleo vegan: Plant-based primal recipes*. Summertown, TN: Book Publishing.

Jönsson, T., Granfeldt, Y., Ahrén, B., Branell, U.-C., Pålsson, G., Hansson, A., . . . Lindeberg, S. (2009). Beneficial effects of a Paleolithic diet on cardiovascular risk factors in type 2 diabetes: A randomized cross-over pilot study. *Cardiovascular Diabetology*, *8*, 35.

Joulwan, M., & Petrucci, K. (2012). *Living Paleo for dummies* (1 ed.). Hoboken, NJ: For Dummies.

Khetan, S. K. (2014). *Endocrine disruptors in the environment* (1st ed.). Hoboken, NJ: Wiley.

Knauff, T. C. (1894). *Athletics for physical culture*. New York, NY: J. S. Tait.

Kresser, C. (2013). *Your personal Paleo code: The 3-step plan to lose weight, reverse disease, and stay fit and healthy for life*. New York, NY: Little, Brown.

Kruse, D. J. (2013). *Epi-Paleo Rx: The prescription for disease reversal and optimal health* (1st ed.). Gallatin, TN: Optimized Life.

Kuipers, R. S., Joordens, J. C. A., & Muskiet, F. A. J. (2012). A multidisciplinary reconstruction of Palaeolithic nutrition that holds promise for the prevention and treatment of diseases of civilisation. *Nutrition Research Reviews*, *25*, 96–129.

Lieberman, D. (2013). *The story of the human body: Evolution, health, and disease*. New York, NY: Pantheon.

Lindeberg, S. (2010). *Food and western disease: Health and nutrition from an evolutionary perspective*. Ames, IA: Wiley-Blackwell.

Lindeberg, S., Jönsson, T., Granfeldt, Y., Borgstrand, E., Soffman, J., Sjöström, K., & Ahrén, B. (2007). A Palaeolithic diet improves glucose tolerance more than a Mediterranean-like diet in individuals with ischaemic heart disease. *Diabetologia*, *50*, 1795–1807.

Longo, V. D., & Panda, S. (2016). Fasting, circadian rhythms, and time-restricted feeding in healthy lifespan. *Cell Metabolism*, *23*, 1048–1059.

Macfadden, B. (1907). Railroad Dust and Soot. *Physical Culture*, *18*, 289–291.

Macfadden, B. (1915). *Vitality supreme*. New York, NY: Physical Culture.

Manheimer, E. W., van Zuuren, E. J., Fedorowicz, Z., & Pijl, H. (2015). Paleolithic nutrition for metabolic syndrome: systematic review and meta-analysis. *American Journal of Clinical Nutrition*, *102*, 922–932.

Masharani, U., Sherchan, P., Schloetter, M., Stratford, S., Xiao, A., Sebastian, A., . . . Frassetto, L. (2015). Metabolic and physiologic effects from consuming a hunter-gatherer (Paleolithic)-type diet in type 2 diabetes. *European Journal of Clinical Nutrition, 69*, 944–948.

Mellberg, C., Sandberg, S., Ryberg, M., Eriksson, M., Brage, S., Larsson, C., . . . Lindahl, B. (2014). Long-term effects of a Palaeolithic-type diet in obese postmenopausal women: A 2-year randomized trial. *European Journal of Clinical Nutrition, 68*, 350–357.

Natterson-Horowitz, B., & Bowers, K. (2012). *Zoobiquity: The astonishing connection between human and animal health.* New York, NY: Knopf.

Nesse, R. M., & Stearns, S. C. (2008). The great opportunity: Evolutionary applications to medicine and public health. *Evolutionary Applications, 1*, 28–48.

Oaten, M., Stevenson, R., & Case, T. (2009). Disgust as a disease-avoidance mechanism. *Psychological Bulletin, 135*, 303–321.

Osterdahl, M., Kocturk, T., Koochek, A., & Wändell, P. E. (2008). Effects of a short-term intervention with a Paleolithic diet in healthy volunteers. *European Journal of Clinical Nutrition, 62*, 682–685.

Otten, J., Mellberg, C., Ryberg, M., Sandberg, S., Kullberg, J., Lindahl, B., . . . Olsson, T. (2016). Strong and persistent effect on liver fat with a Paleolithic diet during a two-year intervention. *International Journal of Obesity (2005), 40*, 747–753.

Otten, J., Stomby, A., Waling, M., Isaksson, A., Tellström, A., Lundin-Olsson, L., . . . Olsson, T. (2016). Benefits of a Paleolithic diet with and without supervised exercise on fat mass, insulin sensitivity, and glycemic control: a randomized controlled trial in individuals with type 2 diabetes. *Diabetes/Metabolism Research and Reviews, 33*, e2828

Pastore, R. L., Brooks, J. T., & Carbone, J. W. (2015). Paleolithic nutrition improves plasma lipid concentrations of hypercholesterolemic adults to a greater extent than traditional heart-healthy dietary recommendations. *Nutrition Research (New York, NY), 35*, 474–479.

Pfister, G. (2009). The role of German turners in American physical education. *International Journal of the History of Sport, 26*, 1893–1925.

Platek, S. M., Geher, G., Heywood, L., Stapell, H., Porter, J. R., & Walters, T. Y. (2011). Walking the walk to teach the talk: Implementing ancestral lifestyle strategies as the newest tool in evolutionary studies. *Evolution: Education and Outreach, 4*, 41–51.

Quinn, N., & Glaspey, J. (2012). *The complete idiot's guide to eating Paleo.* New York, NY: ALPHA.

Ross, C. (2005). *Naked Germany: Health, race and the nation.* New York: Berg.

Ryberg, M., Sandberg, S., Mellberg, C., Stegle, O., Lindahl, B., Larsson, C., . . . Olsson, T. (2013). A Palaeolithic-type diet causes strong tissue-specific effects on ectopic fat deposition in obese postmenopausal women. *Journal of Internal Medicine, 274*, 67–76.

Sandow, E. (1902). *The gospel of strength according to Sandow: A series of talks on the Sandow system of physical culture by its founder.* Melbourne, Australia: T. Shaw Fitchett.

Sanfilippo, D., Staley, B., & Wolf, R. (2012). *Practical Paleo: A customized approach to health and a whole-foods lifestyle.* Las Vegas, NV: Victory Belt.

Schwartz, D., & Stapell, H. (2013). Modern cavemen? Stereotypes and reality of the ancestral health movement. *Journal of Evolution and Health, 1*(1), art. 3.

Sisson, M. (2012). *The primal blueprint: Reprogram your genes for effortless weight loss, vibrant health, and boundless energy.* Malibu, CA: Primal Nutrition.

Starrett, K., Starrett, J., & Cordoza, G. (2016). *Deskbound: Standing up to a sitting world* (1st ed.). Las Vegas, NV: Victory Belt.

Stearns, S. C. (2012). Evolutionary medicine: Its scope, interest and potential. *Proceedings of the Royal Society of London B: Biological Sciences, 279,* 4305–4321.

Stearns, S. C., Nesse, R. M., Govindaraju, D. R., & Ellison, P. T. (2010). Evolutionary perspectives on health and medicine. *Proceedings of the National Academy of Sciences, 107*(Suppl 1), 1691–1695.

Stephenson, N. (2012). *Paleoista: Gain energy, get lean, and feel fabulous with the diet you were born to eat.* New York, NY: Touchstone.

Stewart, M. L. (2000). *For health and beauty: Physical culture for Frenchwomen, 1880s–1930s.* Baltimore, MD: Johns Hopkins University Press.

Stock, J. T. (2008). Are humans still evolving? *EMBO Reports, 9*(Suppl 1), S51–S54.

Strongfort, L. (1920). *Promotion and conservation of health, strength and mental energy.* Newark, NJ: Lionel Strongfort Institute.

Tilden, J. H. (1915). Philosophy of health. *Philosophy of Health, 16*(1), 2.

Voegtlin, W. L. (1975). *The Stone Age diet: Based on in-depth studies of human ecology and the diet of man.* New York, NY: Vantage.

Vries, A. de. (1952). *Primitive man and his food.* Chicago, IL: Chandler.

Weiss, H. B. (1967). *The great American water-cure craze: A history of hydropathy in the United States.* Trenton, NJ: Past Times.

Wolf, R., & Cordain, L. (2010). *The Paleo solution: The original human diet.* Las Vegas, NV: Victory Belt.

Zoe, J. (2015). *Plant-based Paleo: Protein-rich vegan recipes for well-being and vitality.* London, England: Ryland Peters & Small.

Zuk, M. (2013). *Paleofantasy: What evolution really tells us about sex, diet, and how we live* (1st ed.). New York, NY: W. W. Norton.

Zweiniger-Bargielowska, I. (2011). *Managing the body: Beauty, health, and fitness in Britain 1880–1939.* Oxford, England: Oxford University Press.

CHAPTER 11 | From Genetic Evolution to Engineering Optimization

YASER KHALIFA

GENETIC ALGORITHMS (GAS) IS one of the most successful and widely used evolutionary algorithm approaches that pertain to several other evolutionary-inspired heuristic approaches: genetic programming (Koza, 1992), evolutionary programming (Fogel, 1999), and evolutionary strategies (Beyer & Schwefel, 2002). GAs are search and optimization algorithms inspired by biological evolutionary processes. They have been highly successful as techniques for getting computers to automatically solve problems relying on evolutionary heuristics. Since their inception more than 40 years ago, GAs have been used to solve complex computational problems but along with this engineering aspect there has been a growing interest in their theoretical bases and various implementations.

There are a number of different classifications of the existing global optimization techniques described by Zhou (1990). However, Oosthuizen (1989) presents a different classification that would be more appropriate for the case considered in this chapter. It classified global optimization methods into two main categories, namely *volume-oriented* and *path-oriented*. In the volume-oriented method, a search over the whole volume of the search space is conducted (e.g., Monte-Carlo methods; Degroot, 1970). In the path-oriented method, the search starts from an arbitrary or specifically chosen point in the feasible region and then follows one or more paths in the search for a global optimum (e.g., grid search; Times & LaPatra, 1977).

Path-oriented methods are further divided into the *prediction* method and the *exploration* method. The prediction method uses a model of the objective function to predict the steps (e.g., tunneling methods), while the exploration method (e.g., rotating co-ordinates) does not. Evolutionary algorithms, according to the previous classification, fall mainly into the group of path-oriented exploration methods. During the optimization process, however, evolutionary methods usually start the search in a volume-oriented fashion, by considering the whole search space. After some iterations of the search process, the search is then focused on specific volumes of the search space. The most common of these evolutionary algorithms is GA, which has gained much importance in the last few decades.

The purpose of this chapter is to give the reader an overview of two applications of GA: an analog circuit design tool and optimization tool and a music composition tool. Most of the descriptions here refer to what is known in the GA community as *standard* GA.

The field of GA was founded by John Holland in the early 1970s. Holland (1975) emphasizes the ability of simple representations to encode complicated structures and the power of simple transformations to improve that structure. These representations are combined in what is called in biology a *chromosome*. A number of these chromosomes will constitute a *population*. Syntactic operations are then used to alter and improve these coded solutions.

Holland (1975) described GA as a control structure with which these representations and operations could be managed to evolve bit strings that were well developed to the problem to be solved.

As in natural evolution, the problem a GA search faces is one of searching for beneficial chromosomes adaptations to a complicated and changing environment. The knowledge that each iteration of the search gains is embodied within the individual chromosomes that constitute its population.

Genetic Algorithm

The standard GAs introduced by Holland is the basic form from which other improved algorithms have been subsequently developed.

So, what is GA? And how does it work? GA mimics the process of natural selection, reproduction, and evolution. Let's consider a population of humans for example. Each individual of the population carries their biological human characteristics in a chromosome. Our chromosomes consist of genes that identify our individual traits.

In a human population, a new population is created when a number of individuals *select* each other and *mate* to produce children or *offspring*. The selection process depends on the different criterion each has in choosing their partner. The individual who scores high on these criterion is said to be a potential match or a *fit* individual. As per Darwinian survival of the fittest principals, the fit individuals are therefore have better chance of mating and producing offspring that would have their fit genes. Fit genes propagate through new generations of individuals and enhance the overall fitness of these generations in general.

GA mimics the previously described process. For GA, let us make some more precise definitions for some of the terminologies we used earlier.

- **Population**—It is a group of possible solutions, which constitute a subset of all possible solutions to the given problem.
- **Chromosomes**—Is a single individual possible solution.
- **Gene**—A representation of one trait or parameter in a solution.
- **Fitness function**—Is a measure of merit that reflects how good a solution is. This is usually application dependent.
- **Selection**—Is the process of choosing chromosome parents from a current population that would be allowed to mate and produce children (reproduction).
- **Offspring**—Are the children produced by mating selected parents in a current population (reproduction). A group of offspring constitute a new population or a generation.
- **Convergence**—Is the indication that there is very little diversity in the solutions found by the search, and hence no new solutions are likely to be found. This could happen either if the algorithm has found an optimum or acceptable solution, or it got stuck in a local minima.

In a typical GA, starting from an initial randomly generated population, a selection mechanism is used to identify and select members that exhibit better fitness. Those selected chromosomes form the parents. These parents will mate through applying genetic reproduction operators such as *crossover* and *mutation*, which are explained in later sections, to reproduce and generate the subsequent *generation*s. The process continues until either a time limit is reached or until less than a specified error level is achieved. A flowchart of a simple GA is shown in Figure 11.1.

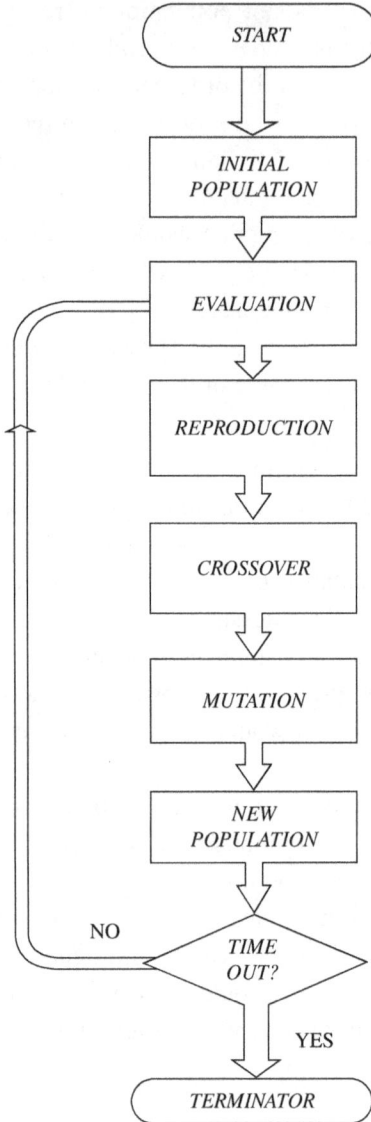

FIGURE 11.1. Simple genetic algorithm flowchart.

Representation

GA works on strings of fixed or variable length. In the following example, a binary-coded decimal representation is used. In general, base-two representations are the most widely used in GA applications. This is mainly because the binary coding offers the maximum number of *schemata* or similarities per bit of information of any coding. A *schema* is a subset of

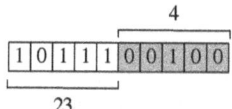

FIGURE 11.2. Chromosome structure with BCD representation.

strings with similarities at certain string positions. It is desirable for a GA search that a maximum number of schemata is available for it to exploit, as illustrated in Figure 11.2. Representation of string characters, or any other special symbols, is equally easy and possible.

Fitness Function

The *fitness* of a chromosome is a measure of its proximity of the solution. The chosen fitness function can be as simple as the inverse of the error encountered, or it can be a complicated function that includes many constraints. The designer is in full control of specifying his or her own requirements and constraints for judging the fitness of the resulting chromosomes.

Genetic Operators

GA operates through a number of operators such as selection, *mutation,* and *crossover* as described by Holland (1975). They resemble natural biological reproduction operators but in an application-specific way. There is a great deal of discussion of parameter setting in the literature (Goldberg, Deb, & Clark, 1992). However, no concrete conclusions have been obtained on which parameter settings are best. The main control parameters are *selection criterion, population size, crossover,* and *mutation rates.* In the sections that follow, descriptions of these parameters are given as well as techniques that enable the GA user to choose reasonable parameter settings.

Selection of Next Generation

In GA, the selection operation is performed on the current population to choose the individuals that will make up the next generation. A number of selection techniques is available. The most widely used technique is the *Roulette wheel* method of selection described by Goldberg et al. (1992).

A selection method in GA operates through exploitation and exploration. It is required that the method of selection should be biased toward

```
FOR l=1 to Population Size
    Cumulative Fitness[l] = Cumulative Fitness + Fitness[l]
    DO WHILE Population < Population Size
        n= random number between 0 and Cumulative Fitness.
        FOR L=0 to Population Size
            IF Cumulative Fitness[L] >= n
                Return Chromosome L
        END FOR
    END WHILE
END FOR
```

FIGURE 11.3. Pseudo code for the Roulette-wheel algorithm.

those members of population that exhibit better performances over others. At the same time, the diversity of solutions should be maintained to search as much of the solution space as possible.

Each fit member of the population is a representative of the volume in the search space that is in its vicinity. *Exploiting* these potential search spaces is one of the tasks that the selection mechanism should perform. However, keeping less than stellar solutions with their genetic contents ensures diversity of genetic contents that also leads to *exploring* wider regions in the solution space.

In the Roulette wheel technique, the selection is performed over the current population with the probability of any individual being chosen is proportional to its own fitness value (Figure 11.3). Individuals with higher fitness values have a better chance of being chosen, and at the same time a small number of individuals with relatively low fitness values will have the chance to contribute in the reproduction of the new *generation*, resulting in new volumes within the search space for the GA to explore. The pseudo code for the Roulette wheel technique is shown in Figure 11.3.

Population Size

The chosen *population* size for each generation has a significant effect on the GA performance. Populations that are too small typically converge prematurely to a local optimum (Goldberg et al., 1992). For the class of problems reported in this chapter, an empirical study has been carried out to enable effective guideline values for population sizes to be determined.

Figure 11.4 shows results obtained. For each case, an average value had been taken over 15 separate runs. Fifteen runs are commonly used in the GA community as a reliable reflection of the performance. However, this also is a highly problem-dependent value.

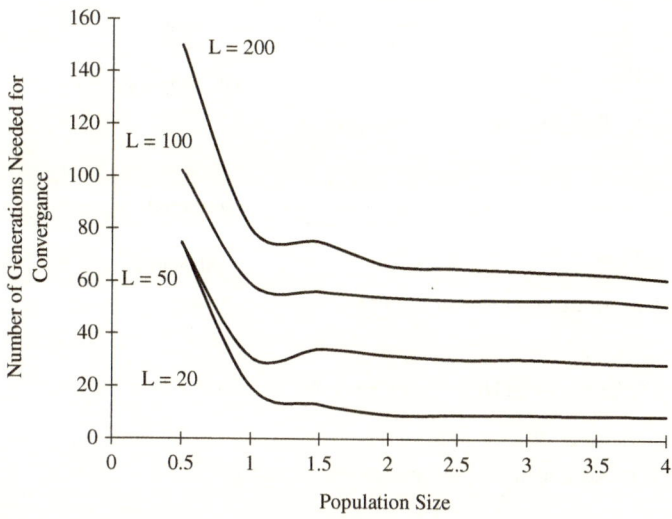

FIGURE 11.4. Population sizes related to the chromosome length (L).

For small populations, research show that increasing the population size results in a valuable reduction in number of generations needed to obtain convergence to a solution or solutions that satisfy the search criterion. However, in an empirical study that looked at the number of generations needed for convergence vs the population size, it was found that these benefits effectively reaches saturation when the population size reaches the chromosome length, L; this is shown in Figure 11.4. The horizontal axis represents multiples of the chromosome size, L. Figure 11.4 shows that for a chromosome with $L = 200$ bit long, about 150 generations were needed to reach convergence when the population size was 0.5L. When the population size was doubled to 1.0L, 80 generations were needed to reach convergence. This is a reduction of about 50% from the previous case. At population size of 2.0L, about 70 generations were needed. The improvement in number of generations needed for convergence is negligible beyond 2.0L.

Crossover Operator

The *crossover* operator is considered to be the most important operator in GAs. It is also a special characteristic of GAs over many other search techniques that are inspired by the process of biological optimization.

In nature, *crossover* occurs when two parents exchange parts of their corresponding chromosomes. In GAs, the same idea forms the basis of the *crossover* operator, where useful parts of different parent chromosomes

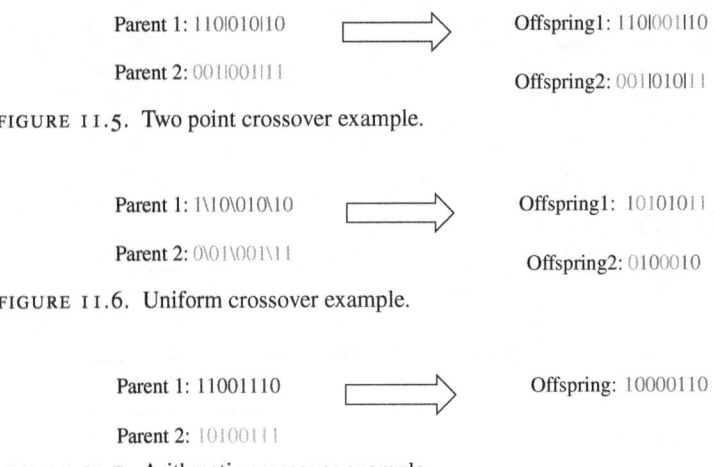

FIGURE 11.5. Two point crossover example.

FIGURE 11.6. Uniform crossover example.

FIGURE 11.7. Arithmetic crossover example.

should be combined to produce an offspring that benefits from the advantageous bit combinations of both parents. Two-point crossover is an operator that randomly selects two crossover points within a chromosome and then interchanges the two-parent chromosomes between these points to produce two new offspring.

Consider the following two parents selected for crossover, Figure 11.5. The "|" symbols indicate the randomly chosen crossover points.

After interchanging the parent chromosomes between the crossover points, the following offspring are produced. Other forms of crossover include uniform and arithmetic, among others. In uniform crossover, each offspring is composed of a fixed ratio of genes from each parent rather than a random segment of the parents, Figure 11.6. The "\" separates different genes within the chromosome. In this particular example, a ratio of 50% is used, so offspring have half of their genes from one parent and the other half from the other parent.

In arithmetic crossover, a logical or arithmetic operation is used to produce the offspring, Figure 11.7. In the example in the following text a logical "AND" operation between the parents is used to produce the offspring.

Mutation

Mutation is a genetic operator that is used to change individual bits in the chromosome at a specified probability. The motivation for using this operator is that it helps to introduce genetic diversity to new generations. While it is desirable in nature to keep the level of mutation low so that an individual produced by the mutation does not differ too much from its

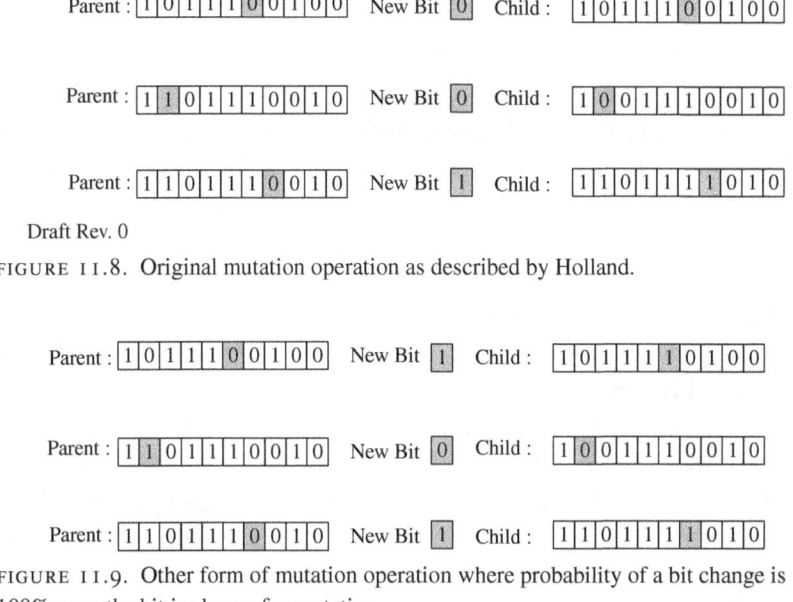

FIGURE 11.8. Original mutation operation as described by Holland.

FIGURE 11.9. Other form of mutation operation where probability of a bit change is 100% once the bit is chosen for mutation.

ancestor, this is not always the case in applications of GA (Grefenstette, 1986). Chosen mutation rates typically can range from 0.001 to 0.05. Rates beyond these limits are not recommended in most GA applications (Goldberg et al., 1992).

Holland defines *mutation* of a certain bit as the substitution of that bit by a random element. This is described in the following text for a chromosome of binary bit strings (Figure 11.8).

In the previously described mutation process, the probability that the new bit will be different from the original bit is 50%. Other mutation operators described in literature would just complement all bits chosen for mutation as in Figure 11.9. A range of different implementations of mutation operators that operate on genes rather than bits (or strings) also exist in literature among others.

Applications of Genetic Algorithms

In this section, three applications of GAs will be shown. For each application, a brief description of the problem is introduced, followed by how GA has been implemented. Lastly, a results section that describes and discusses the results is presented.

Analog Circuits Design Using Genetic Algorithms

Computer-aided design tools for analog circuits consist almost invariably of circuit analysis or simulation packages; no general-purpose analog circuit synthesis tools are available. Conventional numerical optimization techniques can be used to obtain combinations of component values that satisfy a given design specifications. In addition, conventional optimization techniques operate on circuits after having their desired responses approximated by a suitable polynomial, using ideal components with unrestricted values and only operating on predefined circuit structures. The restrictions introduced by predefining the circuit structure waste extra degrees of freedom in the design process and can conceal many novel design structures that can satisfy the target response and at the same time enhance the design process.

In Horrocks and Khalifa (1994), Grimbleby (1997), and Horrocks and Khalifa (1995), evolutionary computation algorithms are applied to fixed structure circuit designs to satisfy certain specifications. As mentioned earlier, the predetermination of the circuit structure required in conventional design techniques is a limiting factor when considering a powerful search technique such as GA.

An efficient measure of performance based on the sensitivity for component variations for the resulting circuits and the incorporation of the parasitic effects associated with components is still needed. This is particularly important in the development of new generations of analog programmable analog arrays design tools. In this section, an efficient GA technique for free structure analog circuit design is described. The technique incorporates a sensitivity measure to compare between potential solutions and incorporates parasitic effects.

The circuit example presented in this section is a low-pass filter circuit design. A typical low pass filter template would be represented as in Figure 11.10. A low-pass filter is a circuit that is designed to allow signals with a frequency lower than a certain cut-off to pass $H_{Low}(w)$ while attenuating those with higher frequencies, $H_{Upp}(w)$. Low-pass filters are widely used in common products such as loudspeakers and phones to eliminate high pitch noise.

GA Implementation: Chromosome Structure

In Holland (1995), Greenberg et al. (1992), and Grefenstette (1986), the predetermination of the circuit structures has helped in reducing

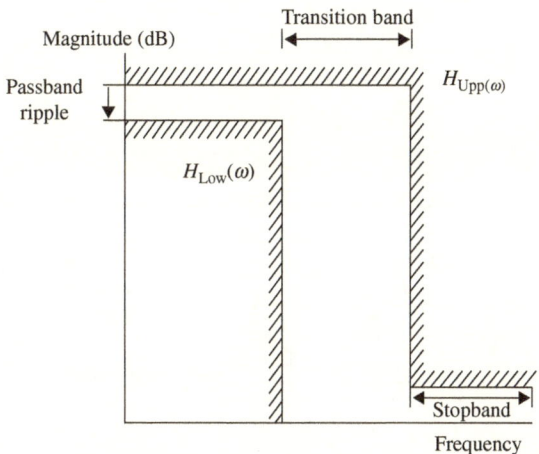

FIGURE 11.10. A typical template of a low pass filter.

the chromosome length. This was achieved by incorporating the order of the chromosome genes into the node connections information. However, in the case of free structure circuit synthesis this technique cannot be used. Koza, Bennett, Andre, Keane, and Dunlap (1997) applied mainly genetic programming to the topological problem as well as sizing problem of an analog circuit. In this chapter, however, we are using GA. In the case of free structure design, the solution space is much larger and the optimization process is composite. It involves the optimization of two different but strongly interconnected problems, the structure as well as the sizing of a circuit. Hence, all specifications of the electrical circuit including the structure and the sizing of all the components should be included in the chromosome representation. The specifications of the electrical circuit include the number of components in the circuit, the type of each component, and a list of connections between the components. Circuit components can include a variety of different types of components, including resistors, capacitors, and inductors.

Given the previous discussion, the chromosome for each circuit comprises a number M of groups of equal bit lengths and an extra group of bits representing the number of nodes in the circuit; see Figure 11.11. Each gene in the chromosome specifies a component and contains four fields (a) type of component (L, C, etc.), (b) a pointer to a menu containing component value, and (c) circuit nodes to which the component is connected;

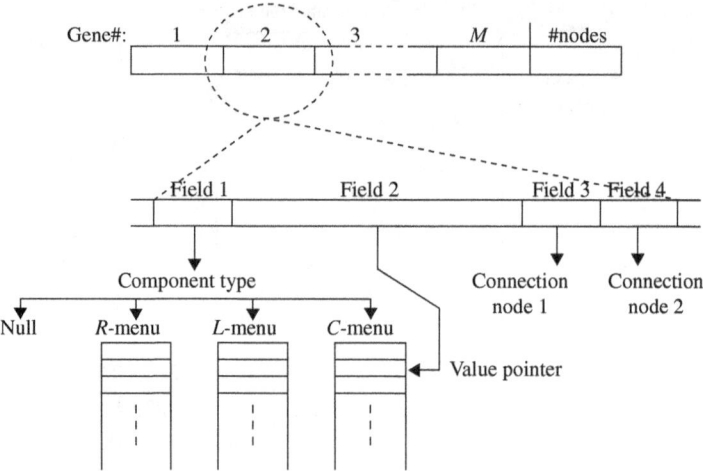

FIGURE 11.11. Proposed chromosome structure.

see Figure 11.11. Two-point crossover rate of 100% and a mutation rate of 0.5% were used for the GA operators.

GA Implementation: Circuit Generation

To start the design process, the user is required to supply the following input data.

(a) $H_{Low}(w)$
(b) $H_{Upp}(w)$
(c) N, the maximum number of electrical nodes. Optional and can be used to limit the search to a certain number of designs.
(d) M, the maximum number of components (a component represents a R/L/C or a null) (optional).

The execution of the program starts by calculating N *(if not provided)*, maximum number of nodes. Then, genes representing elements are then decoded one at a time, splitting them into four fields as explained in the previous section; GA Implementation: Chromosome Structure.

Upon determining type, value, and connection nodes, the component value is inserted in the corresponding nodal admittance matrix location (Khalifa, Khan, & Taha, 2007). When all elements are decoded, contents of the resulting entries in nodal admittance matrix are passed to the calculation function, in which lower–upper decomposition and solutions to the matrix equations are solved. Figure 11.12 shows a pseudocode for the program.

```
FOR population_member = 0 to population_size, DO
    Determine N, number of nodes
    FOR gene_location = 1 to M, DO
        Decode fields 1, 2, 3 and 4
        Descale fields 3 and 4 according to N
        Update NAM
    END FOR
    FOR fequency_test_point = 0 to maximum, DO
        LU decompose NAM
        Calculate response
        Compare with template and find error
    END FOR
END FOR
```

FIGURE 11.12. Pseudocode for a single generation evaluation.

GA Implementation: Fitness Evaluation

The fitness value of each circuit generated, is the reciprocal of the error calculated for that circuit. The error measured is defined as the difference between the specified response constraint, $H_{\text{Upp}}(\omega)$ and $H_{\text{Low}}(\omega)$, denoted by the shading in Figure 11.10, where $H_{\text{Upp}}(\omega)$ is the upper boundary of the frequency response, and $H_{\text{Low}}(\omega)$ is the lower boundary, and the actual response, $H_a(\omega)$, from the design represented by GA chromosomes. The actual frequency response, $H_a(\omega)$, will hence be prescribed by its unique combination of component values. Therefore,

$$H(\omega, R_1, R_2, ..., R_m, C_1, C_2, ..., C_n, L_1, L_2, ..., L_v) \tag{1}$$

For simplicity, all types of components will be declared as k. Thus H becomes

$$H(\omega, k_1, k_2, ..., k_q), \tag{2}$$

where $q = m + n + v$.

Hence, the error function will depend on the region it is applied for. For the passband the error function would be

$$err(\omega, k_1, k_2, ..., k_q) = H_{\text{Low}}(\omega) - H_a(\omega, k_1, k_2, ..., k_q), \tag{3}$$

while in the stopband the error function would be

$$err(\omega, k_1, k_2, ..., k_q) = H_{\text{Upp}}(\omega) - H_a(\omega, k_1, k_2, ..., k_q). \tag{4}$$

The least-squares error criterion E is used as a performance index because of mathematical simplicity and ease of programming. The error is defined as

$$E = \int_l^u err(\omega, k_1, k_2, ..., k_q)^2 \qquad (5)$$

Substituting from (3) and (4) in (5), LSE would be

$$\text{LSE} = \int_{\omega l}^{\omega u} \{H(\omega) - Ha(\omega, k_1, k_2, ..., k_q)\}^2 \qquad (6)$$

where $H(\omega)$ is the boundary response $H_{\text{Upp}}(\omega)$ or $H_{\text{Low}}(\omega)$. Since the calculations must be performed numerically, the integration in (6) will be replaced by summation. Therefore,

$$\text{LSE} = \sum_{i=1}^m \{H(\omega) - Ha(\omega, k_1, k_2, ..., k_q)\}^2$$

where m is equally spaced frequency points assumed over the range of integration. This is chosen because of ease in programming.

Results

The GA used has a crossover rate of 100% and a mutation rate of 5% throughout all runs. The GA was allowed to run for 1,000 generations in each run. However, convergence was reached after an average of 150 generations. Each generation had 50 chromosomes or individuals. Each generation took 10 seconds on average in processing. In these implementations, the desired design template is the starting point of the design process, whereas in conventional designs polynomial approximation for the desired response is used (Temes & LaPatra, 1977). This has the limitation of constraining the feasible solution to sets of design parameters that would satisfy the polynomial approximation. By designing directly from the template specifications, a designer would avoid an extra degree of approximation a polynomial fitting would introduce. At the same time, it provides an extra degree of freedom, which would enlarge the solution space considerably to include any design that would reside within the design template without necessarily being represented by a polynomial.

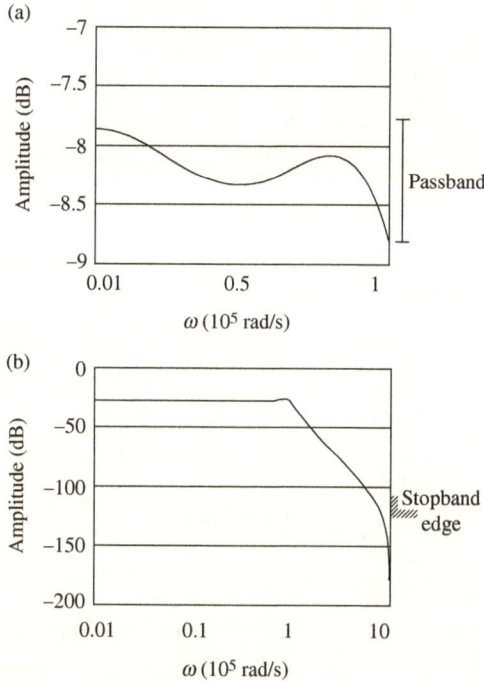

FIGURE 11.13. Frequency response examples for low-pass design template with high quality factor (a) passband and (b) stopband.

Also, the GA search starts from an arbitrary number of points in the solution space and needs not to have any prior knowledge of the feasible region of the required design.

Different low-pass design specifications were considered. One had a passband ripple of 1 dB, minimum attenuation of −100 dB. An average group of five solutions emerged; the frequency response of one of these solutions is plotted in Figure 11.13. The circuit constructed by the GA is shown in Figure 11.14. It is noticed that well-known π and T ladder structures have also been generated by the GA.

A different design specification of 0.5 dB passband ripple and −60 dB stopband attenuation was tested. Figure 11.15 shows the resulting circuit design.

Conclusion

This section presented a circuit design tool using the GA strong search abilities in both the structure and sizing domains of a circuit design problem. It is noticed that well-known π and T ladder structures have also been generated by the GA. That is due to the low sensitivity such structures

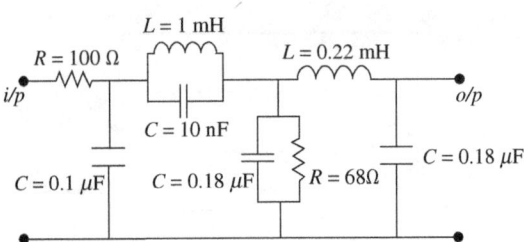

FIGURE 11.14. Genetically designed circuit with a passband ripple of 1 dB and stopband edge at –100 dB.

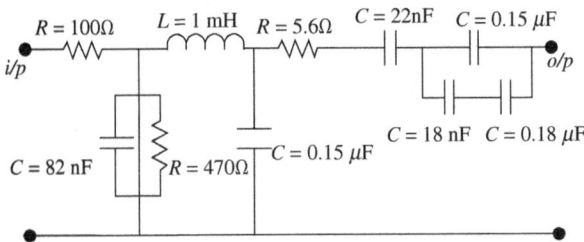

FIGURE 11.15. Genetically designed filter circuit with 0.5 dB passband ripple and –60 dB stopband attenuation.

have, to small changes in component values. This feature was detected by the GA as a result of the sensitivity comparison technique developed but not fully described in this chapter.

Evolutionary Music Composer

The concept of algorithmic music composition has long been attempted in the past; however, due to the nature of music as a creative activity, there is still a need for further work in this area. Gartland-Johnes and Colpey (2003) provide an excellent review of the application of GAs in musical composition. Miranda (2004) discusses different approaches to using evolutionary computation in music. However, most systems listed in literature need a tutor or an external evaluator. Cope (1996) and Rowe (1993) designed systems that created music by analysis of other music. While Cope's system took input from classical compositions, Rowe looked at live input from a performer. Rowe drew a distinction between transformative and generative music composition. Transformative systems transform incoming musical input (e.g., from a human to produce an output). Generative systems use an algorithmic composition to generate the output (e.g., from scratch). While his Cypher system contained elements of both, it was primarily a transformative one—listening to the input from the user, pushing the input through a series of transformations, and then outputting something derivative, although not necessarily reminiscent.

The development of autonomous unsupervised music composers is therefore still very limited but yet has lots of potential. In addition, the concept of using pattern extraction techniques to extract primary patterns, or motives, in established pieces of music has not been extensively explored in the literature. This is somewhat surprising, since composers have made use of motives for composition for centuries. The problem of composing music based on a library of motives is, however, near or perhaps slightly beyond the frontier of current capabilities of artificial intelligence technology. Thus, this area of research spearheads a new direction in automated composition.

The work presented in this section is an attempt in that direction. It presents an autonomous music composition system. The system composes musical pieces based on a library of evolving motifs. The fitness of the generated pieces is based on three evaluation functions: intervals, ratios, and formal grammar. Each of these functions measures the fitness based on an aspect of the musical notes and/or system as explained in the following sections.

Music Background

In this section, some basic fundamentals of music composition are given. Because the piano has a good visual interpretation of music, it will be used for illustration, however these concepts can be transposed to any musical instrument including the human voice. We begin by analyzing the most basic set of notes, called the C major scale, which consists entirely of all the white notes. We will dissect what major scales are, how they are composed, and further our discussion to how they form what are called chords or simultaneously depressed single notes.

In Western music, regardless of the instrument, there are 12 different distinct pitches or tones per octave, which are called keys. A pitch is simply a frequency of sound. Given these pitches a multitude of combinations can be formed to produce "music." However, how can we be assured that a specific combination will be musically pleasing to the ear? Of course the term *musically pleasing* is subjective to the listener, but there must be some fundamental principle underlying the organization of the combination in question.

There is an *interval* that exists between two consecutive pitches, the term *musical interval* refers to a *step* up or down in musical pitch. This is determined by the ratios of the frequencies involved. An octave is a music interval defined by the ratio 2:1 regardless of the starting frequency. From 100 Hz to 200 Hz is an octave, as is the interval from 2000 Hz to 4000 Hz. In music we refer to the interval between two consecutive notes as a *half step*, with two consecutive half steps becoming a *whole step*. This convention is the building block of our major scale.

A scale is a set of musical notes that provides the blueprint of our musical piece. Because our starting point is the musical note C, this major scale will be entitled as such. The major scale consists of a specific sequence of whole steps and half steps: W W H W W W H, where W is a whole step and H is a half-step. A typical musical convention is to number the different notes of the scale corresponding to their sequential order, usually called *roots*.

Using the sequence of the major scale, our C major scale consists of the notes C D E F G A B, returning to note C completing what is known as an *octave*, or a consecutive sequence of eight major scale notes. Numeric values are now assigned where C corresponds to value 1; D, 2; E, 3; and so on. The next C in terms of octaves would restart the count therefore the last value or root would be 7 corresponding to note B. We build on these scales by combining selected roots simultaneously to form what are known as *chords*.

Chords can be any collection of notes, this leads to almost endless possibilities in music; however, for our purposes, we implement the C *major chord* and use its sequence of notes. A major chord consists of the first, third, and fifth root of the major scale; this would mean that we utilize notes C E G.

GA Implementation: Chromosome Structure for Stage 1

In Stage 1, motifs are generated. A table of the 16 best motifs is constructed that is used in Stage 2. These motifs will be used both in their current and transposed locations to generate musical phrases in Stage 2. Figure 11.16 shows the chromosome structure in Stage 1. Each chromosome will contain 16 genes, allowing a maximum of 16 notes per motif. Each motif is limited to a four-quarter-note duration. It could be seen from the chromosome structure in Figure 11.16, that the notes are represented by four bits, resulting in one different possibility. A scaling factor is used to restrict the choices to the 13 available notes in the list.

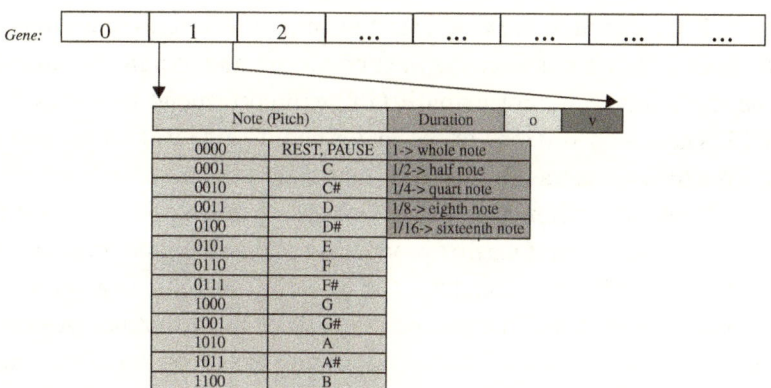

FIGURE 11.16. Chromosome and gene structure for Stage I.

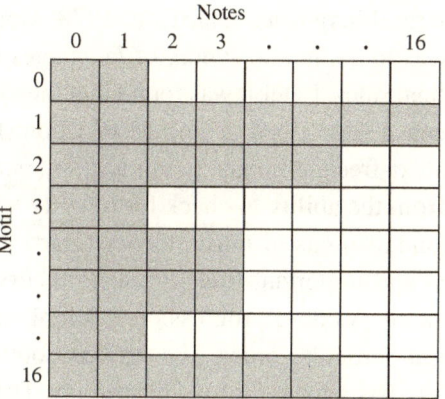

FIGURE 11.17. Motif look-up table generated in Stage I.

At the end of Stage 1, a table of the top 16 motifs is constructed (Figure 11.17). Each row in this look-up table represents a motif. The columns represent the different notes in the motif. Although all motifs generated are one whole note in duration, they could be composed of either one, two, four, six, or eight notes. However, single note motifs are highly discouraged.

GA Implementation: Evaluation Function for Stage 1

Formal Grammar Rule

As previously mentioned the C major chord consists of the first, third, and fifth root of the scale. This will correspond to values 1, 5, and 8. A new genetic reproduction operator, *inversion*, is also introduced in this implementation. We can do an *inversion* of the major chord by using the same notes C E G; this time, however, starting at note E leaving the inversion to be E G C. If we consider the next octave up and assign the value 13 to the repeating C, the inversion E G C will correspond to values 5, 8, and 13. These two versions of the major chords give us two production rules of which we are assured will be musically pleasing. The production rules will take the difference of the values assigning a good fitness if the production rule is met. Our first production rule will be the difference or skip of 4 and 3 ($5 - 1 = 4$ and $8 - 5 = 3$), describing our first version of the major chord C E G. The second rule will be the difference or skip of 3 and 5 ($8 - 5 = 3$ and $13 - 8 = 5$), describing the inversion of the major chord E G C.

Formal grammar theory was introduced by the linguist Noam Chomsky in the 1950s, when he attempted to give a precise characterization of the structure of natural languages (Harrison, 1978; Hopcroft & Ullman, 1979). His goal was to define the syntax of languages using simple and precise mathematical rules. Later it was found that the syntax of programming languages can be described using one of Chomsky's grammatical models called context-free grammars. In this section, our formal grammar can be extended from the ability to check for tonality to the ability to encourage certain fundamentals of music theory (Harrison, 1978). One of the most common and important rules in music theory is the notion of the chord progression. A chord progression is simply a series of chords that are played in a particular order. The various combinations of these progressions are the basis for music we hear today. The most frequently used progressions rely on the first, fourth, and fifth degrees of the major scale. Scale degrees, are the usual nomenclature practiced when relating the name of a note with its corresponding order in which it falls on the diatonic scale.

For example the diatonic major scale of C, would begin at C continuing to D, E, F, G, A, and B. The scale degree of C would be the first, D would be the second, E the third, and so on. These degrees represent only a root tone; subsequent chords would be used in conjunction with that root note, and other notes derived from that root to form a full musical chord. Again, we will only utilize the root and not the corresponding chord.

One of the most common progressions, primarily used in Jazz, is the II–V–I. Observing the diatonic scale of C major, we see that if we start at the II, which is note D, and descend in fifths, we will end at G, which is the V, and continuing in fifths, we end at C which is the I. This pattern in music theory is known as the *circle of fifths*, which describes the relationships among the 12 chromatically distinct keys of music.

Another popular progression used in music is the VII–III–VI, again following the same pattern as the previously mentioned progression, which descends in fifths. In this particular progression we notice that if we end on the VI we can then move to the II if we descend following the circle of fifths theory. The II will then be able transition to the V and finally resolve at I. We now have formed generally pleasing "music" by combining the two progressions, namely, VII–III–VI–II–V–I. This represents a pattern that can be used in producing new production rules.

Dissecting the previously described pattern we notice that each time we descend from a root we subtract four diatonic tones, and when we ascend from a root we add three diatonic tones.

We can apply this 3–4 coupling technique to all roots within the same musical key and find that we will have musically fit pieces. Since we have chosen the key of C, our production rules are simplified due to the lack of sharps and flats in diatonic C major. A formal definition of the grammar we chose follows,

We define a context-free grammar (CFG) thusly: G = {N, Σ, P, S}, where N is the set of nonterminal symbols our grammar accepts, Σ is the set of terminal symbols our grammar accepts, P is a list of production rules in Chomsky normal form, and S is an element of N that represents the start production.

The contents of these sets are:

N: {S, A, B, C, D, E, F, G}
Σ: {a, b, c, d, e, f, g, ε}, where ε is the empty string
P: {
S → beaA | cfbB | dgcC | eadD | fbeE | gcfF | adgG
A → cfbB | dgcC | eadD | fbeE | gcfF | adgG
B → beaA | dgcC | eadD | fbeE | gcfF | adgG
C → beaA | cfbB | eadD | fbeE | gcfF | adgG | ε
D → beaA | cfbB | dgcC | fbeE | gcfF | adgG
E → beaA | cfbB | dgcC | eadD | gcfF | adgG
F → beaA | cfbB | dgcC | eadD | fbeE | adgG
G → beaA | cfbB | dgcC | eadD | fbeE | gcfF
}

This CFG allows for any number of three note tuples, with no particular note tuple repeating. For example, beabea would be rejected, but beafbebea would be accepted. Every three-note tuple follows the same rule: the root note, a note three tones up in the diatonic C scale, and a final note four tones down from the previous note in the diatonic C scale. A given motif will be evaluated note by note according to this grammar, and any motif that cannot be expressed with our CFG will be rejected and given a low fitness value for the next generation to encourage these fundamental progressions in our sound.

Intervals Evaluation Function

wWithin a melody line, there are acceptable and unacceptable jumps between notes. Any jump between two successive notes can be measured as a positive or negative slope.

Certain slopes are acceptable, while others are not. The following types of slopes are adopted:

Step: a difference of one or two half steps. This is an acceptable transition.
Skip: a difference of three or four half steps. This is an acceptable transition.
Acceptable leap: a difference of five, six, or seven half steps. This transition must be resolved properly with a third note (i.e., the third note is a step or a skip from the second note).
Unacceptable leap: a difference greater than seven half steps. This is unacceptable. As observed from the previous information, leaps can be unacceptable in music theory. We model this in GA using penalties within the interval fitness function.

Certain resolutions between notes are pleasant to hear but are not necessary for a "good" melody. These resolutions therefore receive a bonus. Dealing with steps in the chromatic scale, we can define these bonus resolutions as the 12-to-13 and the 6-to-5 resolutions. The 12-to-13 is a much stronger resolution and, therefore, receives a larger weight. It was determined based on empirical assessment that the 12-to-13 resolution have double the bonus of the 6-to-5 one, and that the bonus does not exceed 10% of the total fitness. Thus, the bonuses are calculated by dividing the number of occurrences of each of the two bonus resolutions by the number of allowed resolutions (15 resolutions among 16 different possible note selections):

$$12\text{-to-}13 \text{ bonus} = (\#occurances / 15) * 0.34 \qquad (7)$$

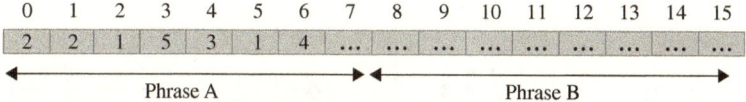

FIGURE 11.18. Chromosome structure for Stage II.

$$6 - \text{to} - 5 \text{ bonus} = (\#\text{occurances}/15) * 0.34 \qquad (8)$$

The total interval fitness:

$$\text{Interval Fitness} = \frac{1}{\text{total\_error}(1 - \text{total}_{\text{bonus}})}$$

GA Implementation: Chromosome Structure Stage 2

A piece of music is composed of a sequence of motives, either same or different ones. In Stage 2, a piece of music composed of two phrases, A and B, is constructed. Each phrase is eight measures, and each measure is a four quarter-note duration motif. The motifs used to compose these 16 measures are selected from the look-up table constructed in Stage 1 (Figure 11.18).

GA Implementation: Stage 2 Evaluation Function

In Stage 2, two evaluation functions are implemented: intervals and ratio. The intervals evaluation function described in the previous section is used to evaluate interval relationships between connecting notes among motifs, that is, between the last note in a motif and the first note in the following motif. The same rules, as described in Stage 1, are used in Stage 2. Other evaluation functions are described in the following discussion.

Ratios Evaluation Function

The basic idea for the ratios section of the fitness function is that a good melody contains a specific ideal ratio of notes, and any deviation from that ideal results in a penalty. There are three categories of notes: the tonal centers that make up the chords within a key; the color notes, which are the remaining notes within a key; and chromatic notes, which are all notes outside a key. Each type of note is given a different weight based on how much a deviation in that portion of the ratio would affect sound

FIGURE 11.19. Sample motif generated in Stage I of the Evolutionary Music Composer.

quality. A proposed ideal ratios sought were presented by Khalifa, Khan, Begovic, Wisdom, and Wheeler (2007) and are as follows: tonal centers make up 60% of the melody, color notes make up 35% of the melody, and chromatic notes make up 5% of the melody. Although these ratios choices could be quite controversial, they are a starting point. Ongoing research is looking into making these ratios editable by the user or music-style dependent.

Results for Stage 1: Motif Selection

The four motifs in Figure 11.19 all resulted from a single running of the program represent a sample of the final most fit motifs from Stage 1. It can be observed that each motif has an identical rhythm consisting of four eighth notes, one quarter note, and two more eighth notes. Summing the durations of the notes yields the correct four quarter-note duration indicated by the time signature $\binom{4}{4}$ at the beginning of each motif.

Using the intervals evaluation algorithm as a reference, we can see why these motifs were chosen to be the elite of the population. Examining motif a, the first three notes are all F♯s, indicating that no penalty will be assigned (a step size of zero). The next note is a G♯ (two half-steps away from F♯). This transition is classified as a step, and no penalty is assigned. The following notes are F♯, G♯, and E (a difference of two, two, and three half-steps, respectively). These transitions are also acceptable; therefore, the intervals evaluation function would not assign any penalty to the motif.

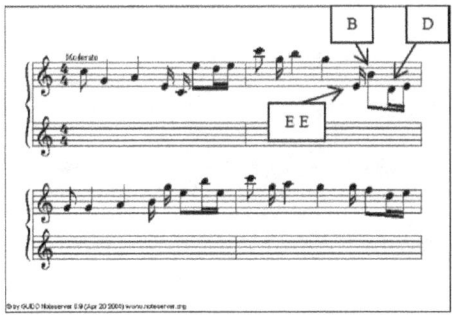

FIGURE 11.20. Sample motifs generated in Stage II of the Evolutionary Music Composer.

When zero error is assigned to a motif, a high fitness value will result. Similar analysis of motifs *b, c,* and *d* yield the same result.

So what is the musical difference between the motifs? Since the notes in each motif are slightly different, the musical "feel" of each motif will vary. Compare motifs *a* and *d* for example. Motif *a* contains four F♯s. They are arranged in such a way that the first beat and a half of the measure are all F♯s, and also the third downbeat (the quarter note).

This repeatedly drives the sound of the F♯ into the listener, resulting in an unconscious comparison of this note to every other note in the measure. This, in turn, will make dissonant notes sound more dissonant, and resolving notes sound more resolved. In the case of motif *d*, the F♯s are arranged in a manner that accents the steady background rhythm of the measure (the repetitive rhythm that your foot taps to when you listen to music). This does not accent the sound of the F♯ as much but rather accents the other rhythms of the measure that occur between the F♯s. A more "primal" feel will result, as opposed to the more "melodic" feel of motif *a*.

For the formal grammar evaluation function, the main musical objective is to implement the 3–4 coupling by analyzing the sequence of notes in the musical piece generated in Figure 11.20, where the piece incorporates almost all the rules and musical parameters that the program has set; however. let us focus on how the 3–4 coupling rule is represented. Taking a look at Figure 11.20 at the end of the second measure, we notice a 16th note E. We know that on the C major scale, speaking numerically, E would be represented by a 3. This would mean that to move three major tones up would leave us at the 6, which is B. From B in accordance with the 3–4 coupling rule, we should then descend four major tones, completing the rule at the 2, or note D. We see a direct implementation of this sequence twice in the entire piece. The third example is in Figure 11.21 and

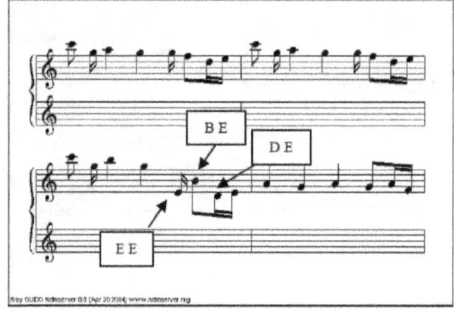

FIGURE 11.21. Sample motifs generated in Stage II of the Evolutionary Music Composer.

is seen at the end of the second to last measure, where the sequence is characterized by the same exact notes as explained in the first instance of the 3–4 coupling rule.

References

Beyer, H. G., & Schwefel, H. P. (2002). Evolution strategies: A comprehensive introduction. *Natural Computing*, *1*, 3–52.
Cope, D. (1996). *Experiments in musical intelligence*. Madison, WI: A-R Editions, 1996.
Degroot, M. H. (1970). *Optimal statistical decisions*. New York, NY: McGraw-Hill.
Fogel, L. J. (1999). *Intelligence through simulated evolution: Forty years of evolutionary programming*. New York, NY: Wiley, 1999.
Gartland-Jones, A., & Copley, P. (2003). What aspects of musical creativity are sympathetic to evolutionary modeling. *Evolutionary Models of Music*, *22*, 43–55.
Goldberg, D. E., Deb, K., & Clark, J. H. (1992). Genetic algorithms, noise, and the sizing of populations. *Complex Systems*, *6*, 333–362.
Grefenstette, J. J. (1986). Optimization of control parameters for genetic algorithms. *IEEE Transactions on Systems, Man, and Cybernetics*, *16*, 122–128.
Grimbleby, J. B. (1997). Automatic synthesis of active electronic networks using algorithms. In *Second International Conference on Genetic Algorithms in Engineering Systems: Innovations and Applications* (pp. 103–107). London, England: Institution of Electrical Engineers.
Harrison, M. (1978). *Introduction to formal language theory*. Reading, MA: Addison-Wesley, 1978
Holland, J. (1975). *Adaptation in natural and artificial systems*. Ann Arbor, MI: University of Michigan Press, 1975.
Hopcroft, J., & Ullman, J. (1979). *Introduction to automata theory, languages and computation*. Reading, MA: Addison-Wesley, 1979.
Horrocks, D. H., & Khalifa, Y. M. A. (1994). Genetically derived filters using preferred value components. In *Colloquim on linear analogue circuits* (pp. 3/1–3/4). Oxford, England: Institution of Electrical Engineers, 1994.

Horrocks, D. H., & Khalifa, Y. M. A. (1995). Genetically evolved FDNR and leapfrog filter using preferred components values. In *Proceeds of the 12th European Conference on Circuit Theory and Design* (pp. 359–362). Istanbul, Turkey: Istanbul Technical University, 1995.

Khalifa, Y. M. A., Khan, B., Begovic, J., Wisdom, A., & Wheeler. A. M. (2007). Autonomous music composition relying on evolutionary formal grammar. *International Journal of Artificial Intelligence and Machine Learning*, 7, 33–39.

Khalifa, Y. M. A., Khan, B., & Taha, F. (2007). Multi-objective optimization tool for a free structure analog circuits design using genetic algorithms (pp. 2527–2534). In *Proceedings of Late Breaking Papers, Genetic and Evolutionary Computation Conference*. London, England: Association for Computing Machinery.

Koza, J. (1992). *Genetic programming: On the programming of computers by means of natural selection*. Cambridge, MA: MIT Press.

Koza, J. R. Bennett, F. H. Andre, D., Keane, M. A., & Dunlap, F. (1997). Automated synthesis of analog electrical circuits by means of genetic programming. *IEEE Transactions on Evolutionary Computation*, 1, 109–128.

Miranda, E. R. (2004). At the crossroads of evolutionary computation and music: Self-programming synthesizers, swarm orchestra and origins of melody. *Evolutionary Computation*, 12, 137–158.

Oosthuizen, G. D. (1989). Machine learning: A mathematical framework for neural network, symbolic and genetics-based learning. In J. D. Schaffer (Ed.), *Proceeding of the 3rd International Conference on Genetic Algorithms* (pp. 385–390). San Mateo, CA: Morgan Kaufmann.

Rowe, R. (1993). *Interactive music systems: Machine listening and composing*. Cambridge, MA: MIT Press, 1993.

Temes G. C., & LaPatra J. W. (1977). *Introduction to circuit synthesis and design*. New York, NY: McGraw-Hill.

Zhou, H. H. (1990). CSM: A computational model of cumulative learning. *Machine Learning*, 5, 383–405.

SECTION 3 | Applied Evolutionary Studies

CHAPTER 12 | The Role of Evolutionary Studies in Education for Sustainable Development

DUSTIN EIRDOSH AND SUSAN HANISCH

THE EARTH HAS ENTERED the Anthropocene, a new geological age in which the human species is now considered the dominant driver of global climate change, biodiversity loss, and thus, sustainability in general (Ellis, 2016). The *wicked problems* of the 21st century are complex, interconnected, controversial, and have no single or simple solution. Global initiatives toward *Education for Sustainable Development* (ESD) have grown from this recognition. Global coordination for ESD initiatives has been largely lead by UNESCO with the ambitious aim to "reorient education and learning so that everyone has the opportunity to acquire the knowledge, skills, values and attitudes that empower them to contribute to sustainable development" (UNESCO, 2014).

The struggles for sustainable development are inherently challenges of human cooperation and collective learning. We will argue that it is critical to engage all students of ESD in studying the *evolution* of these uniquely human dynamics as well as understanding these capacities as *evolutionary processes* in their own right (Wilson, Hayes, Biglan, & Embry, 2014).

Evolutionary studies (EvoS) connects concepts from across disciplines that are critical for understanding the human condition and sustainable development as a whole. The transdisciplinarity inherent in EvoS provides students with important foundational competencies widely recognized across ESD. Despite this potential value, the field of ESD engages the EvoS literature very rarely and selectively (e.g., Krasny & Tidball, 2015, p. 27).

Examining the current global discourse regarding ESD reveals a handful of dialectical tensions that pose both theoretical and practical challenges to developing a coherent and unified approach to curriculum development. ESD seeks to unite our global species while respecting local diversity. ESD seeks to understand both the parts of complex systems as well as the emergent properties of these systems as a whole. ESD seeks to promote multiple levels of cooperation while respecting local autonomy. And in this regard, ESD seeks to promote innovation while respecting tradition. These are complex challenges that have thus far been approached through a patchwork of theoretical frameworks or personal philosophies often disconnected from a cumulative empirical understanding of the human condition.

EvoS provides a flexible toolkit capable of navigating these dialectical tensions with clarity and practical nuance. If human cooperation and collective learning are central to most issues of sustainability, students of ESD should understand how our species developed these capacities and why these abilities are so critical for navigating our changing times. In this way, the linked concepts of *ultimate causation* and *adaptive flexibility* provide a foundational framework for connecting EvoS and ESD content, and a clear direction for developing a global approach to ESD. Sustainable management of complex systems requires students understand the ultimate causation of that complexity, *and* the nature of our species' adaptive flexibility in relation to human values. From these concepts, a range of applied pedagogical directions emerge with the potential to strengthen student learning across both EvoS and ESD. Evolutionary perspectives on sustainable development therefore open new horizons of opportunity for authentic learning for students and educators alike.

Historical Context of EvoS and ESD

In 1946, Julian Huxley became the first elected director of the United Nations Educational, Scientific, and Cultural Organization (UNESCO). On accepting this role, Huxley penned *UNESCO: Its Purpose and Its Philosophy*. For many social science scholars or social justice advocates, this document may read as highly controversial statements about the role of UNESCO to shape "a unified pool of tradition for the human species as a whole" (Huxley, 1946). Huxley was a famed evolutionary biologist, as well as an individual holding complex views on eugenics and humanism. Huxley envisioned the concept of *evolutionary humanism* as the foundational philosophy of UNESCO, arguing its mission was to unify a "world mind" by elevating the well-being of global citizens through a collaborative culture of scientific literacy rooted in secular humanistic values. Huxley had a prescient understanding of *cultural evolution* (Mesoudi, 2011; Muthukrishna & Henrich, 2016) and the role of education in sustainable development, yet his controversial and complex perspectives on eugenics have made him into a pariah not to be referenced positively within ESD. While UNESCO remains the global driver of ESD programming today, the social justice and environmental education traditions informing modern ESD for the last several decades have eliminated virtually any trace of Huxley's influence or original vision. The formal advancement of ESD is, instead, largely recognized as having been influenced by the field of Environmental Education (EE; Monroe, 2012).

To understand the divergent relationship that currently exists between the EvoS and ESD communities in this context, it is important also to refer to the historical relations among the evolutionary, ecological, and complexity sciences themselves, especially in relation to studies of the human condition and EE (Wilson, 2005a). Hagen (1992) describes how an important disciplinary divide between evolution and ecology can be seen starting in the 1960s and escalating into the present. For studies of the human condition, these early tribal divisions across some strains of evolutionists and ecologists only amplified. As the decades have worn on, *ecosystems ecology* (and thus EE) have integrated with complexity sciences on the theoretical end and *social constructivism* on the pedagogical end. Often, social constructivists view biological understandings of our species as something to be explained away by human agency rather than wholly complementary to understanding it (Bandura, 2006). Evolution sciences, during this same period, have built reasonably coherent and well-rounded models of the human condition that integrate ecology, complexity, *and* the values of social constructivism (Gopnik & Wellman, 2012; Raskin, 2008; Wilson, 2005b). However, because evolution scientists studying humans have not historically provided significant leadership in the world of applied educational development, there exists the current disconnect between EvoS literature and the cultural narratives of practice informing ESD.

Critiques and Challenges of ESD

The establishment of ESD as an evidence-driven, globally networked community of practice brings with it challenges for practitioners and researchers alike. The need for rigorous evaluation and integration of the best current knowledge from various disciplines is required to assure valued educational outcomes on a global level. In the context of ESD's history and the steep challenges of the present, it is not surprising that this academic community faces some significant criticism, both from within and outside the field.

A prevailing narrative in ESD literature, exemplified by the Earth Charter (Earth Charter Initiative, 2000), is the need for inducing a shift in values and attitudes, which is considered to lead to changes in action and behavior from the level of the individual to the level of institutions, governments, and transnational organizations. Arbuthnott (2010) argues that the emphasis on transmitting values in ESD is flawed in that a change of attitudes and values among individuals does not directly lead to change in behavior. Often what is needed is targeted design of the environment and of institutions to induce and support behavior change of individuals and groups. This critique seems to have gone unanswered in the field, as the emphasis on specific values and attitudes and an aversion to the concept of *behavior change* as "instrumental" prevails (Hofman, 2015; Kopnina & Meijers, 2014; Tilbury, 2011).

Within the ESD community, it has been noted that current practices and discourse are often limited to the promotion of pedagogical methods that emphasize experiential learning or other very general and common pedagogical practices. ESD falls short of developing and conveying the necessary interdisciplinary perspectives needed for understanding issues of sustainability (Parker, 2010; Tilbury, 2011; Wals, 2012). There is significant lack of consensus as to what ESD really is and how it is different from the field of EE or other sustainability oriented educational approaches, as well as what it should try to achieve (Hofman, 2015; Kopnina & Meijers, 2014; Wals, 2009).

With this lack of consensus comes also a controversy around monitoring and evaluation and a lack of empirical evidence regarding effectiveness of ESD practices to achieve intended learning outcomes. Most publications in the ESD academic literature cover theoretical learning models, such as *transformative learning* (Sterling, 2010) or educational case studies with limited transferability across contexts. The ESD community is aware that the field "remains poorly researched and weakly evidenced" (Tilbury, 2011, p. 9). The current ESD literature thus offers little generalizable explanatory power and, consequently, little guidance for addressing the diversity of wicked problems in sustainable development.

In this regard, Webster (2007) noted that "ESD needs to offer new and coherent stories of what makes a sustainable world," (p. 39) lamenting that ESD was not engaging in developing transferable frameworks—big narratives and mental models—that connect the ecological and social realms and that would give students guidelines for inquiry and discussion. While there currently is a lot of emphasis on interdisciplinary competencies, ESD is vague about how to do this in practice, reverting to emphasis on student-directed knowledge creation but failing to give students the tools and frameworks for effective inquiry (Parker, 2010).

The need for a generalized theory base for sustainability in all its dimensions is thus expressed to the point by Parker (2010):

> In sustainability, the systems that we are trying to change range from the ecological, through the social, to the cultural. . . . How can we gain the understanding to take effective action across these different systems and what kinds of complementary elements might we need in each of these systems in order to generate a positive cycle of change for sustainability? (p. 328)

On the other hand, the ESD community seems to be unable to agree on such a generalized global approach. The prevailing opinion being that such a generalized approach is not possible and even potentially wrong-headed due to the need to appreciate cultural diversity (UNESCO Sustainable Development Goal 4.7) and the contextual diversity of sustainability problems prevailing in any one region (Hofman, 2015; Kopnina & Meijers, 2014; Wals, 2012).

This insufficiency in clarifying the nature of human universals and diversity while striving to be a global program for advancing global sustainability can be regarded as at the basis of multiple unresolved conflicts or dialectics commonly found in ESD discourse.

The Dialectics of ESD Discourse

The Earth Charter describes that "life often involves tensions between important values. This can mean difficult choices. However, we must find ways to harmonize diversity with unity, the exercise of freedom with the common good, short-term objectives with long-term goals" (Earth Charter Initiative, 2000). Despite this important claim, readers of the ESD literature are left with no guidance for making such difficult choices, no practical, tried and tested tools to realize such a harmonization. This section frames four important dialectical values commonly discussed and unresolved within the ESD literature.

The Diversity—Universality Dialectic

Every single human that has ever existed is absolutely unique in so many ways, and yet all humans have a universally common evolutionary lineage. ESD commonly espouses the importance of adopting a human universal perspective while also emphasizing the value and role of embracing human diversity. One of the principles that ESD is advancing is "Humans have universal attributes" (UNESCO, 2012). There exists, however, little clarity on what the term *universal* means specifically or how it should be taught to students and inform collaborative problem-solving. On the other hand, the important role of diversity is often emphasized, such as UNESCO (2012) pointing out the principles "Considering differing views before reaching a decision or judgement" and "Recognizing that economic values, religious values, and societal values compete for importance as people with different interests and backgrounds interact."

However, specific strategies for engaging culturally diverse groups of differing sizes in effective problem-solving and decision-making are rare (e.g., Hansmann et al., 2009) and mostly remain vague, anecdotal, or theoretical (e.g., Coops et al., 2008; Dan, 2016; Hofman, 2015; Wals, 2010; Wals & Schwarzin, 2012). Adding to the confusion of human diversity and universality is the strong distinction often made between Western culture and indigenous cultures. By not employing a coherent framework connecting human universals to local diversity, ESD remains challenged in defining a locally meaningful global approach.

The Reductionism—Holism Dialectic

Ecological and sociocultural systems are almost unfathomably complex, and yet this complexity emerges and evolves from the accumulation of smaller scale interactions. In the ESD literature, a common narrative exists regarding *science* cast as a purely Western, reductionist, rational, materialistic, or mechanistic way of knowing, as brought about by the Enlightenment era (e.g., Jackson, 2011; Sterling, 2014; Stuckey, 2015; Tilbury, 2011; Wals, 2012). This particular vision of science is then often further associated with top–down, transmissive, and rigidly fact-based pedagogical approaches. For many ESD scholars, the enlightenment era contributions to science are considered as key drivers of current ecological and societal problems. This concern gives rise to the concept of "other ways of knowing"—a form of epistemological pluralism embracing modes of thought considered more *holistic, systemic, intuitive, emotional,* or *values-oriented* than the popular narrative of science as purely reductionist and "rationalist." This polarized understanding of science leads to pedagogical methods emphasizing self-directed learning and knowledge (co-)construction, as well as a certain discounting of the scientific method and its accumulated body of knowledge in certain domains.

Similarly, there is a lot of emphasis in ESD discourse on uncertainty in today's complex world: "Rather than setting our academic minds towards minimising uncertainty and maximising predictability, it might be more fruitful to put our energy towards living with uncertainty: seeing it as a given, something that cannot be conquered" (Wals, 2011, p. 183). ESD scholarship often frames a false dichotomy: Striving for zero uncertainty (ascribed to Western scientists) is contrasted by accepting an apparently unmanageable uncertainty. Students of ESD are generally not offered any practical, empirical tools of how to make best decisions in the face of such complex uncertainty.

The Autonomy—Cooperation Dialectic

Universal human rights are predicated on respect for individual autonomy and freedom, and yet achieving global instantiation of human rights clearly requires extensive and effective cooperation across multiple scales of society. The autonomy–cooperation dialectic is found across various themes of ESD. The main pedagogical approaches that ESD distinguishes among are the *transmission-oriented, instrumental modes* as contrasted with *transformation-oriented, emancipatory modes*. The latter "relies more on participation, self-determination, autonomous thinking and knowledge co-creation" (Wals, 2012, p. 23). The ESD community appears to favor this approach to teaching, since there is a general aversion to anything perceived to be a "prescriptive" or top–down approach (Wals, 2012). Sometimes, the endeavor of a *global* ESD effort itself is viewed skeptically as being an instrumental top–down indoctrination (or colonization) of indigenous cultures by the Western world (e.g., Berryman & Sauvé, 2016; Hofman, 2015; Kopnina & Meijers, 2014).

Related to this dialectic is also the conflict between self-interest and mutual interest regarded as opposite sides of a spectrum and expressions of opposing values (e.g., Jickling & Wals, 2012; Preston, 2010; Ruiz, 2010). The prevailing narrative within the Earth Charter and ESD discourse is that to achieve global sustainability, self-interest needs to be overcome by inducing values of cooperation and care (Ruiz, 2010). However, no practical framework is put forward that allows the navigation of nested tensions between autonomy and cooperation inherent at every level of social organization in sustainability problems.

The Tradition—Innovation Dialectic

Cultural traditions bind communities together around shared values and experiences, and yet innovation and change have been never-ending drivers of our shared human story. Western progressives may yearn for change to the modern world while romanticizing the traditions of indigenous cultures. Indigenous cultures face immense opportunity and challenge in their task of navigating tradition and innovation as our globalized world becomes one in which it is harder than ever to remain disconnected. ESD lies at the interface of all of these challenging tensions. The strong distinction made between Western culture and indigenous cultures leads again to the notion of science as a Western phenomenon, following that "both [sustainable development] and ESD are often seen as 'western' concepts" (Wals, 2009, p. 21). Traditional or indigenous cultures are often considered to represent values, traditions, and practices that are seen as inherently more reliable ways of knowing, informing inherently more sustainable ways of living (e.g., Gadotti, 2008; UNESCO, 2009; Pigem, 2007). Tilbury (2011) points out instances of how ESD practices seek to bridge and combine tradition and innovation. To our knowledge, however, there appear to be no clear principles expressed that would guide practitioners toward navigating when and in what context traditional knowledge and practices are adaptive and when innovation or newly derived knowledge and practices will be more effective for achieving sustainable outcomes.

The EvoS Conceptual Toolkit

The dialectical challenges of ESD are deeply interwoven. In viewing these apparently opposing values, humans often intuitively lean toward one or the other end of the spectrum in a given context, where more adaptive solutions might require a complexification of thinking strategies. Dialectical values are important precisely because both ends of the spectrum are valued in different ways and in different contexts. If we don't want to throw the baby out with the bathwater, as the idiom goes, we must become skilled in identifying what is the baby and what is bathwater. In this regard, the value of an evolutionary perspective is that "some questions that appeared central become peripheral and other previously ignored questions become central" (Wilson, 2015a, p. 6). Specifically, an evolutionary analysis emphasizes questions of *why* a given trait exists compared to the many others that could (its *ultimate causation* or *evolutionary function)*, as well as *how* a given trait develops and functions in its current context (its *proximate causation* or *ecological mechanism*).

Ethologist Niko Tinbergen (1963) famously argued that a fully rounded natural explanation for any given trait should explore four questions, two each regarding ultimate and proximate causation. Tinbergen highlights the need for understanding the evolutionary history of a trait as well as its current adaptive functioning (ultimate causation). Additionally, we should understand the developmental trajectory of the trait and the mechanism by which it operates in ecological context (proximate causation). This basic framework serves equally well for understanding pigeons as for people in a holistic, integrative fashion. This includes the broadest conception of human traits of cooperation and collective learning as keystones of our adaptive flexibility. We argue, in line with Waring et al. (2015), that virtually all challenges of environmental, sociocultural, and economic sustainability can be productively framed as challenges of human cooperation, including the human capacity for collective learning. These species-typical capacities are evolved universal dynamics that make us human and represent strengths to build on in some contexts and weaknesses to overcome in others.

Recognizing the *functioning in context* of these behavioral or cultural attributes is a crucial evolutionary insight for adaptively navigating any of the dialectical dynamics of ESD. *Adaptive flexibility* broadly refers to the capacity for biological, behavioral, psychological, or sociocultural changes that enhance relative fitness *or* advance human values in the context of the current environment. Understanding the ultimate causation of certain core human dynamics may support the cultivation of adaptive flexibility at multiple scales of social-ecological organization (Grinde, 2016; Hayes & Ciarrochi, 2015; Waring et al., 2015; Wilson et al., 2014). From this perspective, we consider the linked concepts of *ultimate causation* and *adaptive flexibility* as foundational and global components of the ESD curriculum. These topics are exceedingly more complex than we can discuss. The aim here is only to contextualize the ESD challenges within an EvoS framework.

Understanding Ultimate Causation of Human Dynamics in Sustainable Development

When students or societies try to understand a specific challenge of sustainable development, we often focus strongly on questions of proximate causation, ignoring or only superficially looking at ultimate explanations. Often, there is a perception that ultimate causation is either not necessary or not particularly helpful in finding valued solutions to the wicked problems of sustainable development. However, by not engaging the evolutionary big history of humans as a unifying narrative framework for human diversity, ESD is missing an essential tool for coherent cross-cultural collaboration.

Students of ESD must develop a basic understanding of the evolutionary origins of human cooperation and collective learning as foundational to our species' evolved and pronounced capacity for adaptive flexibility. Exploring how these human dynamics *function* in a given environmental *context* is a central habit of evolutionary thinking vital to the diverse sustainability dilemmas of today's world and for navigating the complex dialectics of ESD. Scholars and practitioners of ESD generally agree about the importance of including local context when looking at global sustainability issues. What is often lacking is a recognition of the value that an evolutionary multilevel understanding of ultimate causation can add to our understandings of specific local contexts.

Local sustainability issues almost always include environmental, technological, behavioral, psychological, or sociocultural contexts that are varyingly similar or different than prior evolutionary or developmental conditions for our species or specific community. For example, cattle herders in Madagascar face intervillage resource issues regarding cattle feed, highly similar to the general dilemmas faced by our common ancestors, yet significantly different than that which their specific community may have memory of dealing with (Eirdosh, 2015). In contrast, the urban environments of cities and the digital world of the 21st century present immense promise for producing sustainability innovations, and yet such environments differ so much from any known human evolutionary developmental environment. In such situations, instances of *mismatch,* negative consequences that result when a trait that evolved in one environment is placed in another environment (Lloyd, Wilson, & Sober, 2011), are to be expected. Further, the challenge of global sustainability is a challenge of planetary-scale altruism, a scale of functional organization never yet accomplished since our earliest human ancestors began to band together. This needed scale of global cooperation will only emerge with ever-cascading and interlocking multilevel cooperation dilemmas resolving at the scales below it (Wilson, 2015a). Understanding the ultimate causation of multilevel organization across the biological and sociological domains, and understanding this in relation to the ultimate causation of our species' biological, psychological, and sociocultural development, provide a vital common foundation of content for all students of ESD.

Understanding ultimate causation guides students of ESD in understanding how evolutionary universals connect with human universals and how these concepts relate to the diversity of local and novel contexts. Understanding the ultimate causation of our human cognitive capacities guides students of ESD to understand and navigate their own and others' information processing and decision-making in the face of complex sustainability problems (Geary & Berch, 2016; Kahneman, 2011). Understanding the ultimate causation of human emotion guides students of ESD in understanding and navigating the dynamics of social-emotional reasoning among stakeholder groups of sustainability problems. Understanding the ultimate causation of multilevel cooperation dynamics across all species, and within our symbolic species specifically, guides students of ESD to effectively navigate the mazes of multilevel interest dilemmas that pervade sustainability studies (Waring et al., 2015). Understanding the ultimate causation of our human capacity for collective learning empowers students with a helpful perspective on local tradition and tools for cultivating innovation and creativity toward valued outcomes (Muthukrishna & Henrich, 2016).

By taking a functional contextual approach to understanding multilevel human dynamics, EvoS in ultimate causation provide a backbone to recognizing, analyzing, and cultivating adaptive flexibility where it is most needed among the grand challenges of sustainability. On this view, students of ESD should be trained to understand the many dialectical tensions of planetary sustainability (see The Dialectics of ESD Discourse previous discussion), and rather than learning to broadly prefer one end of the spectrum over another, students must flexibly seek to cultivate solutions, individually and collectively, that are adaptive in the specific contexts under consideration.

Cultivating Adaptive Flexibility as a Foundation for Sustainable Development

Just as a biologist may explore the functioning of selected traits within a specific ecosystem context, so ESD practitioners can explore how the complex human traits of adaptive flexibility function within a specific individual and community context (Hayes & Ciarrochi, 2015). In this way, the cultivation of adaptive flexibility across scales of social organization can be seen as a primary aim of ESD placed within evolutionary terminology (Eirdosh, 2016). The core challenge remains to identify effective tools of cultivating adaptive flexibility across diverse sociocultural contexts. Understanding the ultimate causation *and* proximate mechanisms of adaptive flexibility, in the broadest sense, frames a highly generalized yet practical foundation for examining important human dynamics across all sustainable development goals. In this way, the basic academic study of adaptive flexibility in humans can itself be seen as a cultural adaptation supporting greater adaptive flexibility and, therefore, sustainability within communities (Eirdosh, 2016).

ESD curriculum designers should therefore view the human dynamics of adaptive flexibility along a spectrum of competency—from the informal training of problem-solving intuitions (Dennett, 2013), to highly formalized research methodologies (Waring et al., 2015). We offer here a sketch of what we regard as foundational connections between the concepts and competencies linking EvoS and ESD. That is, we are arguing for the value of a generalized human evolutionary literacy in supporting the core aims of ESD by moving toward a more common, or shared, understanding of complex human dynamics. We see these dynamics as influential, in varying degrees, in virtually all sustainability problems. For this reason, we do not focus here on addressing specific sustainability challenges or the value of specific evolutionary frameworks but rather on the value of EvoS for teaching a generalized understanding of important human dynamics.

Generalizing the Core Human Dynamics of ESD for Collaborative Curriculum Design

A society able to adapt toward sustaining what it values requires a minimal base of individual citizens who have a basic foundational understanding of the human dynamics that drive sustainability issues. Waring et al. (2015) and Waring, Goff, and Smaldino (2017) emphasize the role that evolutionary theory can and must play in applied studies of sustainability, particularly in understanding human behavior, culture, and multilevel social organization. The challenge remains for curriculum designers to scaffold educational content and experiences that create meaningful connections between the EvoS and ESD communities. We suggest the EvoS literature on human adaptive flexibility can be productively connected to the aims of ESD through focusing on four *core human dynamics*:

Knowledge building: All students should master the skills of discovering, organizing, and testing claims of knowledge; deciding what to believe; and understanding how others decide what to believe.
Social-emotional reasoning: All students should master the skills of identifying the role of values and emotions in individual and social reasoning.
Multilevel cooperation: All students should master the skills of cultivating cooperation toward shared values at multiple levels of social organization.
Collective learning: All students should master the skills of cultivating lifelong learning together across ages, sociocultural differences, geographic distance, disciplinary divides, and ideological disputes.

We view these core human dynamics not as an exhaustive human science synthesis but as a bridge between the evolutionary human sciences and the aims of ESD curriculum design. The four dynamics help frame the biological, emotional, rational, and sociocultural dimensions of human adaptive flexibility in an integrative fashion. Each dynamic connects to a significant literature on the evolutionary history, developmental pathways, current contextual functioning, tools for cultivation towards valued outcomes, and possible future evolution of that respective aspect of our species (see Table 12.1). In this way, each of the core human dynamics represents not only a competency to be developed but also academic content to be engaged.

Scholars across disciplines can and do disagree about myriad details and nuance within each of the identified human dynamics (Greene, 2013; Haidt, 2012; Wilson, 2015a). Productive collaboration between the EvoS and ESD educational communities need not hinge on any particular theoretical or philosophical dispute. Rather, a minimal commitment to building a coherent, transdisciplinary, naturalistic, and functional understanding of these or other core human dynamics could build considerable value for both EvoS and ESD. In the realm of education, focusing on these dynamics offers the beginnings of a *common currency* of scientific self-knowledge

TABLE 12.1. Core Human Dynamics Connected to Education for Sustainable Development Core Competencies, Foundational Evolution Studies References, and Elaboration of the Value Added Regarding Education for Sustainable Development Objectives and Challenges

Core human dynamic (with links to ESD competencies of Schreiber & Siege 2016, p. 102)	Relevant EvoS-related literature	Value added for ESD objectives and challenges
Knowledge building (1,3,6,7,10)	Bain, 2015; Feist, 2006; Gontier, 2006; Gopnik, Meltzoff, & Kuhl, 2000; Raskin, 2008; Kahneman, 2011; Kahneman & Klein, 2009; Lilienfeld, Ammirati, Landfield, Nisbett, Ross, & Gilovich, 2009; Sweller, 2004; Wilson, 1990	Understanding the evolution, development, and functioning in context of the human brain provides context for understanding the cognitive biases that underpin human reasoning in modern complex contexts
Social-emotional reasoning (1,2,3,4,5,6,7)	Biglan, 2015; Greene, 2013; Grinde, 2016; Haidt, 2012; Hayes & Ciarrochi, 2015; Hayes, Strosahl, & Wilson, 2012; Lowe, 2010; Wilson et al., 2014	Understanding the evolution, development, and functioning in context of human emotions and moral reasoning provides context for understanding group dynamics and moral controversies surrounding societal and sustainability issues
Multilevel cooperation (3,4,5,6,7,8,9,10,11)	Biglan, 2015; Boyd & Richerson, 2009; Messner, Guarín, & Haun, 2013; Waring, et al., 2015; Waring & Tremblay, 2016; Wilson et al., 2013; Wilson et al., 2014	Understanding the evolution, development, and functioning in context of human cooperation provides context for understanding the principles for the emergence of cooperation around common goals on multiple levels of social organization

TABLE 12.1. Continued

Collective learning (3,4,5,6,7,8,9,10,11)	Conyers & Wilson, 2015; Feist, 2006; Johnson & Johnson, 2009; Muthukrishna & Henrich, 2016; Tomasello, 2016	Understanding the evolution, development, and functioning in context of collective cumulative learning capacities in humans provides context for understanding the principles for social innovation and collaborative problem-solving

Note: ESD = Education for Sustainable Development. EvoS = Evolutionary Studies.

(Greene, 2013) important for teacher training and collaboration (Conyers & Wilson, 2015). These dynamics pervade the challenges of both sustainability and school development and require a multitude of perspectives to be well understood. Educators of all stripes can benefit from exploring fully rounded evolutionary explanations of these human dynamics (Tinbergen, 1963) and using this basic perspective to engage students in how these capacities emerge, persist, and evolve within modern sustainability issues.

Evolving a Global ESD

> *The key is to envision education as a form of cooperation that requires the same core design principles as most forms of cooperation.*
> —*Kauffman and Wilson (2016, p. 341)*

Both EvoS and ESD share a common goal and purpose of shaping a human society towards sustainability and well-being. Strategically uniting around this purpose presents great opportunity as much as it brings significant challenges. The previously described historical disciplinary and philosophical divides still present perhaps the greatest barrier to sustained productive collaboration and collective learning across EvoS and ESD communities. At the same time, the commitment to viewpoint diversity within ESD and the recognition of challenges within the field itself provide opportunity for shaping productive engagement with EvoS concepts and curricula.

Both fields can benefit from a tighter and more integrated approach connecting theory and practice, research and education. EvoS offers compelling evidence-based narratives of adaptive flexibility rigorously linked to our best collective knowledge of the human condition. ESD is charged with cultivating students and citizens capable of connecting with these science-based cultural narratives and transforming them into valued action.

Social narratives play a key role in the relationship between the human-related sciences and the public understanding of itself. Wilson (2015b) argues that the challenges of the 21st century require a much tighter linkage between science and such cultural narratives of practice in fields of practical importance. This *science-to-narrative chain* represents a valuable new concept for all domains of sustainable development but has special relevance for practitioners of ESD in clarifying the cultural value of science education.

We suggest a strategic focal point for global collaboration across EvoS–ESD communities lies in connecting international teacher training institutes (UNESCO Sustainable Development Goal 4.c) around the previously identified core human dynamics. While ESD has an identity through the UNESCO-led platform and network of scholars and practitioners, much space is open for the EvoS community to advance curricula and projects under the broad umbrella of ESD.

Toward this aim, we have launched a global ESD teacher training course at University of Leipzig in which *ultimate causation* and *adaptive flexibility* are regarded as twin pillars of content and context for the themes of ESD. Specifically, the curriculum aims to strengthen the science-to-narrative chain connecting the human-related sciences to the core global mission of ESD practice. The course provides basic introductory content to strengthen students' understanding of human dynamics and to cultivate competencies in adaptive flexibility (Figure 12.1).

Students systematically explore how these basic human dynamics are functioning across the diversity of specific sustainable development problems. Students can then demonstrate core ESD competencies through developing an introductory-level case study on the human dynamics of a specific sustainable development problem of interest.

The generalized model of global ESD learning outcomes (Figure 12.1) outlines key elements of human adaptive flexibility that can be used to inform the design of single courses, or entire curricula, while easily linking to learning standards across natural and social sciences. The framework also connects to emerging ESD standards and can be linked to the global development/ESD core competencies identified by Schreiber and Siege (2016, p.102; also see Table 12.1). The authors outline 11 core competencies for ESD curriculum design, falling within broader categories of *recognizing,*

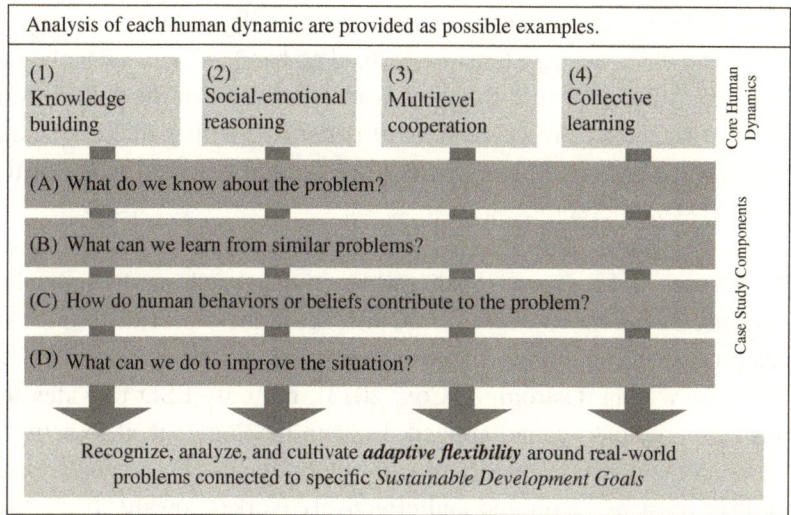

FIGURE 12.1. Global ESD learning outcomes model. Students of sustainable development must be able to *recognize, analyze*, and *act* (*sensu* Schreiber & Siege 2016, p.102) to cultivate adaptive flexibility around real-world problems connected to specific *sustainable development goals* (SDGs). This learning outcomes model connects this central aim of ESD with the evolutionary developmental origins of these competencies as core human dynamics. This is a highly generalized model, representing an educational action-research approach for sustainable development, as well as model of adaptive flexibility itself. Importantly, the core competencies students must work to master themselves as a stakeholder are the same human dynamics they must analyze within other stakeholders of sustainability issues. Specific steps offered for the analysis of each human dynamic are provided as possible examples.

assessing, and *acting*. An evolutionary understanding of adaptive flexibility in humans allows for productive framing of these 11 competencies in terms of the four generalized global ESD core competencies as interconnected human dynamics.

Conclusion

The value of EvoS for ESD can be seen in its strong, cumulative evidence base coming from a wide range of disciplines, its explanatory power across a wide range of phenomena, guidance in asking the right questions, and its elegant simplicity. EvoS offers ESD a uniquely coherent transdisciplinary framework for engaging in applied problem-solving. EvoS helps resolve and dissolve the dialectics often presented in ESD discourse as opposing extremes of a spectrum. Dismissing or actively fighting against one end of a dialectic will not lead to the emergence of the other, as these tensions instead go hand in hand, each playing a functional role in specific contexts. Understanding that the dialectics of ESD exist because of the evolutionary patterns of emergent complexity of our species and world summarizes one central practical value of the EvoS–ESD collaboration.

ESD offers equal value back to the field of EvoS in the form of inclusive global themes and a rigorous real-world testing ground for strengthening the *science-to-narrative chain*. EvoS may describe the critical value of developing shared common purposes and goals for human cooperation in general (Wilson, Ostrom, & Cox, 2013), but only ESD provides the inclusive, participatory, and applied decision-making context to actively cultivate competencies supporting truly global goals. These goals, once identified, must be monitored and effectively worked toward. If the theoretical value of EvoS literature is to become of practical value, the real-world laboratory of ESD classrooms is among the most strategic starting points.

The seemingly steep philosophical divide between EvoS and ESD regarding issues of positivism, relativism, and the value of our cumulative human science knowledge base should not impede cooperation. Rather, the divide can be seen by both EvoS and ESD communities as an entry point to discussing philosophy of science and the nature of sustainability science. Where ESD is an explicitly pluralistic field committed to viewpoint diversity, and in which controversy around approaches and their effectiveness prevails, it may fall on the EvoS academic community to engage and develop resources under the ESD umbrella, rather than waiting for the ESD community to take the lead adopting EvoS frameworks. ESD is intended to be shaped by the global community; therefore, it provides a great opportunity for the EvoS community to frame the applied sustainability value of our approaches. We must additionally work to make our scholarship accessible to teaching communities and education systems through effective curriculum design.

In view of the increasingly urgent need to educate a global citizenship toward an environmentally, socially and economically sustainable future, and the challenges that currently prevail within the programming of ESD, this chapter argues for a call to action. Scholars, educators, and students should work toward a strategic EvoS–ESD collaboration in the arena of international teacher training and classroom collaborations exploring the science of adaptive flexibility in our species and society.

References

Arbuthnott, K. D. (2010). Education for sustainable development beyond attitude change. *International Journal of Sustainability in Higher Education, 10*, 152–163.

Bandura, A. (2006). Toward a psychology of human agency. *Perspectives on Psychological Science, 1*, 164–180.

Bain, B. (2015). Claim testing and critical thinking [Web log post]. *Big History Project*. https://blog.bighistoryproject.com/2015/09/16/claim-testing-and-critical-thinking/

Berryman, T., & Sauvé, L. (2016). Ruling relationships in sustainable development and education for sustainable development. *Journal of Environmental Education, 47*, 104–117.

Biglan, A. (2015). *The nurture effect: How the science of human behavior can improve our lives and our world*. Oakland, CA: New Harbinger.

Boyd, R. T., & Richerson, P. J. (2009). Culture and the evolution of human cooperation. *Philosophical Transactions of the Royal Society B: Biological Sciences, 364*, 3281–3288.

Conyers, M., & Wilson, D. (2015). *Smarter teacher leadership: Neuroscience and the power of purposeful collaboration*. New York, NY: Teachers College Press.

Coops, N. C., Marcus, J., Construt, I., Frank, E., Kellett, R., Mazzi, E., . . . Sipos, Y. (2008). How an entry-level, interdisciplinary sustainability course revealed the benefits and challenges of a university-wide initiative for sustainability education. *International Journal of Sustainability in Higher Education, 9*, 87–98.

Dan, N. L., & Mino, T. (2016). Student diversity augments studying sustainability in higher education. *Journal of Education for Sustainable Development, 10*, 38–53.

Dennett, D. C. (2013). *Intuition pumps and other tools for thinking*. New York, NY: W. W. Norton.

Earth Charter Initiative. (2000). *The earth charter*. Retrieved from http://earthcharter.org/discover/the-earth-charter/

Eirdosh, D. (2015). The tragedy of commonsense project management: Multilevel thinking as a core competency for sustainability scientists. *The Evolution Institute*. Retrieved from https://evolution-institute.org/commentary/the-tragedy-of-commonsense-project-management-multilevel-thinking-as-a-core-competency-for-sustainability-scientists/

Eirdosh, D. (2016). Ultimate causation and evolving social strategies in the anthropocene: A global education perspective. *The Evolution Institute*. Retrieved from https://evolution-institute.org/commentary/ultimate-causation-and-evolving-social-strategies-in-the-anthropocene-a-global-education-perspective/

Ellis, E. C. (2016). Humans: The species that changed the Earth. *The Evolution Institute*. Retrieved from https://evolution-institute.org/focus-article/humans-the-species-that-changed-earth/

Feist, G. J. (2006). *The psychology of science and the origins of the scientific mind*. New Haven, CT: Yale University Press.

Gadotti, M. (2008). What we need to learn to save the planet. *Journal of Education for Sustainable Development, 2*, 21–30.

Geary, D. C., & Berch, D. B. (2016). *Evolutionary perspectives on child development and education*. Cham, Switzerland: Springer.

Gontier, N. (2006). Introduction to evolutionary epistemology, language and culture. In N. Gontier, J. P. van Bendegem, & D. Aerts (Eds.), *Evolutionary epistemology, language and culture. A non-adaptationist, systems theoretical approach* (pp. 1–32). Dordrecht, The Netherlands: Springer.

Gopnik, A., Meltzoff, A. N., & Kuhl, P. K. (2000). *The scientist in the crib*. New York, NY: HarperCollins.

Gopnik, A., & Wellman, H. M. (2012). Reconstructing constructivism: Causal models, Bayesian learning mechanisms and the theory theory. *Psychological Bulletin, 138*, 1085–1108.

Greene, J. (2013). *Moral tribes: Emotion, reason and the gap between us and them*. London, England: Atlantic Books.

Grinde, B. (2016). *The evolution of consciousness: Implications for mental health and quality of life*. Cham, Switzerland: Springer.

Hagen, J. B. (1992). *An entangled bank: The origins of ecosystem ecology*. New Brunswick, NJ: Rutgers University Press.

Haidt, J. (2012). *The righteous mind: Why good people are divided by politics and religion*. New York, NY: Pantheon.

Hansmann, R., Crott, H. W., Mieg, H. A., & Scholz, R. W. (2009). Improving group processes in transdisciplinary case studies for sustainability learning. *International Journal of Sustainability in Higher Education, 10*, 33–42.

Hayes, L., & Ciarrochi, J. (2015). *The thriving adolescent: Using acceptance and commitment therapy and positive psychology to help young people manage emotions, achieve goals, and build positive relationships*. Oakland, CA: Context Press.

Hayes, S. C., Strosahl, K. D., & Wilson, K. G. (2012). *Acceptance and commitment therapy: The process and practice of mindful change* (2nd ed., Vol. 1). New York, NY: Guilford.

Hofman, M. (2015). What is an education for sustainable development supposed to achieve: A question of what, how and why. *Journal of Education for Sustainable Development, 9*, 213–228.

Huxley, J. (1946). *UNESCO: Its purposes and its philosophy*. Preparatory Commission of the United Nations Educational, Scientific and Cultural Organization. London: Mimeo.

Jackson, M. G. (2011). The real challenge of ESD. *Journal of Education for Sustainable Development, 5*, 27–37.

Jickling, B., & Wals, A. E. J. (2012). Debating education for sustainable development 20 years after Rio: A conversation between Bob Jickling and Arjen Wals. *Journal of Education for Sustainable Development, 6*, 49–57.

Johnson, D. W., & Johnson, R. T. (2009). An Educational Psychology Success Story: Social Interdependence Theory and Cooperative Learning. *Educational Researcher, 38*(5), 365–379. http://doi.org/10.3102/0013189X09339057

Kahneman, D. (2011). *Thinking, fast and slow*. New York, NY: Farrar, Straus and Giroux.

Kahneman, D., & Klein, G. (2009). Conditions for intuitive expertise: A failure to disagree. *American Psychologist, 64*, 515–526.

Kauffman, R. A., Jr., & Wilson, D. S. (2016). Beyond academic performance: The effects of an evolution-informed school environment on student performance and well-being. In D. C. Geary & D. B. Berch (Eds.), *Evolutionary perspectives on child development and education* (pp. 307–347). Cham, Switzerland: Springer.

Kopnina, H., & Meijers, F. (2014). Education for sustainable development (ESD). Exploring theoretical and practical challenges. *International Journal of Sustainability in Higher Education, 15*, 188–207.

Krasny, M. E., & Tidball, K. G. (2015). *Civic ecology: Adaptation and transformation from the ground up*. Cambridge, MA: MIT Press.

Lilienfeld, S. O., Ammirati, R., Landfield, K., Nisbett, R., Ross, L., & Gilovich, T. (2009). Giving debiasing away: Can psychological research on correcting cognitive errors promote human welfare? *Perspectives on Psychological Science, 4*, 390–398.

Lloyd, E., Wilson, D. S., & Sober, E. (2011). *Evolutionary mismatch and what to do about it: A basic tutorial*. Wesley Chapel, FL: The Evolution Institute.

Lowe, B. M. (2010). The creation and establishment of moral vocabularies. In S. Hitlin & S. Vaisey (Eds.), *Handbook of the sociology of morality* (pp. 293–312). New York, NY: Springer.

Mesoudi, A. (2011). *Cultural evolution: How Darwinian theory can explain human culture and synthesize the social sciences*. Chicago, IL: University of Chicago Press.

Messner, D., Guarín, A., & Haun, D. (2013). *The behavioural dimensions of international cooperation*. Centre for Global Cooperation Research.

Monroe, M. C. (2012). The co-evolution of ESD and EE. *Journal of Education for Sustainable Development, 6*, 43–47.

Muthukrishna, M., & Henrich, J. (2016). Innovation in the collective brain. *Philosophical Transactions of the Royal Society B: Biological Sciences, 371*, 20150192.

Parker, J. (2010). Competencies for interdisciplinarity in higher education. *International Journal of Sustainability in Higher Education, 11*, 325–338.

Pigem, J. (Ed.). (2007). Faith-based organisations and education for sustainability. *UNESCOCAT*. Retrieved from http://www.arcworld.org/downloads/Barcelona%20Report.pdf

Preston, N. (2010). The why and what of ESD: A rationale for earth charter education (and naming some of its difficulties). *Journal of Education for Sustainable Development, 4*, 187–192.

Raskin, J. D. (2008). The evolution of constructivism. *Journal of Constructivist Psychology, 21*, 1–24.

Ruiz, J. R. (2010). Dangers facing the Earth Charter. *Journal of Education for Sustainable Development, 4*, 181–185.

Schreiber, J.-R., & Siege, H. (2016). *Curriculum framework: Education for sustainable development* (2nd ed.). Bonn, Germany: Engagement Global.

Sterling, S. (2010). Transformative learning and sustainability: Sketching the conceptual ground. *Learning and Teaching in Higher Education, 5*, 17–33.

Sterling, S. (2014). At variance with reality: How to re-think our thinking. *Journal of Sustainability Education, 6*. http://www.jsedimensions.org/wordpress/wp-content/uploads/2014/05/Sterling-Stephen-JSE-May-2014-PDF-Ready.pdf/

Stuckey, P. (2015). Breathing in, breathing out: The biological foundation for sustainable economic and social life. *Journal of Sustainability Education, 9*. http://www.jsedimensions.org/wordpress/wp-content/uploads/2015/03/Weaver-JSE-March-2015-Love-Issue.pdf

Sweller, J. (2004). Instructional design consequences of an analogy between evolution by natural selection and human cognitive architecture. *Instructional Science, 32*, 9–31.

Tilbury, D. (2011). *Education for sustainable development: an expert review of processes and learning*. Paris, France: UNESCO.

Tinbergen, N. (1963). On aims and methods of ethology. *Zeitschrift für Tierpsychologie, 20*, 410–433.

Tomasello, M. (2016). The ontogeny of cultural learning. *Current Opinion in Psychology, 8*, 1–4.

UNESCO (2009). *Bonn declaration*. Retrieved from http://unesdoc.unesco.org/images/0018/001887/188799e.pdf

UNESCO (2012). *Education for sustainable development sourcebook*. Retrieved from http://unesdoc.unesco.org/images/0021/002163/216383e.pdf

UNESCO (2014). Roadmap for implementing the global action programme on education for sustainable development. Retrieved from http://unesdoc.unesco.org/images/0023/002305/230514e.pdf

Wals, A. E. J. (2009). Review of contexts and structures for education for sustainable development 2009. *UNESCO*. Retrieved from http://unesdoc.unesco.org/images/0018/001849/184944e.pdf

Wals, A. E. J. (2010). Mirroring, Gestaltswitching and transformative social learning: Stepping stones for developing sustainability competence. *International Journal of Sustainability in Higher Education, 11*, 380–390.

Wals, A. E. J. (2011). Learning our way to sustainability. *Journal of Education for Sustainable Development, 5*, 177–186.

Wals, A. E. J. (2012). Shaping the education of tomorrow: 2012 full-length report on the UN Decade of Education for Sustainable Development. *UNESCO*. Retrieved from http://unesdoc.unesco.org/images/0021/002164/216472E.pdf

Wals, A. E. J., & Schwarzin, L. (2012). Fostering organizational sustainability through dialogic interaction. *The Learning Organization, 19*, 11–27.

Waring, T. M., Kline, M. A., Brooks, J. S., Goff, S. H., Gowdy, J., Janssen, M. A., . . . Jacquet, J. (2015). A multilevel evolutionary framework for sustainability analysis. *Ecology and Society, 20*(2), art. 34.

Waring, T. M., Goff, S. H., & Smaldino, P. E. (2017). The coevolution of economic institutions and sustainable consumption via cultural group selection. *Ecological Economics, 131*, 524–532.

Waring, T. M., & Tremblay, E. (2016). An evolutionary approach to sustainability science. *Cliodynamics, 7*, 119–167.

Webster, K. (2007). Hidden sources: Understanding natural systems is the key to an evolving and aspirational ESD. *Journal of Education for Sustainable Development, 1*, 37–43.

Wilson, D. S. (1990). Species of thought: A comment on evolutionary epistemology. *Biology and Philosophy, 5*, 37–62.

Wilson, D. S. (2005a). Evolutionary social constructivism. In J. Gottschall & D. S. Wilson (Eds.), *The literary animal: Evolution and the nature of narrative* (pp. 20–37). Evanston, IL: Northwestern University Press.

Wilson, D. S. (2005b). Natural selection and complex systems: a complex interaction. Self-organization and evolution of social systems, 151–165.

Wilson, D. S. (2015a). *Does altruism exist? Culture, genes, and the welfare of others.* New Haven, CT: Yale University Press.

Wilson, D.S. (2015b). The road to ideology: How Friedrich Hayek became a monster. *Evonomics.* Retrieved from http://evonomics.com/the-road-to-ideology-how-friedrich-hayek-became-a-monster/

Wilson, D. S., Hayes, S. C., Biglan, A., & Embry, D. D. (2014). Evolving the future: Toward a science of intentional change. *Behavioral and Brain Sciences, 37,* 395–416.

Wilson, D. S., Ostrom, E., & Cox, M. E. (2013). Generalizing the core design principles for the efficacy of groups. *Journal of Economic Behavior and Organization, 90,* S21–S32.

CHAPTER 13 | Evolutionary Approaches to Health Issues and Behaviors Across the Life Course

DANIEL J. KRUGER AND JESSICA S. KRUGER

THIS CHAPTER PRESENTS AN evolutionary framework for understanding human health issues across the lifespan and variation in health-related behaviors and outcomes. Life history theory in particular provides great insight into health challenges and how health issues relate to the intersections of our evolved genetic heritage and the developmental and current environments we inhabit. This framework would provide a strong foundation to further advance human health conditions in modern societies.

Human health and longevity have improved dramatically in technologically advanced societies due to scientific research and intervention. Advances in medicine, pharmacology, public health, and sanitation have increased average life expectancy by about 40 years in Western industrial nations since the 1840s (National Institute on Aging, 2011). Most of those born in the year 1900 did not live past age 50, now those in many developed nations have life expectancies exceeding 80 years (National Institute on Aging, 2011). The health conditions of living individuals have also improved substantially (National Institute on Aging, 2011), especially among the elderly (Manton, Gu, & Lamb, 2006).

Advances in medical technologies continue to extend the possibilities of saving lives in immediate crisis. However, efforts to promote healthy behaviors and discourage health adverse behaviors struggle with diminishing returns. Control of parasites and infectious diseases and effective treatment of medical emergencies from accidents and other acute causes have altered the scope of pressing health challenges. In the United States and other technologically advanced nations, chronic diseases such as hypertension, type 2 diabetes, and cancer have emerged as major health adversaries. These major medical and public health concerns are notable for their substantial behavioral components in disease trajectory. Diseases resulting from so-called lifestyle causes are the largest challenge to health conditions in modern populations (National Institute on Aging, 2011).

Theoretical Bridges and Disciplinary Divides

Health-related research is broad in scope, diverse in activity, and fragmented theoretically, methodologically, and across disciplines. Progress in understanding and improving human health would be accelerated by establishing a universal and holistic framework for understanding and integrating disparate undertakings. This framework is evolutionary theory, the most powerful explanatory system in the life sciences and the only framework that can unify knowledge in otherwise disparate fields of human research. However, disciplinary divides have shaped and even impeded the adoption of the most powerful theoretical framework in the life sciences. It is difficult to master any one scientific field in a human lifetime, let alone several interrelated areas. Thus, career specialization has affected the scopes of training and practice and the overall complex and convoluted relationship between fields addressing health issues.

Darwinian medicine (Nesse & Williams, 1994; Wilson, 1993) or evolutionary medicine, a term used by those emphasizing that considerable evolutionary insights are continuously generated after the life of Darwin, has made considerable progress in the practical understanding of human physiology and other areas informing medical care and is far more established than the application of evolutionary theory to other health-related areas. Evolutionary approaches to medicine change perspectives in deeply meaningful ways. For example, humans are recognized not just as individuals but also as entire symbiotic ecosystems, as about 90% of the cells in our bodies are nonhuman (Bäckhed, Ley, Sonnenburg, Peterson, & Gordon, 2005; Ley, Peterson, & Gordon, 2006). The microbiome in our digestive systems are critical for proper immune system functioning (Mazmanian, 2005), fat synthesis and storage (Bäckhed et al., 2004), and likely other yet unrecognized contributions.

The divisions between academic and practitioner fields are a result of human constructs. In reality, evolutionary approaches can transcend disciplinary boundaries. Insights from evolutionary anthropologists (e.g., Gurven et al., 2016) may eventually lead to more targeted, less systemic treatments for infection, with the recognition that the management and cultivation of a healthy microbiota is important for human health. Tolerance of the other organisms inhabiting us may be a viable strategy. The recent discovery of the brain's lymphatic system (Louveau et al., 2015; Ransohoff & Englehardt, 2012) demonstrates the transdisciplinary nature of human health dynamics. The convergence of the central nervous system and lymphatic system has implications beyond neuroimmunology and neurodegenerative diseases. There is a growing body of work indicating the interrelationships between environmental pathogens and behavior (e.g., Fincher, Thornhill, Murray, & Schaller, 2008).

At present, there is a high degree of specialization in approaches to human health. Medicine focuses on treating immediate adverse conditions and regulation of physiological homeostatic mechanisms. Evolutionary insights continue to drive advances in medical technology devoted to saving lives in immediate danger (e.g., Anderson, Li, Shao, & Vanden Hoek, 2006; Lampe & Becker, 2007). Primary prevention—that is, preventing exposures to hazards causing disease or injury and changing unhealthy or unsafe behaviors leading to disease or injury—remains the domain of Health Education and Public Health. Although there has been more attention to evolutionary perspectives on medicine, there may be more potential for evolutionary approaches to public health, given that modern disease burdens are largely a product of behavior.

Public health efforts to promote healthy behaviors and discourage health adverse behaviors still struggle in their effectiveness. We may be reaching a limit in the degree of returns from traditional educational and motivational health-promoting practices. There is great potential to advance beyond the current stagnation by integrating evolutionary insights. Given the great potential of an evolutionarily informed public health to shape behavior for health promotion, why does the field lag so far behind Biology, Psychology, and Anthropology in its integration of evolutionary theory?

Health researchers and practitioners may be wary of the application of evolutionary theory to their fields because of the widespread aversion to the eugenics movements that shaped Public Health in the early 20th century. Prominent figures such as H. G. Wells (a prolific writer and one of the founders of science fiction), Margaret Sanger (founder of Planned Parenthood), John Maynard Keynes (one of the most influential economists of the 20th century), and W. K. Kellogg (the namesake of a prominent health-oriented foundation) promoted eugenics programs, which ranged widely from family planning and prenatal care programs for mothers to forced sterilization and state laws banning interracial marriage.

After the genocides committed in Nazi Germany, including programs justified by official policies with arguments loosely based on eugenic principles, eugenics in general was considered unacceptable. In the cultural divide of the 1970s, radicals believed that Sociobiology advocated racism, eugenics, and genocide simply because it held that human behavior was not solely determined by socialization and culture but could also be genetically influenced (Segerstråle, 2000). Thus, it is critical to emphasize the distinction between modern evolutionary approaches to health and the selective breeding programs feared by misinformed critics.

Modern evolutionary health research seeks to improve the health of individuals, rather than establish or maintain some imagined ideal population. Evolution is not teleological, there is no ideal Platonic form of human, and genetic diversity is crucial for selection and resilience in the face of ecological and environmental changes. No individual or group of people is more highly evolved than any other individual or group of people. Everyone who is alive today is descended from a very long line of successful ancestors. Life history theory in particular is a refutation of genetic determinism, as phenotypic plasticity and reaction norms result from the complex interaction between genetic inheritance and developmental environment.

At the same time, individuals and groups are not genetically identical, and some health disparities may partially be a result of differential selection pressures on ancestral populations, which lived in a diverse range of ecologies. Different populations have also experienced different rates of health-relevant environmental changes in recent history. Certain alleles may cause or exacerbate health problems when expressed in modern environments. For example, genes that conferred advantages to heterozygous individuals in resisting malarial parasites also cause sickle cell anemia when homozygous (Kwiatkowski, 2005). Genes for lactose tolerance evolved independently in several human populations that domesticated livestock (Tishkoff et al., 2007), though the likelihood of lactose tolerance differs by ancestry. People who are lactose intolerant have lower levels of calcium, which makes them more susceptible to environmental lead poisoning (Bruening et al., 1999).

Natural and sexual selection did not design us (or any other organism) to maximize health, longevity, or happiness. Our bodies and minds are based on a plan that successfully maximized our ancestors' reproductive success in environments that in many ways are very different from the ones we live in today. Thus, our attempts to optimize health conditions are constrained by the limitations of our biological reality. Life history theory illustrates how organisms must make trade-offs in resource allocations and how environmental conditions shape these allocations (Roff, 1992; Stearns, 1992). Energy used for one purpose cannot be used for another, and consequential trade-offs include those between survival and reproduction, between acquiring reproductive partners and investing in offspring, between current and future offspring, and between size and quantity of offspring (Hill, 1993). In the next section, we will provide an overview of health issues across the human life course, emphasizing evolutionary insights from a life history framework. The final section will review evolutionary approaches to public health, again drawing heavily from life history theory. We note that we do not attempt to cover related areas, such as evolutionary perspectives on mental health issues (e.g., Nesse, 1998), as this would require an entire chapter for adequate review.

Human Life History and Health Issues across the Life Course

The Long Shadow of Pregnancy and Childbirth

From an evolutionary perspective, offspring are crucial as they represent the continuity of one's genetic lineage, and thus mothers invest greatly in a physiologically costly pregnancy. Although pregnancy has typically been conceptualized as an entirely cooperative interaction between a mother and her fetus (Haig, 1993), there is also an inherent conflict or divergence of interests between parent and offspring. Mothers maximize their inclusive fitness by investing equally in each of their offspring, whereas offspring would maximize their inclusive fitness by skewing maternal investment in greater proportions toward themselves (Trivers, 1974). Thus, gestational age and birthweight will be a compromise between maternal and fetal strategies; both fetal and maternal interests are pursued by hormonal regulation (Haig, 1993). Maternal interests are maximized if each offspring receives less investment than would be optimal for its own fitness (Trivers, 1974). Thus, the average population birthweight in humans is lower than the optimal weight for offspring fitness (Blurton Jones, 1978: Karn & Penrose, 1952).

Adverse birth outcomes and demographic disparities persist despite decades of clinical, scientific, and legislative efforts (Hunte, Turner,

Pollack, & Lewis, 2004). Preterm birth is the leading cause of health problems in infants and leads to costs of more than $26 billion annually in the United States (Center for Healthcare Research and Transformation, 2010). Even babies born "late" preterm have greater risk of clinically significant impairments after controlling for many potential prenatal and childhood confounding factors (Talge et al., 2010). The co-varying factors of prematurity and low birthweight are the primary causes of neonatal mortality in developed countries (MacDorman & Mathews &, 2009). The maternal–fetal environment also impacts health outcomes long after birth (Barker, 1998; Gluckman & Hanson, 2005). Adult diseases such as diabetes, obesity, and hypertension are influenced by a mismatch of the phenotype developed by nutritional cues experienced in utero and the adult nutritional environment (Gluckman, & Hanson, 2007).

Birth Outcomes in Part Reflect Variations in Human Life History

Maternal somatic investment in gestating offspring will be contingent on local environmental conditions reflecting the offspring's prospects for survival (Haig, 1993). Because humans typically produce one offspring per reproductive event, maternal manipulations of reproductive frequency and resource allocation are the principle response mechanisms. Thus, the mother biologically manipulates offspring size, body composition, and metabolism based on a selective investment of energy stores (Laskey & Prentice, 1997). In good-quality environments, mothers will have more resources to invest, and outcomes will tend toward the theoretical optimum for offspring fitness, but in more adverse environments maternal and offspring interests will have greater divergence (Wells, 2003).

Maternal fitness will generally be maximized at the expense of the fitness of each individual offspring (Smith & Fretwell, 1974). In marginal environments, reduced somatic investment will lead to low birthweight infants, reducing maternal demands and preserving resources for future offspring (Haig, 1993). In the most severe environments, maternal reproductive investment will be constricted through an inability to conceive (Frisch, 1987), miscarriage early in pregnancy (Wynn, 1987), or stillbirth (Stein, Susser, Saenger, & Marolla, 1975). Maternal survival will be favored at the expense of investment in offspring in adverse circumstances, (Hirschfield & Tinkle, 1975).

Several other factors are related to the prospects of offspring survival and reproduction, including the availability of food and threats from predators. Adverse birth outcomes may partially result from mechanisms that facilitated reproductive success in ancestral environments by evaluating environmental conditions and regulating investment trade-offs. These mechanisms are a legacy from times when mortality rates throughout the lifespan were considerably higher, especially in infancy; they may not promote reproductive success in modern environments and instead may lead to adverse birth outcomes. Features indicating relatively high extrinsic mortality rates, relatively low paternal investment, and the unpredictability of future outcomes may be associated with relatively faster life history strategies, including shifts in investment related to the trade-off between offspring quantity and quality. Somatic investment per offspring may be reduced due to evolved mechanisms, facilitating shorter interbirth intervals and more numerous reproductions in ancestral environments. These mechanisms are designed to increase the chance that at least some offspring will survive and reproduce under conditions of resource scarcity.

Environmental cues consistent with relatively high extrinsic mortality rates may operate both in conjunction with and independently from the influence of traditional socio-economic status indicators. For example, neighborhood physical decay has long been thought to result in mental health problems (Park & Burgess, 1925). In recent decades, the physical deterioration of the human-built environment is gaining recognition as an important influence on health (e.g., Augustin, Glass, James, & Schwartz, 2008). Characteristics of the neighborhood environment may influence birth outcomes through psychosocial and behavioral pathways. The geographic concentration of dilapidated residential structures is associated with greater concentrations of premature and low birthweight births in Flint, Michigan (Kruger, Munsell, & French-Turner, 2011), a city with both high levels of structural deterioration and adverse birth outcomes.

In many species, male influence on the reproductive environment extends beyond conception, in both health-promoting and health-adverse ways. Paternal investment promotes healthy birth outcomes and offspring fitness. In humans, paternal investment of resource provisioning, training in life skills, and defense from threats contributes to the prospects of offspring survival and reproduction (Geary, 2005). The average level of paternal investment is much higher among humans than in other primates (Buss & Schmitt 1993; Geary & Flinn, 2001) and strongly associated with confidence of paternity. In the foraging Aché, children who grow up without an investing father present suffer higher mortality rates (Hill & Hurtado, 1996).

On the other hand, Bruce (1959) noted that pregnant mice miscarried at higher rates when a strange male arrived 24 hours after the females mated.

The Bruce effect occurs in other rodent species (Eleftheriou, Bronson, & Zarrow, 1962; Clulow & Langford, 1971) and in gelada baboons (Roberts, Lu, Bergman, & Beehner, 2012). Research on humans indicates that stepchildren face greatly elevated risk maltreatment (Daly & Wilson, 2001) and are provided less social and resource support (Anderson, Kaplan, Lam, & Lancaster, 1999; Anderson, Kaplan, & Lancaster, 1999; White, 1994) compared to genetic children.

The relative proportions of potentially reproductive males and females in a population influence the average level of paternal investment. The scarcity of sexual access to women in male-biased populations confers additional advantage to women in sexual interactions; men invest more time to wooing, are more romantic in dyadic interactions, and attempt to demonstrate more investment potential (Guttentag & Secord, 1983). When women are scarce, they are more likely to be married (South & Trent, 1988) and tend to marry at younger ages (Kruger, Fitzgerald, & Peterson, 2010). Surplus males intensify male competition for signals of relationship commitment and paternal investment (Pederson, 1991) and raise expectations for paternal care of offspring (Guttentag & Secord, 1983). Women are better able to marry partners higher in socio-economic status than themselves (Lichter, Anderson, & Hayward, 1995), and men who have lower social status and less abundant resources have even greater difficulties getting married (Pollet & Nettle, 2007).

Female-biased populations, where men are scarce, exhibit contrasting patterns including higher divorce rates, more out-of-wedlock births, more single-mother households, higher rates of teenage pregnancies, and lower paternal investment (Barber, 2000, 2004, Guttentag & Secord, 1983; Trent & South, 1989). Men are more likely to have premarital sex and test positive for a sexually transmitted infection (South & Trent, 2010). Women in college campuses with a scarcity of men report poorer treatment by men, who are less willing to commit to relationships and are not to be trusted (Uecker & Regnerus, 2010). Although these women go on fewer traditional dates, they are more likely to be sexually active than women on campuses with greater proportions of men (Uecker & Regnerus, 2010). Women living in female-biased New York City express frustrations with the shortage of single men and that many of those who are available are not good partners because they are simultaneously polygynous or solely interested in short-term sexual relationships ("Single Women," 2007).

Because women in populations with male scarcity face lower prospects for paternal investment, which is historically associated with higher infant and child mortality, they may reduce somatic investment in gestating offspring to conserve maternal resources and speed the production of additional offspring. Truncated maternal investment will result in shorter pregnancies and lighter offspring, increasing the rates of premature gestation births and low birthweight births. Across counties in the United States, male scarcity is associated with higher rates of prematurity and low birthweight. Even when accounting for socio-demographic factors known to predict birth outcomes (e.g., race, income, education), male scarcity uniquely predicts the rate of low birthweight and also indirectly predicts both prematurity and low birthweight through the proportion of families headed by single mothers (Kruger, Clark, & Vanas, 2013).

The proportion of single mother headed households is a demographic indicator inversely related to the level of paternal investment in a population. Fathers are largely absent in interventions intended to improve adverse birth outcomes. The finding described above suggests that interventions increasing the level of paternal involvement and support during pregnancy, especially from unmarried and nonresidential fathers, may generate considerable returns on investment across a broad range of health and social outcomes. Adverse birth outcomes not only predict infant mortality, they also predict a wide range of adverse health outcomes later in life (Bateson, 2004).

An Acquired Taste for Risk

Risk-taking behaviors are related to a wide variety of health concerns. Both boys and girls exhibit risky behavior, yet there are considerable aggregate sex differences in both the quality and magnitude of these behaviors. The observed peak of risky behaviors in young adulthood corresponded with entrance into mating competition (Wilson & Daly, 1992). Males compete for social status and resource control, as these are characteristics valued cross-culturally in intersexual selection (Buss, 1989). In ancestral times, men who controlled more resources partnered with younger women, partnered with more women, and produced offspring earlier (Low, 1998). Measures of male social status and economic power directly relate to reproductive success across a wide variety of societies (Hopcroft, 2006). Even in relatively egalitarian foraging societies, researchers have documented some differentiation of status, such that men with higher status have increased access to mates (Chagnon, 1992; Hill & Hurtado 1996).

Risky behavioral strategies of young males were selected for over time because they tended to aid in mating competition. Male mating effort may peak in young adulthood in part because across our evolutionary history young men often did not yet have partners or offspring to invest in, and they may have been more attractive to females because they had not yet committed their resources to partners or offspring (Hill & Kaplan, 1999). Among Aché foragers, older men tended to produce most of their offspring within long-term relationships, yet younger men fathered most offspring from extrapair sexual affairs (Hill & Hurtado, 1996). The shift in the male allocation of effort from somatic to mating to parenting over the life course helped to explain the pattern of risky behavior underlying the peak in sex differences in mortality from behavioral causes during young adulthood (Kruger & Nesse, 2006).

Energy Intake and Expenditure

Promoting changes in diet and exercise behaviors are two of the central areas in health education efforts. Among all health behaviors, diet has attracted the most attention from the section of the general public interested in the relationships between health and human ancestry. There has been considerable popular literature and discussion based on the notion of a mismatch between the modern food environments and our Paleolithic or Pleistocene ancestors. Some consider agriculture an aspect of this mismatch and promote the importance of plant foods such as nuts, seeds, fruits, tubers, berries and nondomesticated animal products, based on the diets of some modern foraging populations (Elton, 2008). A large commercial and consulting industry has developed based on conceptions of benefits from returning to a "Paleolithic diet."

There are some substantial differences between ancestral and modern populations, including lactase persistence in descendants of pastoralist populations that enable the consumption of milk beyond infancy (Cook, 1978; Holden & Mace, 2002). However, the food environment of human ancestors was heterogenous and not limited to the Pleistocene. Plio-Pleistocene hominins appear to have had remarkable dietary flexibility (Wood & Strait, 2004). The lifestyles of contemporary foraging populations are diverse and exhibit the results of evolution and interaction with other human groups in their ecologies, cultures, and behaviors (Foley, 1995). Similarly, in many modern primates, dietary variability is the norm rather than the exception (for a review, see Elton, 2008).

Because of the inherent range and flexibility of the human diet, there is no need to restrict food types to those consumed by specific foraging populations, used as a model for our Stone Age ancestors (Elton, 2008). Modern humans can eat a vast range of foodstuffs and this broad diet is beneficial, though it would be helpful to eat less overall, eat fewer refined and (predigested) processed foods, eat more fresh fruits and vegetables, and eat meat from grass-fed animals when possible (Milton, 2002). The evolutionary insights gained from comparative behavioral ecology highlight the importance of largely plant-based diets and avoidance of highly processed foods, rather than restricting diets to what has been eaten in particular times and places (Milton, 1999, 2002). There is also evidence that many of the high-carbohydrate foods in modern diets create a high glycemic response and reduce fat oxidation, thus resulting in body fat gain (Brand-Miller, Holt, Pawlak, & McMillan, 2016). Thus, diets based on low-fat, low-glycemic response foods may enhance weight control (Brand-Miller, Holt, Pawlak, & McMillan, 2016). These dietary modifications should be combined with overall increased energy expenditure to maintain energy balance (Milton, 2002).

The impact of dramatic reductions in energy expenditures on the epidemiological transition has received far less public attention. People in modernized societies exercise far less than their foraging counterparts, and the resulting imbalance in energy intake and expenditure has been a major contributor to the modern obesity epidemic (Broyles et al., 2015). In contrast to the modern fad of intense strength training exercises, our modern exercise deficits are in low-impact aerobic exercises such as walking (Katzmarzyk, 2014). Routine exercise was built into daily activities, fetching water, foraging and hunting for food, agricultural labor, and many other activities. Today, industrialization and advanced technology eliminates the necessity for most of this physical activity. However, these technological and cultural changes have not eliminated our physiological need for considerable levels of low-impact aerobic exercise.

The Vulnerable Sex

Being male is the single most prominent demographic risk factor for early mortality in technologically advanced societies (Kruger & Nesse, 2006). We occasionally see a call for attention to men's health and health disparities that emphasizes the lack of a comprehensive framework for understanding men's health issues (Treadwell & Young, 2013). As detailed in this chapter, such a framework exists, yet it has apparently yet to reach the editors of leading public health journals who grasp for explanations rooted in social norms (Treadwell & Young, 2013, p. 5). The women's health movement has made considerable advances in improving health outcomes based on the idea that men and women are different in physiology and in health promotion needs. Perhaps surprisingly, the same insight has not been widely used to examine health issues specific to men.

Males have higher variance in reproductive success compared to females, and male reproductive success may benefit more from greater investments in reproductive competition compared to reproductive success for females. The greater variation and skew in male reproductive success selected for higher investments in mating effort and competition relative to somatic effort (building and maintaining one's body) promoting longevity than for females. Increased male mortality from sexual competition early in life would also decrease selection against senescence in males relative to females. The male biases toward reproduction at the expense of somatic effort, growth at the expense of maintenance, and mating at the expense of parenting result in physiological and behavioral strategies are both riskier than those of women's strategies. Males' riskier strategies will lead to higher levels of mortality from behavioral and most nonbehavioral causes across the lifespan (Kruger & Nesse, 2006).

The cross-cultural consistency in sex differences in mortality requires explanation based in our common evolutionary heritage. Excess male mortality is a result of a trade-off between competitiveness and longevity. However, the degree of excess male mortality in human populations is not genetically determined. The level of male competition for status, resources, and mates drives the level of excess male mortality (Kruger, 2010). Sex differences in mortality rates reflect the interaction of evolved strategies and socio-environmental conditions. The fact that the degree of excess male mortality is not fixed encourages intervention that could produce considerable health and economic benefits.

Evolutionary Insights on Women's Health

An evolutionary perspective also shines new light on issues specific to women's health. Many of these insights relate to women's role as the primary physiological engine of reproduction. Profet (1992) reverse-engineered morning sickness, women's heightened tendencies to be nauseous during pregnancy. Traditional medicine treated pregnancy sickness as an unpleasant symptom or side effect of pregnancy and developed pharmaceuticals to provide relief. Unfortunately, some of these drugs produced severe birth defects in thousands of children (Kim & Scialli, 2011). Instead, Profet (1992) questioned whether pregnancy sickness was a biological adaptation with a useful purpose. She determined that pregnancy sickness met the criteria for an evolutionary adaptation: It is common across all cultures, it develops right when the fetus' basic organs are undergoing formation, and acts to remove potential toxins that could have adverse effects on fetal development.

Premenstrual syndrome (PMS) is another phenomenon affecting large proportions of women worldwide that is traditionally thought of as an unfortunate byproduct of women's reproductive physiology. A recently developed evolutionarily informed model proposes that although PMS is not an adaptation in itself, it is the consequence of a mechanism designed to facilitate women's mate choice (Reiber, 2008, 2009). During the follicular phase when women are fertile, they display several traits and behaviors that make them more attractive to men. After women's fertile phase, these traits diminish and the relative decline in activity, sociability, flirtatiousness, and reduced feelings of desirability and positive affect are experienced as aversive (Reiber, 2008, 2009). Women who may have the most incentives to reproduce, those with ample resources to support additional offspring and those with less time before reproductive cessation, reported feeling better during the follicular (fertile) portion of their cycle (Reiber, 2009). These women felt comparatively worse during the luteal, premenstrual, portion of their cycle. Rather than competing with proximate explanations for PMS based on physiological changes and states during the luteal phase, this evolutionarily informed model provides a more holistic understanding of the phenomenon and indicates the importance of assessing experiences across the cycle as well as the social and resource conditions in women's lives.

Yet another area of evolutionary insight on women's health is on menopause. There is not just one but several partially convergent and partially competing explanations from evolutionary perspectives. These proposals include that menopause evolved because older women could better promote their reproductive success by investment in existing offspring rather than having additional offspring (Williams, 1957), or by investment in grandchildren than by their own continued direct reproduction (Hawkes, O'Connell, & Blurton Jones, 1989), or that women's longevity is a byproduct of reproductively active older men—as well as incidental benefits from kin assistance (Marlowe, 2000), or even that menopause is the cost of the high rate of gamete loss due to female gametes competing to achieve ovulation (Reiber, 2011). There is some degree of evidence for each of these proposals, and perspectives currently vary in the field. This is just one illustration that there may be multiple explanations based on models derived from evolutionary theory, rather than *the* evolutionary account.

Competing hypotheses will be vetted empirically, just as competing models generated by nonevolutionary explanatory frameworks.

Limits to Health Promotion and Life Extension

The decline of health and rise of mortality rates across the latter half of human lifespan raises the inevitable specter of senescence. One may wonder why evolutionary processes occurring over billions of years and millions of generations have not led to perfected complex organisms that can live indefinitely. However, the processes of natural and sexual selection maximize the survival of genes rather than the survival of individuals or species (Dawkins, 1976; Williams, 1957). Building and maintaining a body is in the service of reproduction: our bodies are essentially vehicles for the propagation of genetic information. The effort an individual expends on building and maintaining a body is ultimately for the purpose of reproduction.

Many genes have multiple effects (this is referred to as pleiotropy), which can be both beneficial and hazardous based on the developmental and environmental context. Genes with early benefits but later costs will be selected for because younger individuals have a higher reproductive value (Medawar, 1952). Selection pressure was greater at younger ages because few people survived to old age in ancestral environments, thus early acting beneficial genes spread faster than late acting beneficial genes. The cumulative result of these factors is senescence, a decline of physiological function over time (Williams, 1957). For example, adaptations that extended juvenile characteristics of synaptic activity and plasticity that promoted cognitive capacities in adulthood also increase the risk of Alzheimer's disease (Bufill, Blesa & Agustí, 2013; Raichlen & Alexander, 2014). The relatively higher importance of reproduction at the expense of survival for the sake of longevity also forms the basis for sex differences in mortality rates ultimately created by the processes of sexual selection.

A Life History Framework for Health Behaviors and Health Interventions

A life history framework provides a scientific basis for understanding variation in health-related behaviors and creating broader structural changes promoting healthy lifestyles. In evolutionary terms, particular life history strategies are not inherently "good" or "bad," as a range of strategies have proven successful given the appropriate environmental conditions. As a species, all humans are near the slow end of the life history continuum, so individual variation occurs within the slower range of the continuum (Low, 1998). Public health professionals may value characteristics of slower human life history strategies as good or desirable, as slower life history is associated with greater relationship stability, higher investment in children, lower impulsivity, lower levels of risk taking, and greater regard for social rules (Figueredo et al., 2006). Thus, it would be helpful to create environments promoting slower life histories (see following text).

Quinlan (2007) predicted that parental effort (investment men provide for children) would be lower in environments where parenting cannot improve offspring survival. Using the Standard Cross-Cultural Sample of 186 identified contemporary and historic cultures, Quinlan found that moderate degrees of pathogen load are associated with higher levels of maternal investment. Yet, when pathogen load was very high, maternal investment was considerably reduced. Paternal investment showed a simple inverse relationship to pathogen load and was inversely associated with famine and warfare (Quinlan, 2007).

Mischel and colleagues demonstrated how future-oriented self-control predicts a wide range of outcomes including social competence, educational achievement, and resilience to frustration and stress (e.g., Mischel, Shoda, & Rodriguez, 1989). Individuals' time perspectives (encompassing time horizons, future discounting, planning, etc.) may be central to the psychological representation of life history trade-offs. Future orientation (low future discounting, etc.) reflects a pattern of behavior dominated by a striving for future goals and rewards.

Short time horizons, substantial future discounting, and risky behaviors contribute to a wide variety of health issues, concerns, and outcomes. The future-oriented strategies that health promotion efforts encourage depend on environmental conditions that will be relatively stable over time. Individuals developing in relatively less predictable environments will exhibit riskier, immediate outcome oriented, behavioral strategies because of the historical low probability of reproductive success for more cautious approaches (Hill, Ross, & Low, 1997; Wilson & Daly, 1997). In unstable ancestral environments, the most pressing adaptive problem faced by individuals was avoiding death. Risky behaviors by definition have uncertain outcomes, but, in aggregate, they facilitated early reproduction before death occurred.

The steep discounting of the future by adolescents and young adults could be a rational response to uncertainty (e.g., Gardner, 1993; Wilson

& Daly, 1997), and individuals who develop in relatively uncertain environments will develop riskier behavioral strategies to take advantage of possibly fleeting opportunities (Chisholm, 1999). Those living in chronically risky and uncertain environments (e.g., one with significant family conflict) are more likely to experience earlier menarche, earlier ages of reproduction, and higher reproductive rates (Chisholm, 1999; Kim, Smith, & Palermiti, 1997). In these environments, the most pressing adaptive problem faced by individuals has been avoiding death. Risky behaviors may have facilitated early reproduction before death occurs. In fact, community college students who had shorter lifespan estimates and higher estimates of the unpredictability of the future had a higher frequency of risk-taking (Hill et al., 1997). When mortality rates were low and predictable, individuals may have encountered fewer urgent adaptive problems and a less risky, long-term strategy was optimal. Wilson and Daly (1997) found that neighborhood homicide rates were associated with neighborhood life expectancy and neighborhood income inequality. They argued that risky behaviors such as homicide are a result of steep future discounting, which is a response to environments where the probability of receiving delayed benefits is uncertain or low and the expected benefits of safer courses of action are negligible. They concluded that when competition for resources and social status is more intense, a greater tendency for risk-taking behavioral strategies is evident.

Environmental uncertainty, where personal safety, social support, and access to important resources are unreliable, foster tendencies to discount future health in favor of immediate rewards. Longer-term, risk averse strategies would be more prevalent if perceptions that current efforts will pay off in the future are enhanced by providing greater stability. Longer-term, lower-risk behavioral strategies are expected among those who have experienced reliably stable and supportive environments. Others who experienced environments where personal safety, social support, and resource control are uncertain may be more likely to discount future benefits in favor of more immediate rewards. Children participating in a modification of Mischel et al.'s (1989) classic delayed gratification task were more likely choose immediate yet smaller rewards when the experimenter was unreliable in fulfilling previous promises (Kidd, Palmer, & Aslin, 2013).

In a sample of Americans recruited through the Internet, lower perceived socio-economic standing predicted higher perceived extrinsic mortality risk, which, in turn, predicted the amount of effort they reported investing in their health and safety (Pepper & Nettle, 2014). A demographically representative community health survey showed that those with slower life histories had higher levels of health promoting behaviors, exercise, and consumption of fruits and vegetables and lower levels of health-adverse behaviors: smoking tobacco, alcohol consumption, binge alcohol consumption, and behaviors that create high risk for HIV infection (Kruger & Kruger, 2016). These relationships were statistically significant even when controlling for socio-demographic factors, such as age, gender, and education, which co-vary with life history speed.

A more supportive socio-developmental environment (in terms of perceptions of physical safety, positive socialization, and the helpfulness of others) predicted the strength of future orientation and inversely predicted present orientation in US inner-city middle-school students (Kruger, Reischl, & Zimmerman, 2008). Present orientation predicted both interpersonal aggression and illicit exploitation of resources (property crimes); future orientation also inversely predicted interpersonal aggression uniquely. A similar model of time perspective based on life history theory may also explain and predict substance use behaviors better than conventional models (Richardson & Hardesty, 2012). Time perspective was an important predictor in alcohol, drug, and tobacco use (Keough, Zimbardo, & Boyd, 1999).

Those who argue that risky behavioral strategies are inherited directly have challenged the developmental model of environmental sensitivity. Indeed, twin studies indicate that genes typically account for 50% of the variance in most personality traits (e.g., Jang, McCrae, Angleitner, Riemann, & Livesley, 1998). Researchers have identified, however, a complex interaction between genetic variation and environmental conditions and behavioral tendencies that are contingent on both components (Hill, Ross, & Low, 1997). Some developmental influences on behavioral patterns may be mediated through epigenetics, where gene expression is upregulated or suppressed by methylation or other processes (Juliandi, Abematsu, & Nakashima, 2010; Ma et al., 2010; Sun, Sun, Ming, & Song, 2011). This relatively new area of study will help illuminate some of the physiological mechanisms involved in facultative life history variation.

Environments Shape the Health Phenotype

Health researchers are aware that environments shape health behaviors, for example, that the geographic availability of foods shapes dietary patterns. Greater concentrations of fast-food outlets near one's residence are adversely associated with diet quality (Kruger, Greenberg, Murphy, DiFazio,

& Youra, 2014), whereas households participating in community gardens have higher consumption of fruits and vegetables (Alaimo, Packnett, Miles, & Kruger, 2008). The developmental resource environment may also have lasting impacts on dietary behaviors across the lifespan. In a laboratory study, for students who grew up in a resource-plentiful environment, those who were currently hungry consumed more calories of snack foods than those who were not currently hungry. However, there was no difference in snack food consumption among those who did not grow up in resource-plentiful environments (Hill, Prokosch, DelPriore, Griskevicius, & Kramer, 2016). This mechanism may have promoted survival in harsh and unpredictable ancestral ecologies through the opportunistic acquisition of energy reserves, though it promotes obesity in modern calorie-rich environments.

Research on environmental influences on health initially focused on the negative effects of hazardous agents such as toxic chemicals and radioactivity (Davies, 2013). More recently there has been greater recognition of the health promoting aspects of naturalistic environments. Our health may benefit from natural settings and constructed environments which extensively incorporate natural features. We appear to have an affinity for the natural settings of our ancestral past, preferences that developed over millions of years of evolution (Wilson, 1984). Architects and landscape planners such as Fredric Law Olmstead have recognized the importance of naturalistic elements for well-being, showing an inherent understanding of human biophilia. Landscapes with features necessary for survival and sustenance; fresh running water, plants, and nonhuman animals; defensible space; and moderately dense forestation all promote physical and mental health (Grinde & Patil, 2009). This phenomenon encourages efforts to create more green space in urban and suburban environments and incorporate naturalistic features into our buildings such as open running water and extensive vegetation.

In the next decade, there will be increasing numbers of projects such as community gardens, which provide locally produced fresh produce, better air quality, and additional public green space. Such projects will increasingly take advantage of both horizontal and vertical urban space. Vertical space in urban areas is a largely untapped resource for growing vegetation, and multiple stories increase the potential arable space beyond the building's footprint. Other integrations of the natural environment, from green rooftops to biking and walking paths, will further erode the brutal concrete jungle often featured in the past half century of urban construction.

The mismatch between the modern built environment and ancestral human environments poses other challenges. Those in Public Health are very aware of the current obesity epidemic and other recently emerging health challenges that stem in part from the structure of modern human environments. Educational health promotion efforts have certainly yielded benefits, yet after several decades of health education, many health indicators in the United States are stagnant or declining (e.g., National Center for Health Statistics, 2018). This may be one factor underlying the strong interest in policy change, seen as a theme in several recent Public Health conferences. The motivation to retool public health efforts poses an opportunity for policy advocacy with a basis in evolutionary science. Health advocates will be pleased to know that many such recommendations converge with their own current efforts.

Reshaping the Environment to Promote Health and Longevity

Those residing in European cities and North American locations such as Manhattan enjoy advantages including comprehensive public transportation, more substantial public open spaces and parks, and easily accessible produce and grocery stores, restaurants, shops, and public services. Exercise rates are higher when individuals can access necessities through a combination of walking, biking, and public transportation. In the United States there has been gradual revitalization of urban areas in the past four decades, as educated young professionals have placed greater value on proximity to recreational and cultural opportunities, as well as shorter commutes.

A life history theory framework provides a scientific basis for broader structural changes. Individuals who have experienced environments in which personal safety, social support, and resource control are uncertain may be more likely to discount future outcomes in favor of present ones. Others who experienced more reliably supportive environments where resource control is more certain could be expected to exhibit the reverse pattern as a reflection of experiences that promote longer-term strategies. By facilitating stable access to necessary resources and high perceptions of safety and community cohesion, perceptions that current effort will pay off in the future will be enhanced, and longer-term strategies will emerge.

The creation and extension of public, safe, family-friendly social space may help decrease levels of interpersonal crime and increase perceptions of safety. The architectural properties of defensible space and local recreation and socialization opportunities would contribute to social capital and cohesion. Social isolation is related to morbidity and mortality, with an increased degree of risk comparable to tobacco smoking (House, 2001). Although the human social environment is changing due to Internet communication, changes in the built environment will also have a social impact. Urban residential space may be dense, but it need not be anonymous. Structurally defined neighborhoods or community units with limited access points and shared social space could facilitate socialization for groups of 100 to 200 individuals. It is estimated that a group size of around 150 was most common throughout most of human history, due to cognitive limitations in information processing for social networks (Dunbar, 1992).

Evolutionarily Informed Effective Health Messaging

It will take decades at best to reshape the human environment with a focus on health behaviors. These efforts must be complemented with health education and behavior change interventions designed to be effective in our current environments. Many current health intervention efforts rest on the assumption that if people only had the relevant information and opportunity, they would always make healthy choices (Bentley & Aunger, 2008; Kruger, 2011). Education and availability of opportunities can certainly shape patterns of health behavior, but they are not the only determinants. This chapter identifies several important factors influencing health outcomes that are outside the scope of mainstream public health theory and practice. Adaptive biological responses evolved to maximize reproductive success, not necessarily health (Gluckman, & Hanson, 2007). This includes motivational systems, which enabled our ancestors to successfully survive and reproduce, even if these systems were not designed to optimize mental and physical health. Our world has changed, yet we are left with the legacy of goals and motivations from longer spans of history.

People do not always act in their best long-term health interests; some researchers have recognized the need to understand people's motivations and preferences (Hancock & Garrett, 1995). Appeals to cognitive-level processes—for example, noting the disease risks associated with certain behaviors—have not been overwhelmingly successful (Bentley & Aunger, 2008). Despite the knowledge that ultraviolent radiation is bad for their skin and can cause skin cancer, people still sunbathe because tan skin is associated with health and beauty. Framing sun exposure warnings in terms of accelerated aging of skin may be more effective than messages framed around increased mortality risk (Saad & Peng, 2006). Of course, too little exposure to the sun results in Vitamin D deficiencies, especially in populations experiencing long winter months and without dietary supplementation (Spiro & Buttriss, 2014). Framing messages promoting healthy diets and adequate exercise in terms of attractiveness rather than mortality risk or disease burden may also promote effectiveness.

Likewise, linking smoking to erectile dysfunction in men may be more persuasive than warning against lung cancer (Saad & Peng, 2006). Diarrhea and respiratory infections are the largest source of child mortality in crowded urban environments with inadequate sanitation; infection risk can be reduced by 50% by hand washing with soap (Curtis & Cairncross, 2003). Hand-washing campaigns focusing on disease risks and using fear tactics are not generally effective. Instead, campaigns utilizing disgust, an evolved response for disease avoidance (Curtis, Aunger, & Rabie, 2004), have been more effective at increasing awareness of the importance of hand washing (Curtis, Garbrah-Aidoo, & Scott, 2007).

Conclusion

Evolutionary approaches to understanding health behavior are more recent than for understanding physiology, though the momentum appears to be building as demonstrated by the increase in research literature. Evolutionary principles provide an ultimate explanation for patterns of health outcomes and will gradually enhance the effectiveness of health interventions. Health researchers and practitioners could benefit from an understanding of the basic principles of evolution and how humans have been shaped by natural and sexual selection, even if they are not explicitly testing evolutionary hypotheses. Within evolutionary theory, life history theory in particular holds the promise of promoting understanding of variation in behavioral patterns related to health and why they vary consistent with environmental conditions. The pace at which evolutionary theory becomes the foundation for health research and practice will be influenced by the visibility and utility of evolutionary theory for addressing the health promotion goals of the field. Efforts to actively integrate evolutionary perspectives into health research and promotion are worthwhile both for their potential direct benefit and potential impact on the field.

Hopefully within our lifetimes this will all be widely accepted science. That is, the material covered in this work and elsewhere will be recognized as a unified body of knowledge informed by scientific process, despite wide variation in topic and research methodology. The sciences regarding all aspects of life are organized and integrated by the powerful and pervasive framework of evolutionary theory. There are some who already recognize this unity, even within the fields related to the understanding and promotion of human health. This volume is edited by academics affiliated with the Evolutionary Studies Consortium (EvoS), a group dedicated to promoting evolutionary perspectives across disciplines. Such efforts inherently recognize the integrative nature of evolutionary theory and potential for uniting academic and practical disciplines. In the current state of the literature, environmental conditions, genetic and epigenetic heritage, developmental experiences, psychological processes, and social context are often unnaturally separated in the scientific literature on health behaviors and outcomes.

References

Alaimo, K., Packnett, E., Miles, R., & Kruger, D. J. (2008). Fruit and vegetable intake among urban community gardeners. *Journal of Nutrition Education and Behavior*, *40*, 94–101.

Anderson, K. G., Kaplan, H., Lam, D., & Lancaster, J. (1999). Paternal care by genetic fathers and stepfathers II. *Evolution and Human Behavior, 20*, 433–451.

Anderson, K. G., Kaplan, H., & Lancaster, J. (1999). Paternal care by genetic fathers and stepfathers I". *Evolution and Human Behavior, 20*, 405–431.

Anderson, T. C., Li, C-Q., Shao, Z-H., & Vanden Hoek, T. L. (2006). Transient and partial mitochondrial inhibition for the treatment of postresuscitation injury: Getting it just right. *Critical Care Medicine, 34*, S474–S482.

Augustin, T., Glass, T. A., James, B. D., & Schwartz, B. S. (2008). Neighborhood psychosocial hazards and cardiovascular disease: The Baltimore Memory Study. *American Journal of Public Health, 98*, 1664–1670.

Bäckhed, F., Ding, H., Wang, T., Hooper, L.V., Koh, G.Y., Nagy, A., . . . Gordon, J.I. (2004). The gut microbiota as an environmental factor that regulates fat storage. *Proceedings of the National Academy of Sciences, 101*, 15718–15723.

Bäckhed, F., Ley, R. E., Sonnenburg, L., Peterson, D. A., & Gordon, J. I. (2005). Host-bacterial mutualism in the human intestine. *Science, 307*, 1915–1920.

Barber, N. (2000). On the relationship between country sex ratios and teen pregnancy rates: A replication. *Cross-Cultural Research, 34*, 26–37.

Barber, N. (2004). Reduced female marriage opportunity and history of single parenthood (England, Scotland, U.S.). *Journal of Cross-Cultural Psychology, 35*, 648–651.

Barker, D. (1998). *Mothers, babies and health in later life* (2nd ed.). London, England: Churchill Livingstone.

Bateson, P., Barker, D., Clutton-Brock, T., Deb, D., D'Udine, B., Foley, R. A., . . . Sultan, S. E. (2004). Developmental plasticity and human health. *Nature, 430*, 419–421.

Bentley, G. R., & Aunger, R. (2008). Practical aspects of evolutionary medicine. In S. Elton & P. O'Higgins (Eds.), *Medicine and evolution: Current applications, future prospects* (pp. 217–239). Boca Raton, FL: CRC.

Blurton Jones, N. (1978). Natural selection and birth-weight. *Annals of Human Biology, 5*, 487–489.

Brand-Miller, J. C., Holt, H. S., Pawlak, D. B., & McMillan, J. (2016). Glycemic index and obesity 1,2,3,4. *American Journal of Clinical Nutrition, 76*, 281S–285S.

Broyles, S. T., Denstel, K. D., Church, T. S., Chaput, J.-P., Fogelholm, M., Hu, G., . . . Katzmarzyk, P. T. (2015). The epidemiological transition and the global childhood obesity epidemic. *International Journal of Obesity Supplements, 5*, S3–S8.

Bruce, H. M. (1959). An exteroceptive block to pregnancy in the mouse. *Nature, 184*, 105.

Bruening, K., Kemp, F. W., Simone, N., Holding, Y., Louria, D. B., & Bogden, J. D. (1999). Dietary calcium intakes of urban children at risk of lead poisoning. *Environmental Health Perspectives, 107*, 431–435.

Bufill, E., Blesa, R., & Agustí, J. (2013). Alzheimer's disease: an evolutionary approach. *Journal of Anthropological Sciences, 91*, 135–157.

Buss, D. M. (1989). Sex difference in human mate preferences: Evolutionary hypotheses tested in 37 cultures. *Behavioral and Brain Sciences, 12*, 1–49.

Buss, D. M., & Schmitt, D. P. (1993). Sexual strategies theory: An evolutionary perspective on human mating. *Psychological Review, 100*, 204–232.

Center for Healthcare Research and Transformation. (2010, November 16). *Issues brief on prematurity*. Retrieved from http://www.chrt.org/assets/price-of-care/CHRT-Issue-Brief-November-2010.pdf

Chagnon, N. A. (1992). *Yanomamo* (4th ed.). New York, NY: Harcourt Brace.

Chisholm, J. S. (1999). *Death, hope and sex: Steps to an evolutionary ecology of mind and morality*. Cambridge, England: Cambridge University Press.

Clulow, F. V., & Langford, P. E. (1971). Pregnancy-block in the meadow vole, *Microtus pennsylvanicus*. *Reproduction, 24*, 275–277.

Cook, G. C. (1978). Did persistence of intestinal lactase into adult life originate on the Arabian peninsula? *Man, 13*, 418–427.

Curtis, V. A., Aunger, R., & Rabie, T. (2004). Evidence that disgust evolved to protect from risk of disease. *Proceedings of the Royal Society B: Biological Sciences, 271*, S131–S133.

Curtis, V. A., & Cairncross, S. (2003). Effects of washing hands with soap on diarrhoea risk in the community: A systematic review. *Lancet Infectious Diseases, 3*, 275–281.

Curtis, V.A., Garbrah-Aidoo, N., & Scott, B. (2007). Masters of marketing: Bringing private sector skills to public health partnerships. *American Journal of Public Health, 97*, 634–641.

Daly, M., & Wilson, M. (2001). An assessment of some proposed exceptions to the phenomenon of nepotistic discrimination against stepchildren. *Annales Zoologici Fennici, 38*, 287–296.

Davies, K. (2013). *The rise of the U.S. environmental health movement*. Lanham, MD: Rowman & Littlefield.

Dawkins, R. (1976). *The selfish gene*. New York, NY: Oxford University Press.

Dunbar, R. I. M. (1992). Neocortex size as a constraint on group size in primates. *Journal of Human Evolution, 20*, 469–493.

Eleftheriou, B. E., Bronson, F. H., & Zarrow, M. X. (1962). Interaction of olfactory and other environmental stimuli on implantation in the deer mouse. *Science, 137,* 764.

Elton, S. (2008). Environments, adaptation, and evolutionary medicine: Should we be eating a stone age diet? In S. Elton & P. O'Higgins (Eds.), *Medicine and evolution: Current applications, future prospects* (pp. 9–33). Boca Raton, FL: CRC.

Figueredo, A. J., Vásquez, G., Brumbach, B. H., Schneider, S. M., Sefcek, J. A., Tal, I. R., . . . Jacobs, W. J. (2006). Consilience and life history theory: From genes to brain to reproductive strategy. *Developmental Review, 26,* 243–275.

Fila, S. A., & Smith, C. (2006). Applying the Theory of planned behavior to healthy eating behaviors in urban Native American youth. *International Journal of Behavioral Nutrition and Physical Activity, 3,* 11.

Fincher, C. L., Thornhill, R., Murray, D. R., & Schaller, M. (2008). Pathogen prevalence predicts human cross-cultural variability in individualism/collectivism. *Proceedings of the Royal Society Series B, 275,* 1279–1285.

Foley, R. A. (1995). The adaptive legacy of human evolution: A search for the environment of evolutionary adaptedness. *Evolutionary Anthropology, 4,* 194–203.

Frisch, R. E. (1987). Body fat, menarche, fitness and fertility. *Human Reproduction, 2,* 521–533.

Gardner, W. (1993). A life-span rational-choice theory of risk taking. In N. J. Bell & R. W. Bell (Eds.), *Adolescent risk taking* (pp. 66–83). Newbury Park, CA: SAGE.

Geary, D. C. (2005). Evolution of paternal investment. In D. Buss (Ed.), *The handbook of evolutionary psychology* (pp. 483–505). Hoboken, NJ: Wiley.

Geary, D. C., & Flinn, M. V. (2001). Evolution of human parental behavior and the human family. *Parenting: Science and Practice, 1,* 5–61.

Gluckman, P. D., & Hanson, M. (2005). *The fetal matrix: Evolution, development and disease.* Cambridge, England: Cambridge University Press.

Gluckman, P. D., & Hanson, M. A. (2007). *Mismatch: Why our world no longer fits our bodies.* Oxford, England: Oxford University Press.

Grinde, B., & Patil, G. G. (2009). Biophilia: Does visual contact with nature impact on health and well-being? *International Journal of Environmental Research and Public Health, 6,* 2332–2343.

Guttentag, M., & Secord, P. F. (1983). *Too many women? The sex ratio question.* Beverly Hills, CA: SAGE.

Gurven, M., Trumble, B. C., Stieglitz, J., Blackwell, A. D., Michalik, D. E., Finch, C., & Kaplan, H. (2016) Cardiovascular disease and type 2 diabetes in evolutionary perspective: a critical role for helminths? *Evolution, Medicine, and Public Health, 1,* 338–357.

Haig, D. (1993). Genetic conflicts in human pregnancy. *Quarterly Review of Biology, 68,* 495–532.

Hancock, T., & Garrett, M. (1995). Beyond medicine: Health challenges and strategies in the 21st century. *Futures, 27,* 935–951.

Hawkes, K., O'Connell, J. F., & Blurton Jones, N. G. (1989). Hardworking Hadza grandmothers. In V. Standen & R. A. Foley (Eds.), *Comparative socioecology: the behavioural ecology of humans and other mammals* (pp. 341–366). Oxford, England: Blackwell Scientific.

Hill, E. M., Ross, L. T., & Low, B. S. (1997). The role of future unpredictability in human risk-taking. *Human Nature, 8,* 287–325.

Hill, K. (1993). Life history theory and evolutionary anthropology. *Evolutionary Anthropology, 2*, 78–89.

Hill, K., & Hurtado, M. (1996). *Aché life history: the ecology and demography of a foraging people*. New York, NY: Aldine de Gruyter.

Hill, K., & Kaplan, H. (1999). Life History traits in humans: Theory and empirical studies. *Annual Review of Anthropology, 28*, 397–438.

Hill, S. E., Prokosch, M. L., DelPriore, D. J., Griskevicius, V., & Kramer, A. (2016). Low childhood socioeconomic status promotes eating in the absence of energy need. *Psychological Science, 27*, 354–364.

Hirschfield, M. F., & Tinkle, D. W. (1975). Natural selection and the evolution of reproductive effort. *Proceedings of the National Academy of Sciences U.S.A., 72*, 2227–2231.

Holden, C., & Mace, R. (2002). Pastoralism and the evolution of lactase persistence. In W. R. Leonard & M. H. Crawford (Eds.), *Human biology and pastoral populations* (pp. 280–307). Cambridge, England: Cambridge University Press.

Hopcroft, R. L. (2006). Sex, status and reproductive success in the contemporary U.S. *Evolution and Human Behavior, 27*, 104–120.

House, J. S. (2001). Social isolation kills, but how and why? *Psychosomatic Medicine, 63*, 273–274.

Hunte, H. E., Turner, T. M., Pollack, H. A., & Lewis, E. Y. (2004). A birth records analysis of the Maternal Infant Health Advocate Service program: A paraprofessional intervention aimed at addressing infant mortality in African Americans. *Ethnicity & Disease, 14*, S102–S107.

Jang, K. L., McCrae, R. R., Angleitner, A., Riemann, R., & Livesley, W. J. (1998). Heritability of facet-level traits in a cross-cultural twin sample: Support for a hierarchical model of personality. *Journal of Personality and Social Psychology, 74*, 1556–1565.

Juliandi, B., Abematsu, M., & Nakashima, K. (2010). Epigenetic regulation in neural stem cell differentiation. *Development, Growth & Differentiation, 52*, 493–504.

Karn, M. N., & L. S. Penrose (1952). Birth weight and gestation time in relation to maternal age, parity and infant survival. *Annals of Eugenics, 16, 147–164*.

Keough, K. A., Zimbardo, P. G., & Boyd, J. N. (1999). Who's smoking, drinking, and using drugs? Time perspective as a predictor of substance use. *Basic and Applied Social Psychology, 21*, 149–164.

Kidd, C., Palmeri, H., & Aslin, R. N. (2013). Rational Snacking: Young children's decision-making on the marshmallow task is moderated by beliefs about environmental reliability. *Cognition, 126*, 109–114.

Kim, J. H., & Scialli, A.R. (2011). Thalidomide: The tragedy of birth defects and the effective treatment of disease. *Toxicological Sciences, 122*, 1–6.

Kim, K., Smith, P. K., & Palermiti, A. L. (1997). Conflict in childhood and reproductive development. *Evolution and Human Behavior, 18*, 109–142.

Kruger, D. J. (2010). Socio-demographic factors intensifying male mating competition exacerbate male mortality rates. *Evolutionary Psychology, 8*, 194–204.

Kruger, D. J. (2011). Evolutionary theory in Public Health and the public health of evolutionary theory. *Futures, 43*, 762–770.

Kruger, D. J., Clark, J., & Vanas, S. (2013). Male scarcity is associated with higher prevalence of premature gestation and low birth weight births across the USA. *American Journal of Human Biology, 25*, 225–227.

Kruger, D. J., & Figueredo, A. J. (2010, June). *Health related correlates of life history strategy in a representative community sample*. Oral presentation, Human Behavior and Evolution Society. Eugene, OR.

Kruger, D. J., Greenberg, E., Murphy, J. B, DiFazio, D. A., & Youra, K. R. (2014). Local concentration of fast food outlets is associated with poor nutrition and obesity. *American Journal of Health Promotion, 28*, 340–343.

Kruger, D. J., & Kruger, J. S. (2016). Psychometric assessment of human life history predicts health related behaviors. *Psychological Topics, 25*, 19–28.

Kruger, D. J., Munsell, M. A., & French-Turner, T. M. (2011). Using a life history framework to understand the relationship between neighborhood structural deterioration and adverse birth outcomes. *Journal of Social, Evolutionary, and Cultural Psychology, 5*, 260–274.

Kruger, D. J., & Nesse, R. M. (2006). An evolutionary life-history framework for understanding sex differences in human mortality rates. *Human Nature, 17*, 74–97.

Kruger, D. J., Reischl, T. M., & Zimmerman, M. A. (2008). Time perspective as a mechanism for functional developmental adaptation. *Journal of Social, Evolutionary, and Cultural Psychology, 2*, 1–22.

Kwiatkowski, D. P. (2005). How malaria has affected the human genome and what human genetics can teach us about malaria, *American Journal of Human Genetics, 77*, 171–92.

Lampe, J. W., & Becker, L. B. (2007). Rapid cooling for saving lives: A bioengineering opportunity. *Expert Review of Medical Devices, 4*, 441–446.

Laskey, M. A., & Prentice, A. (1997). Effect of pregnancy on recovery of lactational bone loss. *Lancet, 349*, 1518–1519.

Ley, R. E., Peterson, D. A., & Gordon, J. I. (2006). Ecological and evolutionary forces shaping microbial diversity in the human intestine. *Cell, 124*, 837–848.

Lichter, D. T., Anderson, R. N., & Hayward, M. D. (1995). Marriage markets and marital choice. *Journal of Family Issues, 16*, 412–431.

Louveau, A., Igor Smirnov, I., Keyes, T. J., Eccles, J. D., Rouhani, S. J., Peske, J. D., ... Kipnis, J. (2015). Structural and functional features of central nervous system lymphatic vessels. *Nature, 523*, 337–341.

Low, B. (1998). The evolution of human life histories. In C. Crawford & D. Krebs (Eds.), *Handbook of evolutionary psychology: Issues, ideas, and applications* (pp. 131–161). Mahwah, NJ: Erlbaum.

Ma, D. K., Marchetto, M. C., Guo, J. U., Ming, G. L., Gage, F. H., & Song, H. (2010). Epigenetic choreographers of neurogenesis in the adult mammalian brain. *Nature Neuroscience, 13*, 1338–1344.

MacDorman, M. F., & Mathews, T. J. (2009). The challenge of infant mortality: Have we reached a plateau? *Public Health Reports, 124*, 670–680.

Manton, K. G., Gu, X., & Lamb, V. L. (2006). Change in chronic disability from 1982 to 2004/2005 as measured by long-term changes in function and health in the U.S. elderly population. *Proceedings of the National Academy of Sciences, 103*, 18374–18379.

Marlowe, F. (2000). The patriarch hypothesis: An alternative explanation of menopause. *Human Nature, 11*, 27–42.

Medawar, P. B. (1952). *An unsolved problem of biology*. London, England: H. K. Lewis.

Mazmanian, S. K., Liu, C. H., Tzianabos, A. O., & Kasper, D. L. (2005). An immunomodulatory molecule of symbiotic bacteria directs maturation of the host immune system. *Cell, 122*, 107–118.

Milton, K. (1999). Nutritional characteristics of wild primate foods: Do the diets of our closest living relatives have lessons for us? *Nutrition, 15*, 488–498.

Milton, K. (2002). Hunter-gatherer diets: Wild foods signal relief from diseases of affluence. In P. S. Ungar & M. F. Teaford (Eds.), *Human diet: Its origin and evolution* (pp. 111–122). Westport, CT: Bengin and Garvey.

Mischel, W., Shoda, Y., & Rodriguez, M.L. (1989). Delay of gratification in children. *Science, 244*, 933–938.

National Center for Health Statistics. (2018). *Health, United States, 2017: With special feature on mortality*. Hyattsville, MD.

National Institute on Aging. (2011). *Global health and aging* (NIH Publication 11-7737). Bethesda, MD: Author.

Nesse, R. M. (1998). Emotional disorders in evolutionary perspective. *British Journal of Medical Psychology, 71*, 397–415.

Nesse, R. M., & Williams, G. C. (1994). *Why we get sick: The new science of Darwinian medicine*. New York, NY: Times Books.

Park, R. E., & Burgess, E. W. (1925). *The city*. Chicago, IL: University of Chicago Press.

Pedersen, F. A. (1991). Secular trends in human sex ratios: Their influence on individual and family behavior. *Human Nature, 2*, 271–291.

Pepper, G. V., & Nettle, D. (2014). Perceived extrinsic mortality risk and reported effort in looking after health. *Human Nature, 25*, 378–92.

Pollet, T. V., & Nettle, D. (2007). Driving a hard bargain: Sex ratio and male marriage success in a historical US population. *Biology Letters, 4*, 31–33.

Profet, M. (1992). Pregnancy sickness as adaptation: A deterrent to maternal ingestion of teratogens. In J. H. Barkow, L. Cosmides, & J. Tooby (Eds.), *The adapted mind: Evolutionary psychology and the generation of culture* (pp. 327–365). Oxford, England: Oxford University Press.

Quinlan, R. J. (2007). Human parental effort and environmental risk. *Proceedings of the Royal Society: Series B, 274*, 121–125.

Raichlen, D. A., & Alexander, G. E. (2014). Exercise, APOE genotype, and the evolution of the human lifespan. *Trends in Neurosciences, 37*, 247–255.

Ransohoff, R. M. & Englehardt, B. (2012). The anatomical and cellular basis of immune surveillance in the central nervous system. *Nature Reviews Immunology, 12*, 623–635.

Reiber, C. (2008). An evolutionary model of premenstrual syndrome. *Medical Hypotheses, 70*, 1058–1065.

Reiber, C. (2009). An evolutionary model of premenstrual syndrome. *Journal of Social, Evolutionary, and Cultural Psychology, 3*, 9–28.

Reiber, C. (2011). Female gamete competition: A new evolutionary perspective on menopause. *Journal of Social, Evolutionary, and Cultural Psychology, 4*, 215–240.

Richardson, G. B., & Hardesty, P. (2012). Immediate survival focus: Synthesizing life history theory and dual process models to explain substance use. *Evolutionary Psychology, 10*, 731–749.

Roberts, E. K., Lu, E., Bergman, T. J., & Beehner, J. C. (2012). A Bruce effect in wild geladas. *Science, 335*, 1222–1225.

Roff, D. A. (1992). *The evolution of life histories: Theory and analysis.* New York, NY: Chapman & Hall.

Saad, G., & Peng., A. (2006). Applying Darwinian principles in designing effective intervention strategies: The case of sun tanning. *Psychology & Marketing, 23*, 617–638.

Segerstråle, U. (2000). *Defenders of the truth: The battle for science in the sociobiology debate and beyond.* Oxford, England: Oxford University Press.

Single women: Single minded. (2007, June 28). *Time Out New York, 613*, 28–34.

Smith, C. C., & Fretwell, I. A. (1974). The optimal balance between size and number of offspring. *American Naturalist, 108*, 499–506.

South, S. J., & Trent, K. (1988). Sex ratios and women's roles: A cross-national analysis. *American Journal of Sociology, 93*, 1096–1115.

South, S. J., & Trent, K. (2010). Imbalanced sex ratios, men's sexual behavior, and risk of sexually transmitted infection in China. *Journal of Health and Social Behavior, 5*, 376–390.

Spiro, A., & Buttriss, J. L. (2014). Vitamin D: An overview of vitamin D status and intake in Europe. *Nutrition Bulletin, 39*, 322–350.

Stearns, S. C. (1992). *The evolution of life histories.* Oxford, England: Oxford University Press.

Stein, Z., Susser, M., Saenger, G., & Marolla, F. (1975). *Famine and human development: The Dutch hunger winter of 1944–1945.* New York, NY: Oxford University Press.

Sun, J., Sun, J., Ming, G. L., & Song, H. (2011). Epigenetic regulation of neurogenesis in the adult mammalian brain. *European Journal of Neuroscience, 33*, 1087–1093.

Talge, N. M., Holzman, C., Wang, J., Lucia, V., Gardiner, J., & Breslau, N. (2010). Late-preterm birth and its association with cognitive and socioemotional outcomes at 6 years of age. *Pediatrics, 126*, 1124–1131.

Tishkoff, S. A., Reed, F. A., Ranciaro, A., Voight, B. F., Babbitt, C. C., Silverman, J. S., . . . Deloukas, P. (2007). Convergent adaptation of human lactase persistence in Africa and Europe. *Nature Genetics, 39*, 7–8.

Treadwell, H. M., & Young, A. M. (2013). The right US men's health report: High time to adjust priorities and attack disparities. *American Journal of Public Health, 103*, 5–6.

Trent, K., & South, S. J. (1989). Structural determinants of the divorce rate: A cross-societal analysis. *Journal of Marriage and the Family, 51*, 391–404.

Trivers, R. L. (1974). Parent–offspring conflict. *American Zoologist, 14*, 247–262.

Uecker, J. E., & Regnerus, Mark D. (2010). Bare market: Campus sex ratios, romantic relationships, and sexual behavior. *Sociological Quarterly, 51*, 408–435.

Wells, J. C. (2003). The thrifty phenotype hypothesis: Thrifty offspring or thrifty mother? *Journal of Theoretical Biology, 221*, 143–161.

White, L. (1994). Growing up with single parents and stepparents: Long-term effects on family solidarity. *Journal of Marriage and Family, 56*, 935–948.

Williams, G. C. (1957). Pleiotropy, natural selection, and the evolution of senescence. *Evolution, 11*, 398–411.

Wilson, D. R. (1993). Evolutionary epidemiology: Darwinian theory in the service of medicine and psychiatry. *Acta Biotheoretica, 41*, 205–218.

Wilson, E. O. (1984). *Biophilia.* Cambridge, England: Harvard University Press.

Wilson, M., & Daly, M. (1997). Life expectancy, economic inequality, homicide, and reproductive timing in Chicago neighbourhoods. *British Medical Journal, 314,* 1271–1274.

Wood, B., & Strait, D. (2004). Patterns of resource use in early *Homo* and *Pananthropus. Journal of Human Evolution, 46,* 119–162.

Wynn, A. (1987). Nutrition before conception and the outcome of pregnancy. *Nutrition and Health, 5,* 31–43.

Zimbardo, P., & Boyd, J. (1999). Putting time in perspective: A valid, reliable individual differences metric. *Journal of Personality and Social Psychology, 77,* 1271–1288.

CHAPTER 14 | Integrating Evolutionary Thinking into Medical Education and Curricula

BARBARA NATTERSON-HOROWITZ
AND DANIEL T. BLUMSTEIN

EVOLUTIONARY MEDICINE'S CENTRAL PREMISE is that vulnerabvility to disease, like all other aspects of physiology, is an evolved adaptation. As such, human pathology cannot be fully conceived of and understood without an eco-evolutionary framework. The emerging field seeks to integrate eco-evolutionary principles into biomedical research, medical education, and clinical care to create healthier individuals and populations. Yet, clinical practice and biomedical investigation today do not yet point to ecology and evolutionary biology as foundational sciences in the way engineers point to physics or mathematics. Consequently, ecological and evolutionary concepts rarely appear as central components in the educational curricula of medical, dental, and nursing schools and others that train healthcare providers and investigators.

A central barrier to bringing evolutionary content into educational environments has been the perceived lack of a traditional evidence base to support its benefits to the health fields (Alcock & Schwartz, 2011). A circular problem is created: The perceived absence of evidence about the advantages of evolutionary thinking precludes its deep integration into education and research endeavors. The consequence of its exclusion then contributes to the problem of evidence. "Can evolution save lives?" remains an unanswered question if leaders in biomedical research, clinical care, and public health keep the field at a skeptical "arm's length."

This belief is unwarranted, given the evidence. There are, in fact, several areas in medicine in which evolutionary thinking has already offered the ecological context and biological principles that have resulted in novel hypotheses and even therapeutic approaches. We are seeing new applications emerging from novel, eco-evolutionary approaches to understanding various pathologies. For example, the emergence of antimicrobial resistance over the past several decades now recognized to be, in large part, a byproduct of the indiscriminate and overuse of highly powerful antibiotics. The exposure of a diverse range of pathogens to powerful antimicrobials has promoted the selection of resistant organisms. Infectious disease specialists and others now use evolutionarily informed language and engage evolutionary principles to describe the problem and conceive solutions (Antonovics et al., 2007).

Among clinical oncologists and investigators, human vulnerability to cancer has come to be understood as a consequence of series of evolutionary effects and pressures including life history trade-offs, mismatches, and selection (Aktipis et al., 2015; Boddy, Kokko, Breden, Wilkinson, & Aktipis, 2015; Brown & Aktipis, 2015; Brown, Cunningham, & Gatenby, 2015). The basic science of cancer biology has advanced by the recognition that the vulnerability to cancer emerged with multicellularity itself (Aktipis et al., 2015).

Moreover, the emergence of chemotherapeutic resistance following treatment, for example, is increasingly framed as a predictable result of natural selection acting on a diverse community of cancer cells. The tendency for primary tumors to metastasize is also now being studied using an ecological conceptual framework featuring competition and dispersal.

Finally, mismatch models are emerging to explain modern day epidemics including autoimmune disease, obesity, perhaps even anxiety, depression and more. Mismatch models posit that physiologies which evolved to offer organisms adaptive benefit become maladaptive when environments change. The hygiene hypothesis is an example of a mismatch model which suggests that some autoimmune diseases result from differences between our historically constantly challenged immune systems and the cleaner environments many of us now live in. The hygiene hypothesis compels physicians to consider emergence of pathology through the lens of ancient ancestry. While aspects of the hygiene hypothesis remain controversial, it has sparked the development of novel therapies challenging immune responses which have great promise for treatment (Okada, Kuhn Feillet, & Bach, 2010).

Despite these strong examples from several areas of medicine, the power of eco-evolutionary thinking to improve our understanding of other pathologies and accelerate biomedical research innovation remains significantly underleveraged. First, there is the aforementioned skepticism related to a perceived lack of evidence. But beyond this, even among medical educators who recognize the theoretical advantages of evolutionary thought for physicians, there is uncertainty as to how best apply and teach these principles. Moreover, a roadmap for how evolutionary thought can penetrate and benefit complex healthcare communities is lacking. Addressing these concerns with practical and useable approaches could accelerate the integration of evolutionary principles into medical education, clinical practice, and biomedical investigation. Offering a structured approach that introduces a novel, evolutionarily informed series of questions could expand thinking beyond traditional conceptualizations of disease causation and propagation. Inclusion of such a structure into health curricula would provide a framework for using evolutionary principles to better understand the nature of disease and approaches to developing effective therapies.

A Tinbergean Approach

The novel framework we propose finds its inspiration in the conceptual work of ethologist Niko Tinbergen. In 1963, Tinbergen published "On Aims and Methods in Ethology," a paper in which he offered what he considered to be the four fundamental problems that needed to be asked to understand a specific animal behavior (Tinbergen, 1963). Tinbergen, who went on to win the Nobel Prize in Physiology or Medicine in 1973, along with Karl von Frisch and Konrad Lorenz, for founding the field of ethology, framed these problems as four questions that had to be addressed for a meaningful and complete understanding of a behavioral trait ("The Nobel Prize," n.d.). Tinbergen sought to understand why a behavior occurred by understanding: how it worked, how it developed, how it evolved, and what was its adaptive value (Tinbergen, 1963).

His seismic contribution challenged biologists to move beyond traditional mechanistic and proximate perspectives to more contextualized analyses that also considered developmental, comparative, and adaptive explanations. This approach undoubtedly offered a transformative shift—as his laureate suggests—in how behavior was understood and analyzed. Yet, the applicability of this Tinbergean framework extends far beyond the study of animal behavior to every field in which an evolutionary perspective is relevant, including medicine and other clinical specialties (O'Brien & Gallup, 2011).

A Tinbergean framework offers a structured approach to human health concerns that moves the field beyond its traditional proximate approaches. Tinbergen's questions automatically broaden medical thinking about the nature of disease causation. The four questions expand traditional conceptualizations of cause beyond mechanistic and developmental explanations. They direct thinking toward the context under which the disorder appears, the comparative and historic nature of the disorder, and the adaptive benefits of vulnerability to the disorder.

Applying Tinbergen's Four Questions to Medicine

The introduction of a Tinbergean framework into health science curricula offers a common vocabulary facilitating productive conversations between physicians and evolutionary biologists. Human medicine is largely fixated on proximate causation. Risk factors for disease and lifestyles associated with increased disease risk are conflated with causation. Genetic explanations also offer a decontextualized proximate 'explanation.'

The reflexive and simplistic way in which most clinicians understand and communicate the origins of their patients' disorders reflects this hyper-proximate tendency. Another limitation to physician thinking about non-proximate causes of disease emerges from 1) failure of medical education to emphasize physiology, including vulnerability to pathology, as evolved adaptation. Heightening this limitation is the low awareness among physicians of the spontaneous occurrence of human diseases in nonhuman animals (Natterson-Horowitz & Bowers, 2012). This anthropocentric perspective deprives investigators and clinicians of the information they need to develop intelligent adaptive hypotheses. Finally, human medicine tends to focus on a patient at one point in time instead of considering the presenting symptom or disease in the context of an entire lifetime—from conception (even pre-conception) to the present. Indeed, physician specialties (e.g., pediatrics, adolescent medicine, geriatric medicine) are themselves often determined by this time-stamped form of medicine.

In the remainder of this chapter, we offer such a novel approach to health professions' curricula that emphasizes an integrative eco-evolutionary view of health and disease. There are currently multiple textbooks that consider the application of principles of evolution to medicine (Gluckman, Beedle, & Hanson, 2009; Perlman, 2013; Stearns & Medzhitov, 2015). We do not wish to rehash them here and will focus instead on a novel, clinically oriented structured framework as a guide for curricular integration.

Within this framework evolutionary principles serve as an indispensable common language between collaborating disciplines. As an extension of this framework we also make the case that a broader "change model" is needed before these ideas are fully incorporated into health *professionals'* curricula. By this consideration of the comparative (phylogenetic), developmental, and potential adaptive aspects of disease, Tinbergen's structure pushes medicine beyond its hyper-proximate and mechanistic tendencies.

We present three examples of how high impact human medical and psychiatric conditions can be taught with a Tinbergean approach, and the benefits of doing so.

Congestive Heart Failure

Congestive heart failure (CHF) is one of the leading causes of mortality and morbidity among adults and children around the world (Roger, 2013). CHF due to ventricular systolic dysfunction is diagnosed when the performance of the heart's right and/or left ventricles is impaired resulting in a spectrum of symptoms and signs associated with volume overload, respiratory insufficiency, and underperfusion of the organs of the body. Congestive heart failure can arise as a consequence of many causes of ventricular dysfunction including viral infection of the heart, exposure to some chemotherapeutic agents, abnormalities of the heart's valves, or drug and alcohol abuse. By far the leading cause of congestive heart failure due to impaired systolic dysfunction in the United States is coronary artery disease (Mayo Clinic, 2016; Mosterd & Hoes, 2007).

Regardless of the proximate cause of the ventricular dysfunction, heart failure results in a characteristic pattern of neurohormonal activation leading to volume overload, shortness of breath, exercise intolerance, abnormal heart rhythms, and often death (Mosterd & Hoes, 2007). Until the 1990s, therapies for CHF targeted the underperforming ventricles, "flogging" the weakened myocytes with inotropic agents to make them contract more effectively. Outcomes for patients with CHF were poor at that time (Schocken, Arrieta, Leaverton, & Ross, 1992). Several decades ago, research interest began to shift from ventricular dysfunction itself to the neuroendocrine activation seen in response to reduced cardiac outputs. Interest in neuroendocrine activation centered around a novel question for cardiology: "Was the neuroendocrine activation maladaptive?" Milton Packer(Am J Kidney Dis. 1987 Jul;10(1 Suppl 1):66-73.

Adaptive and maladaptive actions of angiotensin II in patients with severe congestive heart failure.) and others began using the terms 'adaptive' and 'maladtive' to describe the suite of neurohormones released and elevated in patients with systolic dysfunction.

Packer M.

Embedded within the use of words such as 'adaptive' and 'maladaptive' was an evolutionary context. Although the clinicians and investigators involved with this discovery did not explicitly conceive of it as evolutionary medicine, it was precisely that. In fact, this insight transformed the management of congestive heart failure, catalyzed tremendous biomedical innovation, and significantly improved outcomes. It is an example of the power of evolutionary thinking to improve clinical outcomes (Ferrari et al., 1996; Lechat, 2006).

A practical framework for teaching and understanding CHF, structured around the four Tinbergean questions follows.

> When teaching CHF, students can be challenged to consider the following questions:
> 1. MECHANISM:
>   Q: What are the mechanisms underlying the development of CHF. What "causes" CHF?
>   A: Coronary artery disease leading to heart attacks, viruses, valvular abnormalities, chemotherapy, drugs of abuse, etc. There are many ways to respond to this question involving the cellular and subcellular pathways identified in the genesis of CHF. (Mayo Clinic, 2016; Mosterd & Hoes, 2007)?
> 2. ONTOGENY:
>   Q: When in the lifecycle can CHF develop? What are the age-specific factors?
>   A: CHF can occur in any ventricle at any age. The most common cause of ventricular damage in the Westernized world is coronary artery disease, a condition that occurs later in life. However, CHF can occur in infancy, and even a developing ventricle, in utero, can be affected (Mayo Clinic, 2016; Mosterd & Hoes, 2007). Ventricular myocytes are vulnerable from the gestational period through the end of life. Neuroendocrine activation exacerbates heart failure pathology.
> 3. PHYLOGENY:
>   Q: When and how did vulnerability to congestive heart failure evolve?
>   A: Phylogenetic analysis of CHF reveals cases in taxa ranging from nonhuman primates, nonprimate mammals to reptiles, amphibians, and

even single-ventricled fish species (Jacobson, Homer, & Adams, 1991; Mitchell & Tully, 2016; Varki et al., 2009). The vulnerability to ventricular dysfunction is having a myocyte ventricle in the first place. Neuroendocrine systems are seen in all taxa of vertebrates from fish to reptiles to birds to mammals. Activation of the neuroendocrine system is seen in response to many stressors

4. ADAPTIVE VALUE OF VULNERABILITY

Q: What is the adaptive "function" of vulnerability to this disorder?

A: CHF obviously is maladaptive. However, the *vulnerability* to CHF can be understood as having adaptive benefit. The neurohormonal activation seen in CHF is triggered by reduced flow to the kidneys and carotids and is maladaptive for patients with weakened hearts triggering fluid and sodium retention leading to volume overload, shortness of breath, and even death. Yet this represents a highly adaptive response to reduced perfusion to the kidneys and carotids due to dehydration or exsanguination. The conditions under which our ancestors' cardiovascular and renal physiologies evolved, primary life-threatening challenges and the evolution of protective physiologies would have been highly adaptive (Ferrari et al., 1996).

Clinical takeaway: The major improvements in heart failure outcomes over the past two decades reflect the power of evolutionary thinking. Treatment strategies for CHF shifted from targeting the failing ventricle itself to focusing on the maladaptive responses to decreased systemic perfusion. Survival increased by applying functionally informed approach moving beyond treatment based on a mechanistic/proximate explanation (the ventricle is injured) to therapies that reframe the pathology as a maladaptive response, which undoubtedly had adaptive benefits in the past. Since this application, CHF mortality rates have decreased dramatically (Ghali, Cooper, & Ford, 1990; McMurray & Stewart, 2000; Schocken et al., 1992).

Posttraumatic Stress Disorder

Posttraumatic stress disorder (PTSD) is a common and increasingly diagnosed anxiety disorder characterized by crippling symptoms including distress, avoidance, hypervigilance, and insomnia after exposure to a significantly threatening event. Treatments for the disorder involve combinations of talk, behavioral, and pharmacological therapies. Still, and despite these therapies, many individuals with PTSD fail to get better ("What Is PTSD?" 2016).

Some investigators have identified therapeutic benefit by exposing patients to evolutionary-informed explanations to their disorders (Nesse & Williams, 1996). Beyond this, recognition of the highly conserved nature of fear responses to predatory threats coupled with the ubiquity of catecholamine/neuroendocrine responses has allowed for the development of naturalistic animal models for PTSD and related syndromes. Although the clinical and translational application of this work has yet to be fully realized, the increasingly widespread use of catecholamine-blunting therapies in PTSD has certainly been catalyzed by this type of investigation (Beerda, Schilder, Janssen, & Mol, 1996; Yenkosky, Bradshaw, & McCarthy, 2010).

A practical framework for teaching and understanding PTSD, structured around the four Tinbergean questions follows.

> When teaching PTSD, students can be challenged to consider the following questions:
> 1. MECHANISM:
>   Q: What is the mechanism for PTSD?
>   A: The "cause" of PTSD is often identified as the traumatic stressor itself: the witnessed murder, the sexual assault, the near-fatal motor vehicle accident ("What Is PTSD?" 2016). The neurobiology underlying PTSD has been increasingly elucidated.
> 2. ONTOGENY:
>   Q: When in the lifecycle does PTSD originate, present, intensify?
>   A: Vulnerability to PTSD seems to be highest when exposure to the trauma occurs during adolescence and young adult life. Experiences which include trauma or high negative emotional arousal may induce PTSD in vulnerable individuals. The ontogeny of the disorder varies with some individuals triggered by single events while in others ongoing traumatic exposures are necessary.
> 3. PHYLOGENY
>   Q: When and how did vulnerability to PTSD evolve?
>   A: PTSD is believed to exist in several domesticated species including dogs and parrots based on history and behavioral assessments (Beerda et al., 1996; "What Is PTSD?" 2016; Yenkosky et al., 2010). Of course, the internal experiences of animals cannot be understood in the absence of language. However, the conserved nature of response to trauma between species points strongly to PTSD's presence in nonhuman animals. PTSD-like symptoms can

be created through fear-conditioning in rats (Mikics, Baranyi, & Haller, 2008). Notably, PTSD can be conceived of as an inability to 'unlearn' that a stimulus is dangerous. Difference in 'unlearning vs. learning' are beginning to be studied in a variety of taxa (e.g., Brown et al. 2011; Ferrari et al. 2012)

4. ADAPTIVE VALUE OF VULNERABILITY

Q: What is the "function" of vulnerability PTSD

A: While the severe symptoms and avoidant behavior seen in today in PTSD sufferers are maladaptive, embedded with vulnerability to PTSD are adaptive elements. For example, while it is undoubtedly terrifying for a young animal to encounter its predator, rapid one-trial learning is adaptive in many species. Indeed, maintaining hypervigilance in situations where it is not needed is costly (Bonne, Grillon, Vythilingam, Neumeister, & Charney, 2004). Fear-conditioning is a mechanism by which individuals are programmed to calibrate themselves with environmental risks.

Clinical takeaway: Considering PTSD from a broader, more contextualized perspective offers benefits to clinician and investigators. From a scientific perspective, it suggests that understanding the mechanisms underlying 'unlearning' may be a productive approach. And from a patient's perspective, it offers an explanatory narrative for the disorder that depathologizes the response, reframing it as a once adaptive response that is now maladaptive in certain modern contexts.

Metastatic Carcinoma

Cancer is a leading cause of death among adults. Despite the "war on cancer" over the past several decades, rates of cancer deaths remain high, especially among certain populations (American Cancer Society, 2011). Notably, most cancer deaths are the result of metastatic spread of the disease and not from a cancer's localized effects. Traditionally, metastasis has been thought to occur later in the natural history of a malignant tumor (Fidler, 1978). The battle against cancer was conceived of as race to early diagnosis and intensive therapy to "wipe out" the tumor before it could spread (DeVita & Chu, 2008). But this "scorched earth" approach to cancer treatment has not only largely failed to wipe out tumors, it may inadvertently facilitate the metastatic spread of tumors. The widespread use of this approach may have also promoted chemotherapeutic resistance (Boddy et al., 2015; Casás-Selves & DeGregori, 2011; DeVita & Chu, 2008).

Scientists have illustrated this point utilizing infectious disease models to show in-host competition can alter the fitness costs and benefits of resistance in diseases such as cancer. They argue that resistance management requires an evidence-based approach utilizing evolutionary applications for addressing therapeutic limitations and reducing the probability of resistance mutations from drug treatment (Read, Day, & Huijben, 2011).

A practical framework for teaching and understanding metastatic cancer, structured around the four Tinbergean questions follows:

> When teaching about metastatic cancer, students can be challenged to consider the following questions:
> 1. MECHANISM:
>   Q: What are the mechanisms underlying metastasis?
>   A: Traditional proximate explanations for why cancer metastasizes focus on tumor characteristics including size and proximity to vasculature, genetics, and host physiology, (Beerda et al., 1996).
> 2. ONTOGENY
>   Q: When in the lifecycle of the host or within the natural history of disease does vulnerability to metastatic cancer develop?
>   A: Traditional models of metastasis posit that spread occurs over time with a discreet period of localized growth before the appearance of cancer in remote sites (Chambers, Groom, & MacDonald, 2002). Recently, a broader analysis of metastasis suggests that competition for resources between diverse colonies of cancer may promote dispersal (metastasis) following patterns of dispersal modeled by population ecologists (Boddy et al., 2015).
> 3. PHYLOGENY
>   Q: How and when did vulnerability to metastatic cancer evolve?
>   A: Spontaneously occurring cancer occurs across the animal kingdom. Classic metastasis is seen in animals with a vascular network that connects organ systems across the body. There are no published phylogenetic analyses of vulnerability to metastasis. Such an analysis might identify lineages that are more or less prone to metastasis.
> 4. PHYLOGENY
>   Q: What is the purpose/function of vulnerability to metastatic cancer?
>   A: It is impossible to identify an adaptive benefit of metastatic cancer for a host organism. However, the vulnerability to metastases—having a dense and interconnected vascular plexus to effectively provide for an animal's metabolic requirements—has important

functionality for many animal species. Moreover, the foundational property of somatic cells to move toward resources necessary to satisfy their metabolic needs plays an important functional role in animal physiology.

Clinical takeaway: By considering metastatic spread of cancer from a more holistic and ecological perspective, new disruptive and testable hypotheses emerge. In turn, these ecologically informed explanations can yield novel therapeutic approaches aimed at reducing metastases through containment and control rather than annihilation/scorched earth approaches.

Beyond Tinbergen: Cultivating a Culture of Evolutionary Medicine

We selected the above high-impact human pathologies to demonstrate the potential in universally enhancing and reframing our research approach to these diseases using a Tinbergean model. This model can be introduced throughout the health profession curricula as a common language to promote the creation of an evolutionarily informed culture in medicine that considers health and disease in a more holistic and contextualized way. Because this approach is novel and disruptive to traditional approaches, we suggest that a comprehensive change model is required. Table 14.1 illustrates where we are (as of 2016) and articulates a future state vision.

As we have shown, evolutionary thinking has played a significant role in reducing morbidity and mortality in several high-impact human diseases. In some of these cases, evolutionary theory itself has catalyzed these advances. In other cases, evolution's impact has only been identified after the fact (as seen in the case of CHF therapies). The challenge is to create an educational system that facilitates the creation of a community of healthcare providers and biomedical researchers that are aware at the outset of the potential benefits from a Tinbergean approach. By doing so, we hope to accelerate the rate of meaningful changes that lead to beneficial patient and health outcomes.

While the fields of evolutionary biology and biomedical sciences are traditionally perceived to have disparate goals, it is important for educators from across the educational levels—undergraduate to postgraduate—to devise new strategies and collaboratively build a bridge between the

TABLE 14.1. Evolutionary Medicine: Current State and Future State Vision

EVOLUTIONARY MEDICINE	CURRENT STATE (2016)	FUTURE STATE VISION
Context/Culture	Biomedical research tends to be narrowly and mechanistically focused and largely lacks the inclusion of ecologically and evolutionarily-based context, function and history; biomedical research focuses more on immediate, proximate questions primarily intended to cure disease.	A broader, integrated view of health and biomedical and public health research, shared research; ecology and evolutionarily/holistic view of health informed by principles of ecology and evolution that acknowledges context, function and history.
Structures	Research conducted in departmental scientific silos.	Integration of disciplines (including social sciences and humanities) to conduct Evolutionary Medicine, shared research agenda; strategic formation of interdisciplinary teams to answer scientific and medical questions.
Research Processes	Reductionist, proximate studies of disease causation.	Richly interdisciplinary, integrative, and strategically formed-research processes that acknowledges that disease as a context has a function and traits. Define research processes by viewing disease in context of function and history to inform the creation of novel understandings and therapies.
Investigator teams/ Research Education, Training and Career Development	Discipline driven, limited scientific scope.	Interdisciplinary, translational; researchers will appreciate context, function, and history; utilize interdisciplinary evolutionary-medicine-based competencies in future curriculum and training across all educational levels.
Outcomes; Enduring impact	Combat disease following physiological deterioration; treating disease in specific, trait-based view.	Development of new therapies informed by interdisciplinary processes; greater understanding of relationship between life, health and disease in ecological and evolutionary context; contextual, functional and historical understanding of disease states.

disciplines. These efforts are necessarily to formulate a "common language" between them and establish a scientific foundation that facilitates the integration of the two. While clinical training program such as medical schools would be faced with the daunting task of adding evolution education to already-overcrowded curricula, such offerings have yielded positive responses from students who believe the content should be shared with their classmates to supplement the way health questions are asked and answered (Abbott & Abboud, 2006; Antolin et al., 2012; Harris & Malyango, 2005; Hidaka et al., 2015). In fact, since 2015, 20% of exam questions in the biology section of the MCAT (medical school entrance exam in the United States) are on evolution ("What's on the MCAT2015 Exam?" 2015).

In Table 14.2, we propose a change model/template for curricular development and research based on a Tinbergean framework that can be introduced throughout the health professions' curricula to promote the creation of an ecologically and evolutionarily informed approach to solving medical issues.

TABLE 14.2. Example with Retrospective and Prospective Application of the Change Model to Congestive Heart Failure

EVOLUTIONARY MEDICINE	STATE OF CHF IN 1986	CURRENT STATE (2016)	FUTURE VISION
Context/Culture	CHF seen as a "one-organ system" problem. Emphasis on the failing ventricle as the sole issue. Therapies focused on improving contractile state and pumping function of the ventricle.	Marked reduction in CHF mortality related to shift in perspective focusing on a model of adaptation/maladaptation with respect to neuroendocrine and other systems.	Development of effective prevention strategies informed by evolutionary concepts.
Structures	Research conducted in departmental scientific silos.	Comparative research identified conserved CHF-related neuroendocrine activation in animal models.	Collaborative research teams including eco-evolutionary experts along with multiple other relevant teams.

(*continued*)

TABLE 14.2. Continued

EVOLUTIONARY MEDICINE	STATE OF CHF IN 1986	CURRENT STATE (2016)	FUTURE VISION
Research Processes	Reductionist, proximate studies of disease causation.	Research increasingly moving past the proximate issues (how the ventricle was injured) to the maladaptive physiologic responses to the failing ventricle.	Novel hypotheses developing from comparative, phylogenetic, and adaptive models. Research questions emerging from an eco-evolutionary framework.
Investigator teams/ Research Education, Training and Career Development	Discipline driven, limited scientific scope.	Insights into CHF pathophysiology has come from the integration of diverse teams representing cardiology, nephrology, neurology, endocrinology, and animal physiology.	Research teams created with a diversity of fields represented will ask more evolutionarily-informed questions and construct novel hypotheses based on eco-evolutionary framework.
Outcomes, Enduring impact	Combat disease following physiological deterioration; treating disease in specific, trait-based view.	Marked reduction in mortality and morbidity secondary to enhanced accuracy of models of disease leading to effective pharmaceutical and device therapies. Prevention strategies structured around targeting Maladaptive responses.	Development of prevention strategies structured around targeting maladaptive responses.

Note: CHF = congestive heart failure.

Conclusion

There are tools and pathways that can be provided to medical educators and professionals to help them integrate evolutionary concepts into curricula and practice. The widespread application of eco-evolutionary principles across medical and related scientific disciplines can be achieved through a change model we have proposed. On a more granular level, clinician-educators and practitioners can immediately expand their perspectives by applying a Tinbergean lens to the medical and psychiatric conditions.

However, applying a Tinbergean perspective alone is not sufficient to realize the true potential of the field of evolutionary medicine to change health outcomes. Free and open inquiry between disciplines will be necessary to generate novel insights about the essential nature of disease as well as new therapies. Bringing together biomedical researchers with those traditionally not involved in biomedical research—including ecologists and evolutionary biologists—is necessary to create a pipeline for diagnostic and therapeutic discovery and innovation. Essentially, this approach requires more than simply changing the curriculum itself; it requires changing a culture of how biomedical research is conducted and who participates in it.

There are significant barriers to change within disciplines and institutions. An important first step to lowering these barriers is creating a shared vocabulary among participants with very different training. How can this be done?

We recognize that bringing clinicians, investigators, ecologists, evolutionary biologists, and students together is often challenging in highly siloed educational settings. Physicians and biomedical researchers who also have clinical duties are busy and often not co-located with ecologists and evolutionary biologists. This alone makes it difficult to create a shared culture through journal clubs and shared seminars. Some approaches that have worked to promote communication and collaboration include featuring evolutionary biologists as guest speakers at medical Grand Rounds and embedding evolutionary biologists into working rounds with medical students, interns and residents. We have trialed these activities successfully through UCLA's Evolutionary Medicine Program. Additionally, the development of undergraduate minor and graduate programs in evolutionary medicine at UCLA has equipped prehealth students with an evolutionary foundation in understanding health and disease. These examples should be viewed as suggestions, and all require proper and formal evaluation of their efficacy.

Having ecologists and evolutionary biologists give and attend Grand Round lectures is an important opportunity to create a shared mission. Because ecologists and evolutionary biologists use functional and contextual approaches to solve problems on a daily basis, they must share these disciplinary tools with clinicians and investigators. These disciplines offer novel tools for biomedical discovery and by sharing their insights and perspectives we can begin to create a new, shared vocabulary among future collaborators. However, to be most effective at sharing their toolkit, they also must understand the unanswered questions that clinicians have. Joint attendance where the goal is to create a shared vocabulary is essential.

Ecologists and evolutionary biologists have much to learn from current medical education practices. Among those are morning reports and rounds. Here, clinicians identify problems and solutions to difficult cases and, by doing so, teach the next generation of physicians. This didactic approach can also be used to teach ecologists and evolutionary biologists how to think like a doctor, which is an essential trait for truly meaningful collaborations. Specifically, bringing ecological and evolutionary principles to the medical challenges encountered in such diverse fields as cardiology, psychiatry, infectious disease. and oncology will result in advances in investigation and clinical practice and outcomes in these fields.

We believe that meaningful collaborations and ultimately education can be further improved by creating a grant program that requires investigators at an institution to work across traditional disciplinary lines to create collaborative teams of biomedical researchers, ecologists, and evolutionary biologists. We believe that an ecologically and evolutionarily transformed way of approaching research questions can enhance curricula within the health sciences. This, in turn, can generate superior hypotheses resulting in an improved understanding of the nature of health and disease and creating a pipeline for future improved healthcare. Evolution can save lives, if its powerful approaches and principles are integrated meaningfully and usefully into health professions curricula.

References

Abbott, A., & Abboud, G. (2006) Evolutionary medicine: A model for medical school introduction. *Medical Education, 40*, 471.

Aktipis, C. A., Boddy, A. M., Jansen, G., Urszula, H., Hochberg, M. E., Maley, C. C., & Wilkinson, G. S. (2015). Cancer across the tree of life: cooperation and cheating in multicellularity. *Philosophical Transactions of the Royal Society B, 370*, 20140219.

Alcock, J., & Schwartz, M. D. (2011). A clinical perspective in evolutionary medicine: what we wish we had learned in medical school. *Evolution: Education and Outreach, 4*, 547–579.

American Cancer Society. (2011). *Global cancer facts & figures* (2nd ed.). Retrieved from http://www.cancer.org/acs/groups/content/@epidemiologysurveilance/documents/document/acspc-027766.pdf

Antolin, M. F., Jenkins, K. P., Bergstrom, C. T., Crespi, B. J., De, S., Hancock, A., . . . Stearns, S.C. (2012). Evolution and medicine in undergraduate education: a prescription for all biology students. *Evolution, 66*, 1991–2006.

Antonovics, J., Abbate, J. L., Baker, C. H., Daley, D., Hood, M. E., Jenkins, C. E., . . . Vondrasek, J. (2007). Evolution by any other name: Antibiotic resistance and avoidance of the E-word. *PLoS Biology, 5*(2), e30.

Beerda, B., Schilder, M. B. H., Janssen, N. S. & Mol, J. A. (1996). The use of saliva cortisol, urinary cortisol, and catecholamine measurements for a noninvasive assessment of stress responses in dogs. *Hormones and Behavior, 30*, 272–279.

Boddy, A. M., Kokko, H., Breden, F., Wilkinson, G. S., & Aktipis, C. A. (2015). Cancer susceptibility and reproductive trade-offs: a model of the evolution of cancer defences. *Philosophical Transactions of the Royal Society B, 370*, 20140220.

Bonne, O., Grillon, C., Vythilingam, M., Neumeister, A., & Charney, D. S. (2004). Adaptive and maladaptive psychobiological responses to severe psychological stress: implications for the discovery of novel pharmacotherapy. *Neuroscience & Biobehavioral Reviews, 28*, 65–94.

Brown, G. E., Ferrari, M. C. O., Malka, P. H., Oligny, M.-A., Romano, M., & Chivers, D. P. (2011). Growth rate and retention of learned predator cues in juvenile rainbow trout: faster growing fish forget sooner. *Behavioral Ecology and Sociobiology, 65*, 1267–1276.

Brown, J. S., & Aktipis, C. A. (2015). Inclusive fitness effects can select for cancer suppression into old age. *Philosophical Transactions of the Royal Society B, 370*, 20150160.

Brown, J. S., Cunningham, J. J., & Gatenby, R. A. (2015). The multiple facets of Peto's paradox: a life-history model for the evolution of cancer suppression. *Philosophical Transactions of the Royal Society B, 370*, 20140221.

Casás-Selves, M., & DeGregori, J. (2011). How cancer shapes evolution and how evolution shapes cancer. *Evolution: Education and Outreach, 4*, 624–634.

Chambers, A. F., Groom, A. C., & MacDonald, I. C. (2002). Metastasis: dissemination and growth of cancer cells in metastatic sites. *Nature Reviews Cancer, 2*, 563–572.

DeVita, V., Jr., & Chu, E. (2008). A history of cancer chemotherapy. *Cancer Research, 68*, 8643–8653.

Ferrari, M. C. O., Brown, G. E., & Chivers, D. P. (2012). Temperature-mediated changes in rates of predator forgetting in woodfrog tadpoles. *PLoS One, 7*, e51143.

Ferrari, R., Ceconi, C., Curello, S., Ferrari, F., Confortini, R., Pepi, P., & Visioli, O. (1996). Activation of the neuroendocrine response in heart failure: adaptive or maladaptive process? *Cardiovascular Drugs and Therapy, 10*, 623–629.

Fidler, I. J. (1978). Tumor heterogeneity and the biology of cancer invasion and metastasis. *Cancer Research, 38*, 2651–2660.

Ghali, J. K., Cooper, R., & Ford, E. (1990). Trends in hospitalization rates for heart failure in the United States, 1973–1986: evidence for increasing population prevalence. *JAMA Internal Medicine, 150*, 769–773.

Gluckman, P., Beedle, A., & Hanson, M. (2009). *Principles of evolutionary medicine*. Oxford, England: Oxford University Press.

Harris, E. E., & Malyango, A. A. (2005). Evolutionary explanations in medical and health profession courses: Are you answering your students' "why" questions? *BMC Medical Education, 5*, art. 16.

Hidaka, B. H., Asghar, A., Aktipis, A., Nesse, R. M., Wolpaw, T. M., Skursky, N. K., ... Schwartz, M.D. (2015). The status of evolutionary medicine education in North American medical schools. *BMC Medical Education, 15*, art. 38.

Jacobson, E. R., Homer, B., & Adams, W. (1991). Endocarditis and congestive heart failure in a Burmese python (python molurus bivittatus). *Journal of Zoo and Wildlife Medicine, 22*, 245–248.

Lechat, P. (2006). The evolution of heart failure management over recent decades: from CONSENSUS to CIBIS. *European Heart Journal Supplements, 8,* C5–C12.

Mayo Clinic. (2016). Diseases and conditions: Heart failure. Retrieved from http://www.mayoclinic.org/diseases-conditions/heart-failure/basics/definition/con-20029801/

McMurray, J. J., & Stewart, S. (2000). Epidemiology, aetiology, and prognosis of heart failure. *Heart, 83,* 596–602.

Mikics, E., Baranyi, J., & Haller, J. (2008). Rats exposed to traumatic stress bury unfamiliar objects: A novel measure of hyper-vigilance in PTSD models? *Physiology & Behavior, 94,* 341–348.

Mitchell, M. A., & Tully, T. N., Jr. (2016). *Current therapy in exotic pet practice.* Philadelphia, PA: Saunders.

Mosterd, A., & Hoes, A. W. (2007). Clinical epidemiology of heart failure. *Heart, 93,* 1137–1146.

Natterson-Horowitz, B., & Bowers, K. (2012). *Zoobiquity: The astonishing connection between human and animal health.* New York, NY: Vintage.

Nesse, R. M., & Williams, G. C. (1996). *Why we get sick: The new science of Darwinian medicine.* New York, NY: Vintage.

The Nobel Prize in Physiology or Medicine 1973: Karl von Frisch, Konrad Lorenz and Nikolaas Tinbergen. (n.d.). *Nobelprize.org.* Retrieved from http://www.nobelprize.org/nobel_prizes/medicine/laureates/1973/

O'Brien D. T., & Gallup A. C. (2011). Using Tinbergen's four questions (plus one) to facilitate evolution education for human-oriented disciplines. *Evolution: Education and Outreach, 4,* 107–113.

Okada, H., Kuhn, C., Feillet, H., & Bach, J. F. (2010). The "hygiene hypothesis" for autoimmune and allergic diseases: an update. *Clinical and Experimental Immunology, 160,* 1–9.

Perlman, R. L. (2013). *Evolution and medicine.* Oxford, England: Oxford University Press.

Read, A. F., Day, T., & Huijben, S. (2011). The evolution of drug resistance and the curious orthodoxy of aggressive chemotherapy. *PNAS, 108*(Supp 2), 10871–10877.

Roger, V. L. (2013). Epidemiology of heart failure. *Circulation Research, 113,* 646–659.

Schocken, D. D., Arrieta, M. I., Leaverton, P. E., & Ross, E. A. (1992). Prevalence and mortality rate of congestive heart failure in the United States. *Journal of the American College of Cardiology, 20,* 301–306.

Stearns, S. C., & Medzhitov, R. (2015). *Evolutionary medicine.* Sunderland, MA: Sinauer.

Tinbergen, N. (1963). On aims and methods of Ethology. *Zeitschrift für Tierpsychologie, 4,* 410–433.

Varki, N., Anderson, D., Herndon, J. G., Pham, T., Gregg, C. J., Cheriyan, M., . . . Varki, A. (2009). Heart disease is common in humans and chimpanzees, but is caused by different pathological processes. *Evolutionary Applications, 2,* 101–112.

What is PTSD? (2016). *National Center for PTSD, US Department of Veterans Affairs.* Retrieved from http://www.ptsd.va.gov/public/PTSD-overview/basics/what-is-ptsd.asp

What's on the MCAT2015 exam? (2015). *Association of American Medical Colleges.* Retrieved from https://aamc-orange.global.ssl.fastly.net/production/media/filer_public/f7/e5/f7e57fb2-44fa-4c00-83dd-c17cee034c47/mcat2015-content.pdf

Yenkosky, J. P., Bradshaw, G. A., & McCarthy, E. (2010). Session 120: Post-traumatic stress disorder among parrots in captivity: Treatment considerations. In *Proceedings of the Association of Avian Veterinarians: Waves of wisdom in San Diego.* Weatherford, TX: Association of Avian Veterinarians

CHAPTER 15 | How Evolutionary Studies Enables People to Think Outside the Box

GORDON G. GALLUP JR., JENNIFER A. STOLZ,
REBECCA L. BURCH, AND JENNIFER A. BREMSER

EVOLUTIONARY THEORY NOT ONLY organizes and integrates a lot of diverse information about human behavior; it represents a means of generating many new provocative and testable predictions about what people would be expected to do under a variety of contemporary circumstances. In this chapter, we illustrate how evolutionary theory enables people to think about things in new and often radically different ways.

Examples include a number of essays we have prepared for this book that provide answers to the following questions. Why was the idea that sex is necessary for procreation a long time in coming? Why is semen sampling a putative mate choice mechanism? Why may pubic hair removal usher in a new generation of pedophiles? Why is the shape of the human penis something that may change the way you think about sex? Why have all foods been genetically engineered? How has artificial illumination created a continuous summer? Why does bottle-feeding put mothers at risk? Why is oral sex an artifact of personal hygiene? Why was Neal Armstrong wrong about the moon? Why does artificial insemination increase the risk of preeclampsia? Why is the risk of conception higher as a consequence of rape? Why does circumcision have adaptive but unanticipated consequences? Why has human evolution not caught up with easy credit? Why are smart phones dumbing us down? And finally, why do smart people gravitate toward evolutionary studies?

We think the answers to these questions will illustrate how evolutionary theory can be used to think outside the box in ways that may seem radically different and in ways that may have important implications for generating solutions to practical problems. And, importantly, this approach to asking and answering questions across a broad array of areas by invoking evolutionary principles is central to the mission of an Evolutionary Studies (EvoS) education.

How Long Have Humans Known about the Connection between Sex and Reproduction?

Too often people make the mistake of thinking that what they know has always been true, and the corollary is that what they do is what humans have always done. Our discussion about sex and reproduction and, elsewhere in this chapter, about pubic hair removal, oral sex, and artificial illumination illustrates just how easy it is to fall victim to these kinds of mistaken assumptions.

Most people take the connection between sex and procreation as a given and so obvious as to be self-evident. However, from a more informed evolutionary perspective, there are plenty of reasons why this connection was not always as obvious as it might seem. Throughout human evolutionary history, engaging in sex was a necessary but certainly not sufficient condition for reproduction. The fact is you do not conceive every time you have sex.

Some people have sex on a regular basis for years before they reproduce, and some people have sex all their lives but remain childless. Although the correlation between sex and procreation is necessarily greater than zero, it is much closer to zero than 1.0. If you and your partner engage in heterosexual intercourse three times a week, week after week, year after year, and decade after decade, by the time you reach 65 you will have had sex approximately 150 times per year for 40 or more years. That works out to roughly 6,000 sexual encounters. If you wind up producing two children, which is typical for many people these days, that gives you a conception to copulation ratio of 1 baby for every 3,000 instances of sexual intercourse. Under such conditions the numerical value of the correlation between sex and reproduction would be so close to zero that it would be trivial, at best. To be due to something other than chance and be deemed statistically significant (i.e., $p < 0.05$), you would have to get pregnant on average of at least once out of every 20 times you had sex. Even factoring in the contemporary availability of various contraceptive measures, the human reproductive rate does not even come close to that. Even if you had as many as five children, you would have had sex an average of 1,200 times for each child. Thus, the bottom line is that the connection between sex and reproduction fails to even approach statistical significance. Consistent with the fact that many people are surprised when they get pregnant, among humans, conception as a consequence of heterosexual intercourse is clearly the exception rather than the rule.

If you are a woman it is also important to realize that even without the use of contraceptives, there are extended periods of time when sex and conception are biologically uncoupled. The likelihood of getting pregnant while you are menstruating is very remote. Impregnation itself promotes hormonal changes that preclude reimpregnation during pregnancy, and breastfeeding leads to lactational anovulation following the birth of a child, which is a hormonal condition that puts the menstrual cycle and ovulation on hold and functions as an evolved birth spacing mechanism (Kennedy & Visness, 1992). Add to this the fact that during human evolutionary history the typical healthy woman spent most of her reproductive life pregnant or breastfeeding. Finally, once a woman reaches menopause, as early as her late thirties or as late as her mid-forties, the capacity to conceive drops to zero.

There are also sex differences in a person's reproductive life span: the amount of time during your life that you are reproductively viable. Assume for the sake of illustration that men begin to produce viable sperm and women begin to release viable eggs at 15 and that nowadays men and women have a life expectancy of 75. Healthy males are often capable of fathering children into their seventies and therefore can be reproductively viable for up to 80% of their lives. Women, in contrast usually reach menopause by age 45 or earlier, and therefore are reproductively viable for only 30 years or less (40%) of their lives.

There is another reason why sex will not suffice when it comes to reproduction. Timing is of paramount importance. For impregnation to occur, insemination must be synchronized with ovulation. There is a very narrow temporal window that constrains when conception will occur. Because sperm viability is short-lived (Gallup, Finn, & Sammis, 2009), insemination must typically occur within a period of plus or minus 12 hours of ovulation for conception to occur. Although sperm in the female reproductive tract may stay alive longer than that, they lose their capacity to achieve capacitation, which is required for fertilization. Combine this with the fact that women no longer produce easily discernable ovulatory cues, and the probability that a random sexual encounter will eventuate in conception is even more remote.

The other aspect of reproduction that obscures the connection between sex and reproduction is that it is not obvious when conception occurs. Conception itself does not typically produce any feedback or sensory cues. Short of modern-day pregnancy test kits, women often had no reason to suspect they were pregnant until they missed one or more periods of menstruation and/or experienced unusual changes in breast sensitivity. Sometimes outward signs of pregnancy do not occur for months and, on rare occasions, not at all. There are documented cases of obese women who have gotten pregnant but have not realized it until they went into labor (Wessel & Buscher, 2002).

Bear in mind also that for 99% of human evolutionary history no one had ever seen a human egg or knew anything about spermatophores. Moreover, consistent with the idea that the connection between sex and reproduction emerged only recently, there are remote groups of primitive people that until recently saw no connection between having sex and having babies (Malinowski, 1929).

What was lacking for most of human evolutionary history was a control group. To demonstrate that sex is a necessary condition for having babies, you need a group of females who are having sex on a regular basis and a comparable group of reproductively viable females who are not. There are reasons to suspect that remaining celibate was simply not an option for most women during human evolutionary history, bearing in mind that human evolutionary history and recorded human history are two very different periods. Recorded human history is but a split second of human evolutionary history. The news is filled with reports of dire consequences (e.g., physical abuse, rape, murder) for women refusing sex in the modern age, and it is very likely that these consequences plus the ramifications associated with abandonment (e.g., no provision for shelter, food, or care for previous offspring in the absence of any protective social programs) would have been magnified in our evolutionary history.

One interpretation is that it was not until the domestication of animals with the opportunity to segregate animals into same sex and mixed sex groups that it became apparent that sex and reproduction were related. The repeated observations of birth following acts of sex between male and female animals with shorter gestational periods than humans would also have provided tangible and more salient evidence than could have been gleaned by examining the human condition alone.

Semen Sampling as an Evolved Mate Choice Mechanism

Evolved mate choice mechanisms abound, and, directly or indirectly, they all put a premium on picking a healthy reproductively viable partner (Gallup & Frederick, 2012). One of the most important ways to gain genetic representation in future generations is to pair your genes with someone who is reproductively viable. Some of these mate choice mechanisms focus on morphological cues that correlate with health and fertility such as bilateral symmetry, low waist to hip ratios in women, and high shoulder to hip ratios in men (see Gallup, Frederick, & Pipitone 2008, for a review). Other cues are more subtle and more likely to operate below the radar.

Take kissing as a case in point. It will come as no surprise that among humans, romantic kissing is part of a hard-wired courtship strategy. What was unanticipated, however, was the discovery that the majority of men and women have found themselves attracted to a member of the opposite sex on one or more occasions, only to discover that after they kiss that person for the first time they were no longer interested (Hughes, Harrison, & Gallup, 2007). Therefore, the first kiss can literally be a make or break proposition. At the moment of the first kiss, there is a very complicated exchange of subtle cues, including postural adjustments, mouth and body odor cues, tactile cues, and even salivary cues. There may be mechanisms operating at a relatively unconscious level that process this information and allow the participants to make a preliminary assessment of the other person's health, fertility, and genetic compatibility in ways that bear on their value as a potential mate.

In many instances, courtship in the ancestral as well as the contemporary environment also involved being inseminated by different prospective mates. Semen is a very complicated concoction of sex hormones, neurotransmitters, endorphins, placental hormones, immune-suppressants, and other ingredients that have evolved to promote the reproductive best interests of males (Burch & Gallup, 2006). In addition to sharing most features in common, there is growing evidence that each man may have his own "semen signature" of sorts. That is, the biochemistry of one man's semen may be unique to that particular man. Some of the best evidence for a semen signature is evidence that shows that a change in paternity is a risk factor for preeclampsia (Davis & Gallup, 2006).

In parallel with kissing, it has been suggested that semen sampling among prospective suitors may also be an evolved female mate choice mechanism (Gallup & Reynolds, 2014). In other words, the composition of any particular man's seminal fluid may contain cues to his health, fertility, and genetic compatibility. As a result, just like kissing, there may be unconscious mechanisms that are activated upon being inseminated and operate to process seminal fluid for cues that bear on whether that man would be a good match.

Gallup and Reynolds (2014) argue that a peculiar condition called "seminal plasma hypersensitivity" may represent the extreme negative end of this continuum. Seminal plasma hypersensitivity resembles an allergic reaction to semen, where genital or even topical exposure to semen leads to swelling, irritation, and even burning and painful side effects in some cases. But unlike an allergic reaction, for many patients this reaction appears specific to the semen of a particular man, rather than a generalized reaction to semen per se. That is, numerous patients who have been inseminated by a number of different men only report an adverse reaction to the semen of a particular male. Although few follow-up studies have been conducted, case histories of couples who have experienced this reaction but persevere in an attempt to get pregnant include histories of spontaneous abortions, neonatal deaths, and an inability to conceive (for details, see Gallup & Reynolds, 2014).

Semen sampling as a mate choice mechanism is a testable hypothesis. A clear prediction that would follow from this hypothesis is that following an instance of insemination a female might either wake up the next morning with feelings of regret and remorse for what she did the night before, or she could wake up head over heels in love.

How Pubic Hair Removal May Usher in a New Generation of Pedophiles

Although fashionable in many quarters these days, pubic hair removal was not an option during most of human evolutionary history. Just because you go to great pains to remove your pubic hair, do not make the mistake of assuming that it was always the case that people were preoccupied with this peculiar practice. Get yourself a copy of an issue of Playboy Magazine published in the 1980s or earlier, and you may be shocked (may be even mortified) to discover that almost without exception all of the nude female models, including the centerfold, had generous amounts of unfettered, unconcealed, and untrimmed pubic hair.

Before we begin to examine some of the implications of pubic hair removal, it would be instructive to go back in time. Why do humans have pubic hair in the first place?

Pubic hair is not an embarrassing accident. Pubic hair, along with underarm hair, chest hair, and facial hair is sexual hair. Sexual hair is body hair that emerges with the onset of puberty and, as such, serves to advertise sexual maturation and fertility. In the ancestral environment when people lacked clothes, it would have been obvious in a glance which females were fertile and which were prepubescent. Sexual hair also functions as an odor trap that captures and distributes sex pheromones that also convey fertility cues. Pubic hair is distinctive because it is curly and coarse. Located at the base of sexual hair follicles in both men and women are apocrine glands. Unlike sweat glands, which are distributed over most of the body and emit water and salt that functions to promote cooling, apocrine glands function differently. Apocrine glands secrete a more complicated oily concoction of hormone-dependent body odor cues that travel up the shank of the pubic hair follicles and around all of the tight kinky curls, to take advantage of the greater surface area of pubic hair follicles. As the chemicals become airborne they broadcast subtle fertility cues to members of the opposite sex. Not surprisingly, men rate vaginal odor cues as more attractive when women are in the fertile phase of their menstrual cycle (Doty, Ford, Preti, & Huggins, 1975), and men also rate female underarm odor cues as more attractive when they are in the fertile phase of their menstrual cycle (Singh & Bronstad, 2001).

Until very recently, pubic hair removal was not an option. Why? The answer is simple. For most of human evolutionary history, we did not have scissors, razors, lasers, or bikini wax. We did not even have tweezers that could be used to pull those pesky pubic hair follicles out by the roots. Therefore, during human evolution men who were attracted to females without pubic hair would have been pedophiles (Gallup & Frederick, 2010) and those who mated preferentially with women sporting pubic hair would have left more descendants. Prior to pubic hair removal, the principal distinction between the genitals of a child and a mature woman would have been the presence or absence of pubic hair.

There is growing evidence that some male sexual preferences and even fetishes may be a byproduct of early imprinting-like sexual experiences that occur during the onset of puberty (Gallup, 1996). A growing proportion of adolescent males in the Western world now rely almost exclusively on the Internet to access pornography for purposes of achieving sufficient sexual arousal to masturbate and experience sexual gratification. Since it has become increasingly popular nowadays for women to remove their pubic hair, almost all of the contemporary heterosexual video pornography and the Internet sites that feature these films exclusively depict naked women without any pubic hair. Therefore, this may be the occasion for pubic hair removal to usher in a new generation of male pedophiles imprinted on hairless female genitalia. Indeed, as a testable implication of this hypothesis, we would predict that a growing number of adolescent males nowadays would show evidence of increased blood flow to the penis in response to seeing pictures of nude prepubescent girls.

How the Shape of the Human Penis May Change the Way You Think about Sex

The penis evolved as an internal fertilization device. The human penis, with a relatively larger glans and more pronounced coronal ridge than is found in many other primates, may also function to compete with rival male semen and displace seminal fluid from other males in the vagina by forcing it back over/under the glans. During intercourse the effect of repeated thrusting would be to draw out and displace foreign semen away from the cervix. As a consequence, if a female copulated with more than one male within a short period of time this would allow subsequent males to "scoop out" semen deposited by others before ejaculating (Baker & Bellis, 1995). Gallup et al. (2003) simulated sexual encounters using artificial models and found that the displacement of simulated semen was robust across different prosthetic humanlike phalluses (but only those with a pronounced glans), different artificial vaginas, different semen recipes, and different semen viscosities.

Ancestral males would have been intact (uncircumcised). If the penis functions as a semen displacement device, as just discussed, seminal fluid left by other males in the vagina would collect on the penile shaft of subsequent males. This could easily result in rival male ejaculate being trapped under the foreskin. Depending on the intercopulatory interval, it would be possible for uncircumcised males to transfer displaced seminal fluid unwittingly trapped under their foreskin to their next sexual partner.

Imagine the following scenario: An uncircumcised man (Male B) has sex with a woman (Female A) who recently had sex with another man (Male A). In the process of thrusting his penis back and forth in her vagina, some of Male A's semen is forced under Male B's frenulum, collects behind his coronal ridge and is displaced from the area proximate to the cervix. After Male B ejaculates and substitutes his semen for that of Male A, as he withdraws from the vagina some of Male A's semen will be present on the shaft of his penis and behind his coronal ridge. As his erection subsides, the penis withdraws under the foreskin, raising the possibility that some of Male A's semen could be captured underneath the foreskin in the process. Were Male B to then have sex with Female B several hours later, it is possible that some of the displaced semen from Male A would still be present under his foreskin and thus may be unwittingly transmitted to Female B who, in turn, could then be impregnated by Male A's sperm. Were Male B circumcised, this would be a far less likely outcome because the residual foreign sperm on his penis would not be afforded the protection of the foreskin from desiccation, light, and cooling and would likely perish during the interim separating sexual encounters. This is called the "Piggybacking hypothesis" (Gallup & Burch, 2004) and is also known as fertilization by proxy in other species (Haubruge, Arnaud, Mignon, & Gage, 1999; Tigreros, South, Fedina, & Lewis, 2009).

If your sex partner is an uncircumcised male, you are at risk of being inseminated by the man who had sex with your partner's previous partner. Or to put it another way, if your partner is not circumcised you could be having sex with the men who had sex with your partner's previous partner.

If the foreskin makes the human penis a vector for fertilization by proxy, why is the foreskin still there? The prepuce, or foreskin, has been present in primates for at least 65 million years and is likely to have existed for at least 100 million years based on its preponderance in mammals (Martin, 1990). We assume that during human evolutionary history the incidence of self-cuckoldry was not high enough to offset the advantages of the foreskin, which affords protection of the glans. Interestingly, the prepuce is normally at least partially adhered to the penile shaft throughout the childhood, with separation occurring for most males before age 13 and by age 17 at the latest (Cold & Taylor, 1999). This may indicate the importance of the protective properties during childhood and the separation for proper sexual functioning by puberty. Indeed the innervation of the prepuce (some 50,000 nerve endings) could greatly assist in rapid stimulation and sexual response in males that could aid in reproduction. The separation of the prepuce and the mucosal lining has been shown to lubricate the penis and aid in erection and penetration (Fleiss, Hodges, & Van Howe, 1998).

This discussion generates an interesting question: have the adaptive problems posed by the existence of piggybacking semen from rival males led to adaptations that incapacitate foreign sperm? The immunological properties of the prepuce do not become activated until the prepuce is retractable (Cold & Taylor, 1999), which coincides with puberty and the beginning of sexual activity. It appears that the immunological response would be important for sexual health/prevention of urinary tract infections.

Although research on the immunological properties of the foreskin or prepuce has yet to be conducted in earnest, Fleiss et al. (1998) argue that the prepuce performs important immunological functions but only after the foreskin separates from the penile shaft.

Even with the immunological functions of the foreskin, researchers have shown that circumcision can lead to at least some protection from various sexually transmitted diseases (STDs). Gray et al. (2007) found that circumcised men were at reduced risk for HIV acquisition, as did Baeten et al. (2005), Gray et al. (2000), Lavreys et al. (1999), Reynolds et al. (2004), and Weiss, Quigley, and Hayes (2000). When examining human papilloma virus (HPV), there are a number of conflicting studies, most likely due to the various strains of HPV and various genital sites being examined, but many have found a decreased risk in circumcised males (Auvert et al., 2009; Giuliano et al., 2009; Nielson, Schiaffino, Dunne, Salemi, & Giuliano, 2009). Van Buskirk et al. (2011) found that whereas the likelihood of HPV acquisition did not differ by circumcision status, uncircumcised men were more likely than circumcised men to have infections detected at multiple genital sites, which may have implications for HPV transmission.

Some observational studies suggested that male circumcision significantly decreased the incidence of herpes simplex 2 (HSV-2; genital herpes) infection (Auvert et al., 2001; Weiss et al., 2001). Similarly, observational studies showed that male circumcision decreased the incidence of syphilis (Bwayo et al., 1994; Cook, Koutsky, & Holmes, 1994). Tobian et al. (2009) ultimately found that male circumcision significantly reduced the incidence of HIV infection, incidence of HSV-2 infection and the prevalence of HPV infection but did not affect syphilis infection. However, Weiss Thomas, Munabi, and Hayes (2006) in a large scale meta-analysis of male circumcision and ulcerative sexually transmitted infections strongly indicated that circumcised men are at lower risk of chancroid and syphilis and showed a weaker but still significant association with HSV-2.

In short, circumcision, while decreasing sensation, significantly reduces the probability of acquiring STDs or transporting displaced ejaculate to other partners. A useful diagram that identifies the different features of the human penis discussed in this section and elsewhere in this chapter (e.g., glans, coronal ridge, frenulum, smegma) can be found at this link: http://www.luckymojo.com/faqs/altsex/penis.html.

Think Again, All Food Is Genetically Engineered

Although many people nowadays are concerned about the prospect of eating so-called genetically modified food (GMOs) from an evolutionary perspective such concerns are unfounded.

Make no mistake about it, all animals and plants have been genetically modified and engineered by their evolutionary history (i.e., natural selection). With the domestication of animals and plants, humans unwittingly became involved in this process. For example, during domestication chickens have been selectively bred for egg production, cows for milk production, pigs for meat production, and different crops have been selectively bred for better yields, higher protein, more fructose, and resistance to mold and fungus. Indeed, it would be difficult to find any foods in contemporary supermarkets that have not been affected by this kind of beneficent genetic engineering.

It can even be argued that humans have unwittingly been involved in genetically engineering each other. When it comes to mate choice, women who hold out for more attractive, intelligent and stronger men, men who excel at competing with other men for scarce resources, and men who are "tall, dark, and handsome" are participating in the process of unknowingly attempting to feature and fine tune such traits among their future descendants through selective breeding. This is known as the "sexy son hypothesis" (Fisher, 1930).

What upsets a lot of people is not domestication and selective breeding per se but attempts to more directly intervene into the process by manipulating and inserting genes. This boils down to a matter of selection based on random genetic mutations, which is what characterizes natural selection, as opposed to a process based on attempts using biotechnology to promote a particular feature such as meat that is high in iron and protein but low in adverse health-threatening components such as fat and low-density lipoprotein cholesterol.

Not only is concern about genetically engineered food unfounded from an evolutionary perspective, there is emerging evidence that such concerns have been promoted for political and financial purposes. Over 100 Nobel Laureates recently issued a letter blaming Greenpeace for ignoring the evidence in favor of GMOs and deliberately trying to scare people in an effort to raise money for their organization (Achenbach, 2016). Here is an excerpt from that letter.

Scientific and regulatory agencies around the world have repeatedly and consistently found crops and foods improved through biotechnology to be as safe as, if not safer than those derived from any other method of production. There has never been a single confirmed case of a negative health outcome for humans or animals from their consumption. Their environmental impacts have been shown repeatedly to be less damaging to the environment, and a boon to global diversity.

Neil Armstrong Got It Wrong

When the first person set foot on the moon it was not simply "a small step for man, and a giant leap for mankind." It was far more profound than that. The moment Neil Armstrong stepped off the space capsule on to the moon, it represented an instance of coming face to face with our maker.

An evolutionary enigma is how life on planet Earth, which started in the ocean, ever managed to evolve to bridge the gap between the aquatic domain and eventually acquire all the necessary adaptations required to exploit the terrestrial domain and take up life on land. The problem applies to both plants and animals, but the gap seems greater for animals. Consider just a brief laundry list of some of the requirements. In contrast to water, mobility on land requires a new and diverse set of both morphological and behavioral adaptations. The same is true for exacting oxygen from the environment to support metabolism, not to mention thermoregulation, where the thermal inertia of water is a buffer against dramatic day-to-day or even hour-to-hour fluctuations of temperature on land. Sexual reproduction is also affected, as evidenced by the fact that many aquatic creatures, such as spawning fish simply dump their genetic material into the environment in a way that some of the sperm from the male will come in contact with the eggs of the female to achieve fertilization. Add to that adaptations to conserve internal fluids and minimize water loss, as well as adaptations to exploit and subsist on new sources of food, and it should be apparent that one random mutation or even a bunch of accidental mutations would not suffice to enable aquatic organisms to bridge the gap.

The moon may have been the solution to this problem. The prevailing view is that the moon was once part of planet Earth. The origin of the moon is thought to be a result of a cataclysmic asteroid or even planet-to-planet collision where a huge piece of the planet was torn off and went into orbit around what was left of planet earth. Geologists who have examined moon rocks and other material astronauts brought back from the moon feel that what we now know about the composition of the moon is consistent with this theory (e.g., Wiechert et al., 2001). But, whatever the origin of the moon might have been, the moon may be the key to the emergence of life on land. The moon provided the opportunity by which these changes could have occurred.

As the moon orbits the earth, it exerts a small gravitational pull on the surface of the planet, and this is reflected in tidal movements that characterize large bodies of water. These tidal movements create tidal zones where the ocean meets the land that are defined by the difference between high tide and low tide. The existence of regularly recurring tidal changes at the interface between land and the ocean has been a pervasive and highly consistent feature of the planet for billions of years.

These tidal zones along with persistent tidal pools represented an opportunity for sea animals and plants to gradually grow accustomed to the otherwise stark differences between land and water and would have given aquatic creatures the opportunity to gradually acquire the necessary mutations and adaptive changes required to eventually make a relatively clean break with the aquatic world and take up life on land.

In all likelihood, we would not be here if it was not for the moon.

Why the Risk of Conception Is Higher as a Result of Rape

Although it may seem counterintuitive, there is growing evidence that nonconsensual sexual encounters are associated with a higher likelihood of conception than are consensual sexual encounters. Gottschall and Gottschall (2003) report a conception rate of 6.42% in nonconsensual sexual encounters. Indeed, when adjusted for intrauterine device (IUD) and oral contraceptive use, this percentage increases to 7.98%. The US national rape-related pregnancy rate is 5.0% per rape among victims of reproductive age (aged 12–45); among adult women (over 18 years) an estimated 32,101 pregnancies result from rape each year (Holmes, Resnick, Kilpatrick, & Best, 1996; see also Stewart & Trussell, 2000).[1]

---

[1] These estimates do not include women under the age of 18 or women who do not report the assault, thereby creating conservative estimates, at best (Holmes et al., 1996; Stewart & Trussell, 2000). Indeed, 60% of all rapes involve girls younger than 18 years of age (Holmes et al., 1996). A national study in the Netherlands with a sample of 6,000 rape survivors found a pregnancy rate of 7.0% (de Haas, van Berlo, Bakker, & Vanwesenbeeck, 2012). Chu and Tung (2005) found a 5.8% pregnancy rate among rape survivors in Hong Kong. Ohayi et al. (2015) found a lower incidence of 3.3% in Nigeria, but the age of this sample averaged just 13 years and therefore may contain females who were not yet reproductively viable (the average age of menarche in Nigeria was 13.94 years; Abioye-Kuteyi et al., 1997).

In contrast, the pregnancy rate for consensual, unprotected intercourse is calculated at 3.1% (Wilcox, Dunson, Weinberg, Trussell, & Baird, 2001). The risk of becoming pregnant as a result of being raped is therefore considerably higher than as a consequence of having consensual sexual intercourse according to Gottschall and Gottschall (2003) and Wilcox et al. (2001). While these differences are already dramatic, the section How Long Have Humans Known about the Connection between Sex and Reproduction? in this chapter makes it clear that the estimate by Wilcox et al. (2001) is an overestimation of conception rates as it does not take into account periods of pregnancy, lactational anovulation (the hormonal suppression of ovulation during breastfeeding), or secondary amenorrhea (loss of menstruation due to low fat stores).

The data showing an increased likelihood of conception following rape as compared to consensual intercourse are all the more compelling when one examines the behavior of fertile females. Women are more sensitive to perceived sexual coerciveness when they are in their fertile phase (Garver-Apgar et al., 2007) and appear to avoid situations that may increase the risk for rape. Because females appear to behave in ways that minimize conception as a byproduct of being raped (Chavanne & Gallup, 1998; Petralia & Gallup, 2002), the increased risk of conception is even more anomalous. In fact, Broder and Hohmann (2003) replicated the study of Chavanne and Gallup (1998) and found that females specifically reduced risky behaviors but not nonrisky behaviors when they were most fertile. Taken together, these data make it even more incongruous that rape would result in higher pregnancy rates.

The key to understanding this phenomenon may be contained in the title of Jochle's article; "Coitus-Induced Ovulation." Induced ovulation is a well-documented phenomenon in other species and has traditionally been thought to be the result of the tactile stimulation of coitus. However, research on induced ovulation shows that it is the seminal fluid that induces ovulation, specifically a stimulator of luteinizing hormone (LH; Adams & Ratto, 2013). These ovulation induction factors (OIFs; bioactive prohormone forms) are absorbed through the female reproductive track into the bloodstream and have been found in several species (Bogle, Ratto, & Adams, 2011; Chen, Yuen, & Pan, 1985; Pan et al., 1992; Ratto, Huanca, Singh, & Adamset, 2006). Jochle (1973) reviews induced ovulation in other species as well as the controversy concerning its existence in humans. Jochle describes the compounds in human seminal fluid that may be responsible for ovarian changes as including LH, prostaglandins, and other compounds from the prostate.

Gallup, Burch, and colleagues have argued that human seminal fluid evolved to manipulate female psychology and physiology (Burch & Gallup, 2006, Gallup, Burch, & Petricone, 2012; Gallup, Burch, & Platek, 2002). The following ovulatory hormones are present in human semen: estrone (Ney, 1986), estradiol (Luboshitzky, Shen-Orr, & Herer 2002; Ney, 1986), follicle-stimulating hormone, LH (Ney, 1986), and LH-releasing hormone (Chan & Tang, 1983). Of these, LH-releasing hormone is the most obvious OIF, acting in physiologically similar ways to the previously listed OIFs. Human males also have five times more LH in their semen than circulating blood levels in women. Therefore, induced ovulation in humans could be due to LH, LH-releasing hormone, another ovulation-inducing factor, or all in combination.

This hypothesis could be tested in two ways. The first approach would be to compare semen assays using semen collected from rape kits with semen obtained from sperm donation. The second, and more compelling, method would be have males provide semen samples while watching pornographic videotapes that contained either consensual or nonconsensual elements. You could then conduct semen assays to determine whether semen collected while watching nonconsensual pornography contained higher levels of LH and follicle-stimulating hormone FSH (Gallup et al., 2012).

Burch and Gallup also argue that these semen parameters may change as a function of the context in which the ejaculation occurs. Baker and Bellis (1993) have shown that sperm counts increase exponentially when males were unable to monitor their partners. There are currently no data on whether semen chemistry shifts with jealous or coercive contexts, but future research may determine whether the chemistry of ejaculate released during rape differs from that released during a consensual encounter. There may also be individual differences in addition to contextual differences. Dekker, Tubbergen, Valk, Althuisius, and Lachmeijer (1998) found that semen parameters vary between individuals; this could implicate another mechanism by which certain males may be more capable of inducing ovulation. There may also be a combination of these mechanisms, where sexually disadvantaged males with few mating opportunities are more fertile.

Artificial Light Throws Shade on Natural Processes

Humans, like all other living creatures on the planet evolved in the presence of the natural light–dark cycle of day and night. Indeed, for much of our existence, the rhythm of our day-to-day life would have been tied to the daily and yearly (i.e., seasonal) relation of the Earth to the Sun. Although our sense of taste, touch, smell, and hearing would not have been affected much when the sun went below the horizon, our visual world would have drastically changed after sunset. Having no light by which to see would have impacted our ability to navigate the environment, to find food and water, and to see and evade predation, for example. As such, individuals with a penchant for sleeping, or at least remaining quiet and still, during this time likely would have been more successful than those who attempted to continue on as during the day (Stear, 2005).

Recently, however, through the invention of artificial illumination, humans have mastered the ability to create an essentially endless day. No longer must activity stop and quiet time begin when the sun goes down. The visual world no longer provides the same constraints on human behavior as it once did. What are the consequences of this drastic change? First, as noted by Kanazawa and Perina (2009), higher IQ individuals tend to stay up later than their lower IQ counterparts. This extra waking time could be filled with other "unnatural" behaviors that are more likely to be carried out by higher IQ individuals, such as educational pursuits, sexual activity, and drug use.

Although an important difference in general scheduling between higher and lower IQ individuals has become apparent in the light of the artificial day, there is growing evidence that artificial illumination has had an impact on all humans. It is known that we sleep about an hour less per day than we did just 100 years ago (Nadelson, 2001), and probably two to three hours less than we did prior to the advent of artificial lighting. Sleep is increasingly been shown to be important for both physical and mental health and has been tied to ailments such as certain cancers, gastro-intestinal distress, sleep disturbances, cardiac issues, as well as many psychiatric disorders (e.g., depression and anxiety; Sigurdson & Ayas, 2007). An important and easily highlighted case of individuals affected by artificial illumination is the shift worker. These individuals often work exclusively during the night or sometimes have to alternate between day and night shifts. Their general well-being is worse than that of daytime workers, and they are more likely to be involved in both workplace accidents and accidents outside of the workplace, and they are more prone to develop cancer than are day workers (Schernhammer, Kroenke, Laden, & Hankinson, 2006).

Even if you still work during the day and sleep at night, artificial illumination has created the equivalent of a continuous summer. People who sleep 8 hours a day are exposed to 16 hours of light every day, which used to be only typical of the summer but is now true whether its spring, summer, winter, or fall. Humans also show patterns of seasonal breeding sensitive to a latitudinal gradient that peak during the summer, as evidenced by the fact that more full-term babies are born in the spring than during other seasons of the year for babies born in areas north of the equator (Martinez-Bakker, Bakker, King, & Rohani, 2014). By emancipating ourselves from seasonal changes in the photoperiod and living under lighting conditions that used to be restricted to the summer, we may have unwittingly created some of the conditions that may be conducive to patterns of sexual hyperactivity.

Another recent, almost exclusively human (and domesticated animal) problem is obesity, and this is negatively associated with sleep duration (Knutson, Spiegel, Penev, & Van Cauter, 2007). Lack of sleep, per se, likely sets off a chain of physiological responses, one of which is to store fat. Another contributing possibility, however, derives not just from the effects of lack of sleep but also from the increased opportunity to eat and the decreased likelihood of exercise during wakeful periods spent in artificial light. That is, it is likely to be the case that if we are spending more hours awake than our ancestors spent, we are spending some of them eating calorie-laden food that our ancestors did not eat (e.g., snacking in front of the TV or computer). Artificial lighting opens up more opportunity for increased hours in a lit environment but would not necessarily allow us to spend more time outdoors exercising.

It is important to note that humans are not the only species that have been affected by the development of artificial illumination. Essentially, no ecosystem that exists in proximity to man-made lighting is exempt from the consequences. Artificial light has been harnessed in green houses to grow food and plants in places (and/or during times) where it normally would not be possible to do so. Moreover, even in cases in which artificial light is not purposefully used to extend or modify the growing life of a plant, the presence of artificial light for human use has had drastic effects on wildlife surrounding it. For example, the presence of well-lit skyscrapers has been shown to be extremely detrimental to birds. Migration patterns can be altered by the presence of extra light, and many birds have become disoriented and killed when flying into lit buildings at night (Loss, Will, Loss, & Marra, 2014).

Therefore, although the advent of around-the-clock lighting has been a boon to humans, it is not without its problems. Artificial lighting has contributed to an increasing number of problems in not just human life, but all life, and the outcomes of these new environmental pressures will continue to be experienced for as long as the light shines.

The most obvious advice is to take advantage of the benefits of artificial illumination but to do so armed with the knowledge that artificial lighting has the power to be disruptive as well. By being mindful of this, it is possible to take measures to ensure that conditions conducive to sleep, most notably adequate darkness for an ample period of time, are created and maintained. This applies not only to fixed light sources, as recent research has demonstrated that the list emanating from handheld devices such as tablets and cell phones can be detrimental to sleep as well (Chang, Aeschbach, Duffy, & Czeisler, 2015).

Why Does Artificial Insemination Increase the Risk of Preeclampsia?

Artificial insemination, just like artificial sweeteners, was not a possibility over the course of human evolutionary history. But sometimes an evolutionarily recent phenomenon, such as artificial insemination, can trigger a mechanism more deeply rooted in the conditions that prevailed over our ancestors' existence. Just as artificial sweeteners can temporarily satisfy our cravings for sweet foods, artificial insemination may at a biological level, simulate the conditions of rape. Consider that the victim's single exposure to the perpetrator's semen characterizes many, but not all, instances of rape. Incidentally, a single exposure to donor semen characterizes many, but not all, instances of artificial insemination. Therefore, in terms of the frequency of exposure to semen, artificial insemination unwittingly resembles the conditions that would have been consistent with rape.

Today, and throughout ancestral conditions, rape represents a serious threat to a woman's reproductive resources. Rape precludes consent, a deliberate decision to invest her reproductive resources in a particular mate. Without such consent, a woman forgoes the important variable of mate choice. In addition, most instances of rape would preclude any future contribution from the perpetrator in terms of child-rearing, provisioning, or protection of the woman or for her offspring. Given that rape results in conception more often than consensual sex (Gottschall & Gottschall, 2003), adaptations to avoid rape and prevent pregnancies from occurring from rape would have been at a premium over the course of our evolutionary history. The lack of paternal investment poses risks to both mother and her offspring. Without the added protection provided by a committed male, the women would be more vulnerable to predation. Without provisioning afforded by the male, it would be more difficult to provide critical resources for her and her offspring's survival. In addition, the time and resources the mother invests in her offspring are at the expense of time that could be invested in securing a committed mate.

Investment from the father has been shown to lead to more positive outcomes for the offspring as well. For example, the risk of preterm delivery and low birthweight is reduced when the father is involved (Multale, Creed, Maresh, & Hunt, 1991; Oakley, 1985; Ramsey, Abell & Baker, 1986; Turner, Grindstaff, & Phillips, 1990). The mere presence of the father's name on the birth certificate decreases infant mortality in the first year of life by 2.3 times (Gaudino, Jenkins, & Rouchat, 1999).

Clearly, the lack of paternal investment is associated with less positive outcomes. Accordingly, the female may have evolved biological mechanisms to distinguish a committed pair bond from an uncommitted one. Davis and Gallup (2006) have argued that semen familiarity may be one such mechanism. Frequent insemination by the same male over a period of time would have been a proxy for a committed pair bond and, consequently, an index of the likelihood of receiving benefits that go along with such bonds.

Preeclampsia is a disorder of pregnancy that occurs between the 14th and 16th week of pregnancy and characterized by high maternal hypertension and excess protein in the urine. The modern methods of blood pressure regulating medications, pregnancy induction, and/or Caesarian section were not available options during most of our evolutionary history; therefore, preeclampsia could be considered a leading cause of infant and maternal mortality (Duley, 2009). Although the exact pathogenesis of preeclampsia is not known, the disease is characterized by a failure to achieve the second phase of implantation occurring at the end of the first trimester. This second phase of implantation is a unique feature of great ape and human evolution that evolved to provide increased blood flow important for fetal brain development (Robillard, Hulsey, Dekker, & Chaouat, 2003).

Preeclampsia is more common in nulliparous women (those who have never given birth before) and mothers of multiples, in which the demand for maternal resources is greater than in singleton pregnancies (Coonrod, Hickok, Zhu, Easterling, & Daling, 1995). However, there is a third class of risk factors for preeclampsia that is associated with semen familiarity.

Women who use barrier methods of contraception (e.g., condoms) are twice as likely to develop preeclampsia, even after controlling for a number of other potentially confounding factors including marital status, smoking, and family history of the disease (Klonoff-Cohen, Savitz, Cefalo, & McCann, 1989). Similarly, women who engaged in oral sex with their partners prior to pregnancy had significantly lower levels of preeclampsia (Koelman et al., 2000).

Another ostensible measure of exposure to, and therefore familiarity with, their partners' semen is length of cohabitation prior to pregnancy. Other things being equal, women with a prolonged period of cohabitation with their partner are less likely to develop preeclampsia (Marti & Herman, 1977). A change in paternity is another obvious sign of a change in semen familiarity. Women who have previously given birth (multiparous women) are generally at a reduced risk of preeclampsia. The protective effect of parity, however, is lost with a change in paternity (Trupin, Simon, & Eskenazi, 1996; Robillard et Al., 2003).

Artificial insemination from a sperm donor is another category of unfamiliar semen that elevates the risk of preeclampsia, but it is important to note that the increased risk of preeclampsia is diminished when women are artificially inseminated with their committed partner's sperm (Hoy, Venn, Halliday, Kovacs, & Waalwyk, 1999; Smith, Walker, Tessier, & Millar, 1997). Why does artificial insemination increase the likelihood of preeclampsia? In short, along with the other previously described, conditions, it simulates a reproductive situation in which investment from the father is uncertain or unlikely.

Semen familiarity may enhance the conditions required to support and maintain pregnancy, and preeclampsia may develop in the absence of repeated insemination of the father. Human semen contains the cytokines interkeukins 6 and 8, which are related to mechanisms that regulate implantation and fetal maturation (Maegawa et al., 2002). In addition, exposure to the prostaglandins present in semen may lessen the immune response of mother to fetus (Gallup & Burch, 2006). Robertson and colleagues have found that transforming growth factor-β and other cytokines and prostaglandins culminate in improved endometrial receptivity to the implanting embryo. Cytokines have embryotropic properties and also contribute directly to the optimal development of the early embryo (Robertson, 2005). T-reg cell-inducing cytokine TGF and male alloantigens present in seminal fluid delivered at coitus are sufficient to induce a state of "active immune tolerance" to paternal alloantigen. This seminal fluid priming also impacts fetal survival in later gestation (Robertson, Guerin, Bromfield, Branson, Ahlstrom, & Care, 2009).

Semen unfamiliarity, in contrast, inadvertently simulates the difficult ancestral conditions that went hand in hand with a lack of paternal investment. Thus, the use of evolutionarily recent medical advances such as artificial insemination by women trying to conceive may unwittingly increase the risk of preeclampsia.

Why Oral Sex in Humans Is an Artifact of Personal Hygiene

Because of the widespread practice of cunnilingus and fellatio and the recent trend toward analingus (oral stimulation of the anus), many people assume that oral sex was a common feature of human sexual history. However, as detailed by Gallup et al. (2012), it is clear that oral sex was not prevalent during most of human evolution. Oral sex has only recently become a widespread practice because of effective and widely available methods of personal hygiene. Soap, disinfectants, cleaning agents, deodorants, perfumes, colognes, douches, toilet paper, and other means of odor-masking and odor/germ removal that are taken for granted and widely employed today were not available during 99% of human evolutionary history.

Hundreds of thousands of years ago, the typical genitalia would have been a source of foul odor, producing bacteria and other contaminants that would trigger hard-wired disgust reactions and revulsion. Contained in the vagina of a sexually active woman would have been decomposing seminal fluid from prior sexual encounters, residual menstrual flow, and other vaginal secretions. Given the proximity of the vagina to the urethra and the anus, the contents of the vagina would contain trace amounts of urine and fecal material as well. If that was not enough to be a sufficient deterrent to oral sex, the cleavage of the buttocks of both sexes would have served as a virtual magnet/channel for transferring the collection of sweat and dirt to the genitals as a consequence of the human propensity to assume seating postures on the ground and other less than hygienic places. Anyone who has spent the day in a swimsuit at a beach can attest to the sand's ability to find its way into the human body. All of this would have combined to create a place where most people would not want to put their mouths. Much the same would have been true for the typical penis of a sexually active man as a consequence of the collection and retention of comparable materials captured underneath the foreskin. In addition, the production and collection of smegma underneath the foreskin would trap various bacteria and viruses (Cold & Taylor, 1999; Rawls, Laurel, Melnick, Glicksman, & Kaufman, 1968).

Combined with this lack of hygiene would have been an abundance of pubic hair. Pubic hair has long been thought of as an odor trap for sweat and other secretions (Baker & Bellis, 1995), but given the lack of hygiene, the lack of clothing, and the need to sit and lay on the ground, the hair in the pubic region would also entrap various pests, parasites, and detritus, made all the less inviting by the combination of these materials with the aforementioned bodily secretions. Indeed, consider that humans lost their body hair approximately 1.2 million years ago (Rogers, Iltus, & Wooding, 2004). Prior to that, humans would have been excessively hairy (resembling fur) by modern standards, and this hair would have trapped numerous parasites, food items, and other materials. This is not to say that humans were completely hairless after the "loss" of body hair 1.2 million years ago; they still would have possessed body hair (just not entirely covered with hair), and this hair would have been concentrated on the head and in the underarm regions and the pubic area. It is also important to note that the amount of body hair would have varied with climate, with those in colder environments possessing the genetic variants for more vestigial hair growth (Rogers et al., 2004). In addition, humans did not develop the clothing that would have shielded their genitalia from dirt, mud, insects and other items until 70,000 years ago. In total, this creates a long history of humans with extensive odor traps and little barrier between their skin and various invaders that would have functioned as a deterrent to oral sex.

If you take exception to this hypothesis, we invite you to return to the ancestral mode and refrain from bathing, douching, using toilet paper, brushing your teeth, using deodorants and soap for a month and then see how many people you can get to engage in oral sex with you.

Bottle-Feeding, Postpartum Depression, and the Risk of Autism in Your Next Child

For most women, the decision to bottle-feed an infant is determined by a host of personal, social, and cultural factors. The demands of modern life push women back to work, just weeks after giving birth. The United States is the only industrialized nation that does not provide paid leave to new mothers so that they can recover from pregnancy and birth, bond with their child and meet the around-the-clock demands of feeding and caring for a newborn infant.

The financial strains from unpaid leave means that for many families, returning to work is the only option. Although there are laws in place to ensure women have clean spaces and time to pump breast milk for their children, the rate of women still breastfeeding at one-year postpartum is low. For example, of children born in the United States in 2012, only approximately 30% were still receiving any breastfeeding at 12 months of age (Centers for Disease Control, 2012).

What else might affect a woman's decision concerning breastfeeding? Women who do choose to breastfeed may incur social costs from having to do so in public. Still other women may opt not to breastfeed over concerns about their physical appearance and the way in which breastfeeding may change breast shape and volume.

It is important to recognize that there are many cases in which bottle-feeding is not driven by choice, but by necessity. Many preterm infants being treated in the neonatal intensive care unit must be bottle-fed, as well as infants born to incarcerated women and infants whose mothers are ill, require certain medications, or, in the worst case, infants whose mothers have died. In these cases bottle-feeding provides a way of ensuring survival for infants that may have otherwise perished in the ancestral environment.

What are the consequences of never breastfeeding or terminating breastfeeding early in the infant's development? The absence or cessation of breastfeeding is associated with a number of health outcomes that illustrate the emerging field of evolutionary medicine and the concept of an evolutionary mismatch between our modern environment and the conditions that used to prevail (Gallup, Pipitone, Carrone, & Ledholm, 2010). First, formula is nutritionally inferior to breast milk, and the safety of formula depends on access to clean water, clean bottles, and proper-mixing, all of which may present challenges to people living in developing countries. Other studies have found a link between feeding with formula and a number of negative health outcomes (Heinig & Dewey, 1996).

The costs associated with formula feeding are not born solely by the infant. Bottle-feeding mothers have an increased risk of obesity (Chapman & Perez-Escamilla, 1999) and developing breast cancers during premenopause, when the cancer is typically more aggressive (Steube, Willet, Xue, & Michels, 2009).

The application of evolutionary theory can offer a unique perspective on the unintended consequences the modern convenience of bottle-feeding entails. First, we consider what set of conditions the absence or abrupt cessation of breastfeeding was associated with over the course of human evolutionary history. Probabilistically, the cessation of breastfeeding would be occasioned by the loss of a child. In addition, the cessation of breastfeeding would be followed by hormonal shifts in women, because the stimulus that drove milk production has been removed. In this vein, at a basic biological level bottle-feeding with formula unwittingly simulates child loss. In support of this perspective, Gallup et al. (2010) found that compared to women who breastfed their babies, bottle-feeding mothers were more likely to experience symptoms of postpartum depression. The effect remained robust even after controlling for a number of other factors known to influence depressive symptoms including age, income, education, and the mother's relationship to the father. In addition, bottle-feeding mothers expressed a stronger desire to hold their babies, which parallels a common behavior in nonhuman primates in response to child loss. Primate mothers often persist in carrying the corpse of their infant for prolonged periods of time.

The effects of failing to recognize what insight an evolutionary perspective on bottle-feeding provides for mothers already experiencing signs of postpartum depression is another issue to consider. Many women benefit from treatment with antidepressants. However, the Federal Drug Administration does not recommend the use of antidepressants by postpartum mothers because of the potential transfer of these medications (and their potentially negative effects) to the infant through breast milk (Gjerdigen, 2003). Although most antidepressants are considered safe and only transfer a dose proportionate to 1% of the amount ingested by the mother, many healthcare professionals suggest discontinuing breastfeeding while taking antidepressants. This flawed logic may exacerbate, rather than ameliorate the symptoms of postpartum depression given that formula feeding simulates the conditions of child loss, which is a strong stimulus in its own right for depression.

For women who must spend extended amounts of time away from their infants, pumping breast milk may provide a viable alternative to offer both the nutritional and some of the psychological benefits of breastfeeding to mothers and infants. However, an evolutionary informed logic would further suggest that to maximize the benefit, women pumping milk who attempt to mimic the actual experience of breastfeeding may be at an advantage compared to those who do not. It is here that evolutionary theory can impart practical guidance to women pumping breast milk. For example, the use of sensory cues, such as the baby's scent, sounds, and images may elicit a stronger milk let down response and greater yields than pumping in a sterile environment. Indeed, the ideal situation would be to pump with as many infant-related cues present as possible.

Finally, the benefits of breastfeeding are not limited to mother and infant but may even impact the mother's subsequent offspring. Breastfeeding is associated with lactational anovulation, which is an evolved mechanism that promotes birth spacing (Howie & McNeilly, 1982). Because the demands of pregnancy and lactation drain the mother's protein and energetic stores, early reimpregnation may have negative consequences for mothers and their unborn infants because there has not been ample time to recoup the physical resources required for pregnancy. Consistent with this, recent research suggests inadequate birth spacing is a risk factor for autism. Children conceived within a year or two of their next oldest siblings are three times more likely to develop autism (Chapman & Perez-Escamilla, 1999) than those born further apart. Women who formula-feed their infants trigger mechanisms that make early reimpregnation more likely and, therefore, may inadvertently increase the risk of autism in subsequent children. Limited spacing between births poses other health problems as well including preterm birth, low birthweight, and small size for gestational age. Formula-feeding may therefore unintentionally contribute to the increased risk of these consequences as well (Gallup, Spaulding, & Aboul-Seoud, 2016).

The application of evolutionary theory provides a unique lens to analyze the impact of modern conveniences that put us out of phase with our evolutionary history, such as bottle-feeding, and a context in which to examine their unintended consequences. Although the decision to bottle-feed may be necessitated by the social, financial, and cultural influences associated with modern society, it is important to realize that this decision inadvertently increases the risk of postpartum depression, has health consequences for the mother, and may have even more far-reaching implications for the mother's current and subsequent offspring's health.

Why Circumcision Has Adaptive but Unintended Consequences

The practice of circumcision has largely been perpetuated for hygienic and religious reasons. However, recent evidence shows that surgically removing the foreskin from the penis may promote a number of unanticipated but nonetheless adaptive consequences.

As detailed in our section on human penis morphology, the semen displacement properties of the human penis affords it the potential to piggyback rival male semen from one vagina to the next and lead to self-cuckoldry or what is also termed *fertilization by proxy* (see Gallup & Burch, 2004). Thus, the foreskin would appear to be a vector for self-cuckoldry. It is interesting to note in this regard that the piggybacking hypothesis has recently inspired an intriguing novel entitled *The Hitchhiker's Child* by Robin Baker (2013).

Among circumcised men, the absence of the foreskin leaves any rival male semen behind the coronal ridge and on the shaft of the penis exposed where it is subject to the adverse effects of desiccation, cooling, and light during the period separating one sexual encounter from the next, which would greatly diminish or even eliminate the likelihood of self-cuckoldry.

Not only does circumcision make men less likely to be cuckolded by transferring rival male semen to their partner, removing the foreskin reduces the diameter of the shaft of the penis immediately posterior to the glans and coronal ridge creating a greater area for displaced sperm to collect and rendering the circumcised penis a supernormal semen displacement device. As evidence for this effect, the coronal ridge among circumcised males becomes more fluted and accentuated and thereby appears better suited to scoop or displace larger volumes of rival male semen back away from the cervix and from the vagina of a recently inseminated female. Thus, other things being equal, we would predict as a testable consequence of this analysis that the incidence of nonpaternity among circumcised men ought to be lower than among their intact uncircumcised counterparts.

As further evidence for this effect of circumcision, women who have experienced sexual intercourse with both circumcised and intact men often rate intercourse with the former as lasting longer but being more uncomfortable. Because of a loss of sensitivity in the glans, circumcised males engage in deeper and more vigorous thrusting, and, as a consequence, the circumcised penis not only functions as a more effective semen displacement device, it is more likely to displace and deplete vaginal lubrication as well (O'Hara & O'Hara, 1999).

A corollary advantage that accrues to circumcised males involves a lower likelihood of being infected by STDs, as well as being less likely to transmit STDs. That is, the semen displacement properties of the intact penis not only puts men at risk of unwittingly piggybacking rival male semen from one vagina to the next by capturing semen under the foreskin, it also renders them more likely to piggyback STDs captured under the foreskin from one vagina to the next. Consistent with this analysis, evidence reviewed in our section on human penis morphology shows that women with uncircumcised partners are at greater risk of contracting a number of STDs. By the same token, intact males are more likely themselves to become infected by STDs. For instance, as a result of the greater risk of transmitting and being infected by HPV, uncircumcised males are also at greater risk of developing cancer of the glans (Larke, Thomas, Silva, & Weiss, 2011) in ways that parallel the increased risk of cervical cancer among women who contract HPV. For similar reasons, circumcised men are also at a reduced risk of contracting urinary tract infections (Shapiro, 1999).

It is important to acknowledge, however, that there is a dark side to circumcision. Based on case studies of men who have opted for voluntary circumcision as adults, the results show that, following recovery from surgery, they experience a loss of sensitivity of the glans and report less enjoyment from sex (Money & Davidson, 1983). Indeed, many of these men report that based on what they now know they would not have undergone circumcision in the first place.

But even the dark side to circumcision contains a silver lining. Because circumcision reduces the sensitivity of the glans, circumcised males have more staying power. The advantage of greater staying power is that periods of prolonged intercourse not only increase the effectiveness of displacing rival male semen, it also increases the likelihood of female orgasm. The reason this may be adaptive is because of growing consensus that the vaginal and intrauterine contractions that accompany female orgasm function to promote sperm mobility and transport and thereby increase the chances that some of the sperm will reach the fallopian tubes and fertilize an egg (Gallup, Ampel, Wedberg, & Pogosjan, 2014).

Thus, there are six unintended reasons why circumcision may be adaptive. Circumcised men are (a) less likely to cuckold themselves by unwitting piggybacking rival male semen from one vagina to the next, (b) less likely to be cuckolded by other males because the circumcised penis is a more effective semen displacement device, (c) less likely to transmit STDs, (d) less likely to contract STDs, (e) less likely to develop urinary tract infections, and (f) more likely to bring their partners to orgasm and achieve impregnation.

Getting a Mate on Easy Credit

Human offspring are extremely altricial (i.e., helpless at birth and remain parent-dependent for extended periods of time). Indeed, the needs of offspring provided the primary impetus for human adult pair bonding. For most of human evolutionary history successful infant- and child-rearing necessitated two healthy individuals, one to act as the primary caregiver—typically the mother, who is capable of nursing the offspring—and one to provide for the other needs of the family unit (i.e., shelter, security, food, etc.). Ideal mate selection on the part of the female, therefore, would involve choosing a male who could contribute not only good genes to her offspring but also ample resources to the family, which is known as the provisioning hypothesis (Lovejoy, 1981). The focus of this section is to explore how societal and economic evolution has complicated the process of finding a mate with good resources.

Throughout much of our evolutionary history, determining whether a potential mate could provide was likely not a difficult feat. Individuals who appeared well-nourished were likely successful at procuring food, individuals with shelter likely discovered or somehow fashioned that shelter out of available resources, and individuals whose bodies appeared fit and healthy likely offered a strong chance of being able to ward off predators and parasites. Therefore, the appearance of resources signaled success in obtaining resources. Females today are still attracted to high-providing males, as evidenced by the fact that males with high occupational status are able to attract and marry attractive females (e.g., Daly & Wilson, 1983; Dunn & Searle, 2010). The picture today, however, is not as clear due to the availability of easy credit and the impact that has on the ability of males to gather the things that signal a strong ability to provision. Indeed, we argue that credit is the resource analog to cosmetic changes in physical appearance and is at least equally powerful in obscuring and even distorting important factors in mate choice.

Whereas in our ancestral history it was almost guaranteed to be the case that an individual who had possession of a shelter was the owner of that shelter (either by finding it, building it, or overpowering the previous owner), that is not as clearly the case today. The advent of mortgages means it is possible to live in a residence for a long period of time before one is officially the owner. It can be argued that the ability to obtain a mortgage is dependent on earning potential, etc. and therefore provides a good proxy for access to resources, but the recent housing bubble and real estate crash compromises that argument.

Home ownership is likely the largest purchase in most individuals' lives, but the issue of credit transcends it. Credit extends beyond mortgages and can be used to "buy" cars, clothing, vacations, jewelry and even food. Therefore, it is not entirely clear when one sees an individual in possession of something whether that individual actually owns the item, is purchasing it on credit, or is even renting or leasing it. Moreover, the type of credit that one obtains can vary tremendously. Highly secure low-interest rates loans or mortgages, which in some cases can actually be a sign of good financial resources, exist alongside payday loans, which require paying a huge ransom to get money over a relatively short-term period. Unfortunately, there is no easy way to determine which of these sources (or any of the ones in-between) were used to gather resources. That information is known only to the loan holder and the individual with the loan, yet both sources of money could be used to purchase the same goods (to a certain extent). Therefore, when a date picks you up in an expensive, brand-new sports car, you have no clear way of knowing whether he owns that car or whether he is paying an incredibly high interest rate and borrowing far more money than is reasonable to pay for it (Dunn & Searle, 2010). Those two scenarios would obviously have very different implications for the likely success of being able to provide resources in the future.

Thus, in today's world of plastic currency and "have it today, pay for it tomorrow," the obvious signals of resource procurement have been obscured in a way similar to cosmetic augmentation to the human body. Just as it is wise to question whether one is seeing signs of physical beauty or Maybelline, it is wise to ask whether one is enjoying a meal paid for by cash or MasterCard.

Most women are commitment skeptical, because feigning good intentions in an attempt to gain sexual favors was a common male strategy during our evolutionary history. However, the fact that women fall for the appearance of good genes based on easy/inflated credit is evidence for how evolution has yet to catch up to contemporary changes and how these changes may be outstripping our biological ability to keep pace. Imagine the consequences of falling for a man with lots of material possessions, only to discover after you marry him that he has $120,000 in college loans and $75,000 in credit card debt. You and he would have to work your fingers to the bone for the first 10 or more years of your marriage to pay for his previous philandering and life in the fast lane. Let's face it: This might not be conducive to a happy and productive marriage.

Smartphones and the Evolution of Dumb Humans

Recent history has seen a spate of technological developments, perhaps none so ubiquitous as the smartphone. Although it allows us the opportunity to be within "vocal reach" almost all the time, at this point, the "telephone" is seemingly an afterthought. Instead, from the current perspective, the major impact from these devices comes from the fact that we are essentially carrying a computer, with access to anything on the Web from just about all locations on the planet, all the time, and a device that allows us to communicate via written, rather than spoken language. Whereas the intelligence behind the development of these devices is incredible, we argue that many of the likely impacts created by the devices may be detrimental to humans from a cognitive, as well as a social, perspective.

A recent study by Barr, Pennycook, Stolz, and Fugelsang (2015) found that individuals who had a penchant to rely more on intuitive methods to solve problems were more likely to turn to their smartphones to recruit answers to questions than were their more analytic counterparts. Across problems as varied as calculating a restaurant tip to answering a question, they were more likely to access their smartphones' calculator or Web browser and spent more of their day using the device than did more analytic individuals. Further, Ferguson, McLean, and Risko (2015) have shown that, regardless of thinking style, individuals are less likely to rely on their own knowledge and problem-solving ability to answer a question when they have Internet access than when they do not. As with any other skill, practice and use leads to better ability, and thinking is no exception. Recently, in the developed nations, at least, the first generations of children have been born who will not know a time when this technology was not available. In addition to phones, for many more well-to-do children, a good deal of playtime probably involves access to a tablet device. This likely means the maturation of the first generations of children for whom boredom is not an issue and for whom watching a movie or program on a tablet will supplant picking up a comic book or novel. Although it seemingly provides great opportunity to have a wealth of information at one's fingertips, it may be the case that boredom jump starts creativity. Recent studies point to negative health consequences for children with excessive access to electronic media (e.g., Strasburger & Jordan, 2010), and it will be interesting to examine the long-term cognitive consequences as these children mature.

In addition to the cognitive deficiencies that might be associated with prevalent smartphone and tablet use, there are also likely to be social costs. For many children today, texting and messaging seem to replace face-to-face communication—both with peers and with adults. Our own academic experience indicates that with the development of handheld email, visits to office hours have decreased and electronic messages have increased in frequency. This is important because spoken language often conveys richer information than electronic messaging. Face-to-face communication involves a lot more than verbal messages; nonverbal messages often carry the show. Things like sarcasm and irony, and even frustration and exasperation cannot often be expressed well in email or text, although both could be expressed with facial expressions, postural changes, or even changes in pitch or tone of voice on a telephone. Indeed, "Poe's law" has been proposed to account for instances in which a parody of an extreme view is misinterpreted as support for that view when no concrete signals to the author's true intent are available (Poe, 2005). It will be important to ask whether individuals who grow up with electronic messaging as a primary means of communication fail to develop the skills necessary to interpret and send these more subtle social cues.

Finally, the issue of vocal and/or face-to-face communication as compared to electronic communication is also critical in the development of romantic relationships. Recent research has shown that an individual's voice carries important cues to fitness (Gallup & Frederick, 2012). Thus, relying too heavily on electronic communication might set up a situation in which a good deal of time or resources is spent getting to "know" someone you've never spoken to, thereby setting up a "deaf date." Moreover, not only are the vocal cues absent in electronic messaging, but there is also a time lag (albeit short) and a backspace key that allows the message sender time to quickly reflect and polish the message before sending it on. The potential for inauthenticity is magnified greatly in a condition where an individual is getting the bulk of his or her information about the other person from a dating website accessed on a phone, which might contain not only doctored pictures but also doctored intentions and doctored thoughts.

Thus, although the technology underlying the advent of the smartphone undoubtedly reflects a very high collective human IQ, its use might very well degrade the selective pressures that contributed to the evolution of that high IQ in the first place.

Why Smart People Gravitate Toward Evolutionary Studies

According to the Savanah principle of human intelligence (Kanazawa, 2004), solutions to many of the recurring adaptive problems in the ancestral environment were incorporated into what are now represented by genetic/hard-wired solutions. In contrast, unique and novel problems that arise in the contemporary environment are solved by general intelligence as typically measured by standardized IQ tests. In support of the existence of these two domain specific types of intelligence, Kanazawa has conducted a series of studies that show that people who score high on standardized intelligence tests often excel in seemingly counterintuitive and paradoxical ways at solving atypical problems (see Kanazawa, 2012, for an overview). Examples include evidence showing that people with liberal political views have average IQ scores that exceed those with more conservative views, and these differences may contribute to the striking political differences in problem-solving when it comes to such things as economic plans, healthcare, vaccination, global warming, and the availability of weapons. Intelligent people are also more likely to be night owls, more likely to prefer classical music, more likely to be homosexual, and more likely to be atheists. Intelligent people make more money, make better scientists, and become better astronauts, but it is important to realize that these are all evolutionarily novel pursuits.

The same is true for education. According to the Savanah principle, intelligent people do better in school because school is evolutionarily novel. Attending lectures, reading books, and answering written essay questions had little or no counterpart to what people used to do.

While the same is generally true for what gets taught in school, in our view, some topics are more novel than others. Courses in child development and abnormal psychology can be used to illustrate this point. Issues related to raising children were a pervasive source of recurrent problems during human evolution as were dealing with strange, unusual, irrational, and/or delusional people. You can also make a case that cooking was a precursor to the discipline of chemistry, and an interest in the moon and the stars was a harbinger to astronomy. Although contemporary physics with an emphasis on subatomic particles seems devoid of any evolutionary precursors, in the process of growing up everyone had to learn to master different aspects of intuitive physics associated with adapting to such things as leverage, gravity, and momentum. The treatment of all of these topics has origins in human history that predates formal education.

In our view, one of the few subject matters taught in colleges and universities these days that is not anchored directly or indirectly in adaptive problems embedded in human history is evolution. People did not need to know anything about natural selection, inclusive fitness, or the Hardy Weinberg Law to solve day-to-day problems of evading predators, competing for scarce resources, and caring for offspring. It follows, therefore, that if intelligent people are drawn to novel problems, then intelligent people and those who can think outside the box may be more prone to cut their teeth on problems related to the impact of evolution on human behavior.

At this juncture, it is interesting to note that Dawkins (1976) feels that intelligent life emerges when it works out the reasons for its own existence. Whereas even now most people still fail to realize that their existence or even their behavior may be driven by underlying evolved mechanisms, some people with a penchant for studying evolution have become the only creatures on this planet to seriously and scientifically grapple with the reasons for their own existence.

But there is a dark side to human intelligence. Evolution is not based on the survival of the fittest; it is based on the perpetuation of genes. Therefore, when it comes to competing with one another for genetic representation in future generations, intelligent people may unwittingly be in the process of undermining their evolutionary best interests. Intelligent people are more likely to use contraceptives and more likely to remain childless, and Kanazawa (2012) argues that therefore intelligent people make the worst kind of parents simply because they are less likely to have children. It was evolutionarily novel to remain childless by choice.

Although some people claim that taking courses about evolution may improve performance in other domains (Kauffman & Wilson, 2016), according to our hypothesis, it remains a distinct possibility that smarter students may simply be more likely to take courses in evolution in the first place, which would confound any apparent advantage of taking such courses.

Conclusion

Evolutionary theory can be used to integrate and organize an unprecedented amount of information about human behavior. It also has heuristic value in the sense that it prompts people to do important research that otherwise might never be done. Finally, as we have attempted to emphasize in this chapter, evolutionary theory enables people to think about human behavior and human existence in ways that are outside the box. That is, evolutionary theory can be used to generate not only new but sometimes radically different ways of thinking about behavior that often have both testable implications and important practical applications.

The entire idea of EvoS, a full-on academic curriculum that integrates evolutionary ideas across all academic areas, has the capacity to serve as a powerful model for education writ large. Evolutionary principles, as shown in this chapter, can help shed light on a broad array of phenomena. The EvoS curriculum is the kind of curriculum that can clearly help improve how we educate the next generation of thinkers. EvoS is an approach to education that can be thought of as part of a great synthesis in the fields of teaching and learning.

References

Abioye-Kuteyi, E. A., Ojofeitimi, E. O., Aina, O. I., Kio, F., Aluko, Y., & Mosuro, O. (1997). The influence of socioeconomic and nutritional status on menarche in Nigerian school girls. *Nutrition and Health, 11*(3), 185–195.

Achenbach, J. (2016, June 30). 107 Nobel laureates sign letter blasting Greenpeace over GMOs. Washington Post. Retrieved from https://www.washingtonpost.com

Adams, G. P., & Ratto, M. H. (2013). Ovulation-inducing factor in seminal plasma: A review. *Animal Reproduction Science, 136*, 148–156.

Auvert, B., Sobngwi-Tambekou, J., Cutler, E., Nieuwoudt, M., Lissouba, P., Puren, A., & Taljaard, D. (2009). Effect of male circumcision on the prevalence of high-risk human papillomavirus in young men: results of a randomized controlled trial conducted in Orange Farm, South Africa. *Journal of Infectious Diseases, 199*, 14–19.

Auvert, B., Buve, A., Lagarde, E., Kahindo, M., Chege, J., Rutenberg, N., . . . Weiss, H. A. (2001). Male circumcision and HIV infection in four cities in sub-Saharan Africa. *AIDS, 15*(Suppl 4), S31–S40.

Baeten, J. M., Richardson, B. A., Lavreys, L., Rakwar, J. P., Mandaliya, K., Bwayo, J. J., & Kreiss, J. K. (2005). Female-to-male infectivity of HIV-1 among circumcised and uncircumcised Kenyan men. *Journal of Infectious Diseases, 191*, 546–553.

Baker, R. (2013). *The hitchhiker's child*. Booklocker.com

Baker, R. L., & Bellis, M. A. (1993). Human sperm competition: Ejaculate adjustment by males and the function of masturbation. *Animal Behavior, 46*, 861–885.

Baker, R. R., & Bellis, M. A. (1995). Human sperm competition: Copulation. masturbation, and infidelity London, England: Chapman and Hill.

Cold, C. J., & Taylor, J. R. (1999). The prepuce. *BJU International, 83*(S1), 34–44.

Barr, N., Pennycook, G., Stolz, J. A., & Fugelsang, J. A. (2015). The brain in your pocket: Evidence that smartphones are used to supplant thinking. *Computers in Human Behavior, 48*, 473–480.

Bogle, O. A., Ratto, M. H., & Adams, G. P. (2011). Evidence for the conservation of biological activity of ovulation-inducing factor in seminal plasma. *Reproduction, 142,* 277–283.

Bröder, A., & Hohmann, N. (2003). Variations in risk taking behavior over the menstrual cycle: An improved replication. *Evolution and Human Behavior, 24,* 391–398.

Burch, R. L., & Gallup, G. G., Jr. (2006). The psychobiology of semen: Female infidelity paternal uncertainty In S. M. Platek & T. Shackelford (Eds.), *Female infidelity and paternal uncertainty: Evolutionary perspectives on male anti-cuckoldry tactics* (pp. 141–172). Cambridge, England: Cambridge University Press.

Bwayo, J., Plummer, F., Omari, M., Mutere, A., Moses, S., Ndinya-Achola, J., . . . Kreiss, J. (1994). Human immunodeficiency virus infection in long- distance truck drivers in east Africa. *Archives of Internal Medicine, 154,* 1391–1396.

Center for Disease Control (2012). The CDC breastfeeding report card. Retrieved from www.cdc.gov/breastfeeding/data/

Chan, S., & Tang, C. (1983) Immunoreactive LHRH-like factors in human seminal plasma. *Archives of Andrology, 10,* 29–32.

Chang, A. M., Aeschbach, D., Duffy, J. F., & Czeisler, C. A. (2015). Evening use of light-emitting eReaders negatively affects sleep, circadian timing, and next-morning alertness. *PNAS, 112,* 1232–1237.

Chapman, D. J., & Perez-Escamilla, R. (1999). Identification of risk factors for delayed onset of lactation. *Journal of the American Dietic Association, 99,* 450–454.

Chavanne, T. J., & Gallup, G. G. (1998). Variation in risk taking behavior among female college students as a function of the menstrual cycle. *Evolution and Human Behavior, 19,* 27–32.

Chen, B., Yuen, Z., & Pan, G. (1985). Semen induced ovulation in the bactrian camel (*Camelus bactrianus*). *Journal of Reproduction and Fertility, 47,* 335–339.

Chu, L. C., & Tung, W K. (2005). The clinical outcome of 137 rape victims in Hong Kong. *Hong Kong Medical Journal, 11,* 391–396.

Cold, C. J., & Taylor, J. R. (1999). The prepuce. *BJU International, 83,* 34–44.

Coonrod, D. V., Hickok, D. E., Zhu, k., Easterling, T. R., & Daling, J.R. (1995). Risk factors for preeclampsia in twin pregnancies: A population-based cohort study. *Obstetrics and Gynecology, 5,* 645–650.

Cook, L. S., Koutsky, L. A., & Holmes, K. K. (1994). Circumcision and sexually transmitted diseases. *American Journal of Public Health, 84,* 197–201

Daly, M., & Wilson, M. (1983). *Sex, evolution and behaviour* (2nd ed.). Belmont, CA. Wadsworth.

Davis, J. A., & Gallup, G. G., Jr. (2006). Preeclampsia and other pregnancy complications as an adaptive response to unfamiliar semen. In S. Platek & T. Shackelford, (Eds.), *Female infidelity and paternal uncertainty: Evolutionary perspectives on male anti-cuckoldry tactics* (pp. 191–204). Cambridge, England: Cambridge University Press.

Dawkins, R. (1976). *The selfish gene.* Oxford, England: Oxford University Press.

de Haas, S., van Berlo, W., Bakker, F., & Vanwesenbeeck, I. (2012). Prevalence and characteristics of sexual violence in the Netherlands, the risk of revictimization and pregnancy: Results from a national population survey. *Violence and Victims, 27,* 592–608.

Dekker, G. A., Tubbergen, P., Valk, M., Althuisius, S. M., & Lachmeijer, A. M. A. (1998). Change in paternity: A risk factor for preeclampsia in multiparous women. *American Journal of Obstetrics and Gynecology, 178*(1 Pt 2), S120.

Doty, R. L., Form, M., Preti, G., & Huggins, G. R. (1975). Changes in the intensity and pleasantness of human vaginal odors during the menstrual cycle. *Science, 190,* 1316–1318.

Duley, L. (2009). The global impact of preeclampsia and eclampsia. *Seminars in Perinatology, 33,* 130–137.

Dunn, M. J., & Searle, R. (2010). Effect of manipulated prestige-car ownership on both sex attractiveness ratings. *British Journal of Psychology, 101,* 69–80.

Ferguson, A. M., McLean, D., & Risko, E. F. (2015). Answers at your fingertips: Access to the Internet influences willingness to answer questions. *Consciousness and Cognition, 37,* 91–102.

Fisher, R. (1930). *The genetical theory of natural selection.* Oxford, England: Oxford University Press.

Fleiss, P. M., Hodges, F. M., & Van Howe, R. S. (1998). Immunological functions of the human prepuce. *Sexually transmitted infections, 74*(5), 364.

Gallup, G. G., Jr. (1996). Attitudes toward homosexuals and evolutionary theory: The role of evidence. *Ethology and Sociobiology, 17,* 281–284.

Gallup, G. G., Jr., Ampel, B. C., Wedberg, N., & Pogosjan, A. (2014). Do orgasms give women feedback about mate choice? *Evolutionary Psychology, 12,* 957–977.

Gallup, G. G., Jr., & Burch, R. L. (2004). Semen displacement as a sperm competition strategy in humans. *Evolutionary Psychology, 2,* 12–23.

Gallup, G. G., Jr. & Burch, R. L. (2006). The semen displacement hypothesis: Semen hydraulics, double mating, adaptations to self-semen displacement, and the IPC proclivity model. In S. Platek & T. Shackelford, (Eds.), *Female infidelity and paternal uncertainty: Evolutionary perspectives on male anti-cuckoldry tactics* (pp. 129–140). Cambridge, England: Cambridge University Press.

Gallup Jr, G. G., Burch, R. L., Zappieri, M. L., Parvez, R. A., Stockwell, M. L., & Davis, J. A. (2003). The human penis as a semen displacement device. *Evolution and Human Behavior, 24*(4), 277–289.

Gallup. G. G., Jr., Burch, R. L., & Petricone, L. (2012). Sexual conflict, infidelity, and semen chemistry. In A. Goetz & T. Shackelford (Eds.), *The Oxford handbook of sexual conflict in humans* (pp. 217–232). New York, NY: Oxford University Press.

Gallup, G. G., Jr., Burch, R. L., & Platek, S. M. (2002). Does semen contain antidepressant properties? *Archives of Sexual Behavior, 39,* 289–291.

Gallup, G. G., Jr., Finn, M. M., & Sammis, B. (2009). On the origin of descended scrotal testicles: The activation hypothesis. *Evolutionary Psychology, 7,* 517–524.

Gallup, G. G., Jr., & Frederick, D. A. (2010). The science of sex appeal: An evolutionary perspective. *Review of General Psychology, 14,* 240–250.

Gallup, G. G., Jr., Frederick, M. J., & Pipitone, R. N. (2008). Morphology and behavior: Phrenology revisited. *Review of General Psychology, 12,* 297–304.

Gallup, G. G., Jr., Pipitone, N. R., Carrone, K.J. & Leadholm, K.L. (2010). Bottle feeding simulates child loss: Postpartum depression and evolutionary medicine. *Medical Hypotheses, 74,* 174–176.

Gallup, G. G., Jr., & Reynolds, C. J. (2014). Evolutionary medicine: Semen sampling and seminal plasma hypersensitivity. *Evolutionary Psychology, 12*, 245–250.

Gallup, G. G., Jr., Spaulding, K. N., & Aboul-Seoud, F. (2016). Bottle feeding: The impact on post-partum depression, birth spacing and autism. In A. Alvergne, J. Crispin, & C. Faurie (Eds.), *Evolutionary thinking in medicine: From research to policy and practice* (pp. 47–57). New York, NY: Springer.

Garver-Apgar, C. E., Gangestad, S. W., & Simpson, J. A. (2007). Women's perceptions of men's sexual coerciveness change across the menstrual cycle. *Acta Psychologica Sinica, 39*, 536–540.

Gaudino, J. A., Jr., Jenkins, B., & Rouchat, R. W. (1999). No father's names: A risk factor for infant mortality in the state of Georgia. *Social Science and Medicine, 48*, 253–265.

Gjerdingen, D. (2003). The effectiveness of various postpartum depression treatments and the impact of antidepressant drugs on nursing infants. *Journal of the American Board of Family Medicine, 16*, 372–382.

Giuliano, A. R., Lazcano, E., Villa, L. L., Flores, R., Salmeron, J., Lee, J.-H. . . . Quiterio, M. (2009). Circumcision and sexual behavior: factors independently associated with human papillomavirus detection among men in the HIM study. *International Journal of Cancer, 124*, 1251–1257.

Gottschall, J. A., & Gottschall, T. A. (2003). Are per-incident rape-pregnancy rates higher than per-incident consensual pregnancy rates?. *Human Nature, 14*, 1–20.

Gray, R. H., Kiwanuka, N., Quinn, T.C., Sewankambo, N. K., Serwadda, D., Mangen, F. W., . . . Wawer, M. J. (2000). Male circumcision and HIV acquisition and transmission: cohort studies in Rakai, Uganda. *AIDS, 14*, 2371–81.

Gray, R. H., Kigozi, G., Serwadda, D., Makumbi, F., Watya, S., Nalugoda, F., . . . Sewankambo, N. K. (2007). Male circumcision for HIV prevention in men in Rakai, Uganda: a randomised trial. *The Lancet, 369*(9562), 657–666.

Haubruge, E., Arnaud, L., Mignon, J., & Gage, M. G. (1999). Fertilization by proxy: Rival sperm removal and translocation in a beetle. *Proceedings of the Royal Society of London B: Biological Sciences, 266*, 1183–1187.

Heinig, M. G., & Dewey, K. G. (1996). Health advantages of breastfeeding for infants: A critical review. *Nutrition Research Reviews, 9*, 89–110.

Holmes, M. M., Resnick, H. S., Kilpatrick, D. G., & Best, C. L. (1996). Rape-related pregnancy estimates and descriptive characteristics from a national sample of women. *American Journal of Obstetrics and Gynecology, 175*, 320–325.

Howie, P. W., & McNeilly, A. S. (1982). Effect of breast-feeding patterns on human birth intervals. *Journal of Reproduction and Fertility, 65*, 545–557.

Hoy, J., Venn, A. Halliday, J., Kovacs, G., & Waalwyk, K. (1999). Perinatal and obstetric outcomes of donor insemination using cryopreserved semen in Victoria, Australia. *Human Reproduction, 14*, 1760–1764.

Hughes, S. M., Harrison, M. A., & Gallup, G. G., Jr. (2007). Sex differences in romantic kissing among college students: An evolutionary perspective. *Evolutionary Psychology, 5*, 612–631.

Jöchle, W. (1973). Coitus-induced ovulation. *Contraception, 7*, 523–564.

Kanazawa, S. (2004). General intelligence as a domain-specific adaptation. *Psychological Review, 111*, 512–523.

Kanazawa, S. (2012). *The intelligence paradox: Why the intelligent choice isn't always the smart one*. Hoboken, NJ: Wiley.

Kanazawa, S., & Perina, K. (2009). Why night owls are more intelligent. *Personality and Individual Differences, 47*, 685–690.

Kauffman, R. A. Jr., & Wilson, D. S. (2016). Beyond academic performance: The effects of an evolution-informed school environment on student performance and well-being. In D. C. Geary & D. B. Berch (Eds.) *Evolutionary perspectives on child development and education* (pp. 307–347). Cham, Switzerland, Springer.

Kennedy, K. I., & Visness, C. M. (1992). Contraceptive efficacy of lactational amenorrhoea. *Lancet, 339*, 227–30.

Klonoff-Choen, H. S., Savitz, D. A., Cefalo, R. C., & McCann, M. F. (1989). An epidemiologic study of contraception and preeclampsia. *Journal of the American Medical Association, 262*(22), 3143–3147.

Knutson, K. L., Spiegel, K., Penev, P., & Van Cauter, E. (2007). The metabolic consequences of sleep deprivation. *Sleep Medicine Reviews, 11*, 163–178.

Koelman, C. A., Coumans, A. B., Nijman, H. W., Doxiadis, I. I., Dekker, G. A., & Claas, F. H. (2000). Correlation between oral sex and low incidence of preeclampsia: A role for soluble HLA in seminal fluid? *Journal of Reproductive Immunology, 46*, 155–166.

Larke, N. L., Thomas, S. L., Silva, I. S., & Weiss, H. A. (2011). Male circumcision and penile Cancer: A systematic review and meta-analysis. *Cancer Cause and Control, 8*, 1097–1110.

Lavreys, L., Rakwar, J. P., Thompson, M. L., Jackson, D. J., Mandaliya, K., Chohan, B. H., . . . Kreiss, J. K. (1999). Effect of circumcision on incidence of human immunodeficiency virus type 1 and other sexually transmitted diseases: A prospective cohort study of trucking company employees in Kenya. *Journal of Infectious Diseases, 180*, 330–36.

Lavreys, L., Rakwar, J. P., Thompson, M. L., Jackson, D. J., Mandaliya, K., Chohan, B. H., . . . Kreiss, J. K. (1999). Effect of circumcision on incidence of human immunodeficiency virus type 1 and other sexually transmitted diseases: A prospective cohort study of trucking company employees in Kenya. *Journal of Infectious Diseases, 180*, 330–336.

Loss, S. R., Will, T., Loss, S. S., & Marra, P. P. (2014). Bird-building collisions in the United States: Estimates of annual mortality and species vulnerability. *The Condor: Ornithological Applications, 116*, 8–23.

Lovejoy, C. O. (1981). The origin of man. *Science, 211*, 341–350.

Luboshitzky, R., Shen-Orr, Z., & Herer, P. (2002). Seminal plasma melatonin and gonadal steroids concentrations in normal men. *Archives of Andrology, 48*, 225–232.

Maegawa, M., Kamada, M., Irahara, M., Yamamoto, S., Yoshikawa, S., Kasai, Ohmoto, Y., . . . Aono, T., (2002). A repertoire of human cytokines in human seminal plasma. *Journal of Reproductive Immunology, 54*, 33–42.

Malinowski, B. (1929). *The sexual life of savages in north-western Melanesia*. London, England: Routledge.

Marti, J. J., & Herrmann, U. (1977). Immunogestosis: A new etiologic concept of "essential" EPH gestosis, with special consideration of the primigravid patient: Preliminary report of a clinical study. *American Journal of Obstetrics and Gynecology, 128,* 489–493.

Martin, R. D. (1990). *Primate origins and evolution: a phylogenetic reconstruction.* Princeton, NJ: Princeton University Press.

Martinez-Bakker, M., Bakker, K. M., King, A. A., & Rohani, P. (2014). Human birth seasonality: Latitudinal gradient and interplay with childhood disease dynamics. *Proceedings of the Royal Society: B, 281,* 2013–2438.

Money, J., & Davison, J. (1983). Adult penile circumcision: Erotosexual and cosmetic sequelae. *Journal of Sex Research, 19,* 289–292.

Multale, T., Creed, F., Mareshm M., & Hunt, L. (1991). Life events and low birth weight-analysis by preterm and small for gestational age. *British Journal of Obstetrics and Gynecology, 98,* 166–172.

Nadelson, C. C. (2001). *Sleep disorders.* Philadelphia, PA. Chelsea House.

Ney, P. G. (1986). The intravaginal absorption of male generated hormones and their possible effect on female behaviour. *Medical Hypotheses, 20,* 221–231.

Nielson, C. M., Schiaffino, M. K., Dunne, E. F., Salemi, J. L., & Giuliano, A. R. (2009) Associations between male anogenital human papillomavirus infection and circumcision by anatomic site sampled and lifetime number of female sex partners. *Journal of Infectious Diseases, 199,* 7–13.

Oakley, A. (1985). Social support and the outcome in pregnancy: The soft way to increase birth weight? *Social Science and Medicine, 21,* 1259–1268.

O'Hara, K., & O'Hara, J. (1999). The effect of male circumcision on the sexual enjoyment of the female partner. *British Journal of Urology, International, 83,* 79–84.

Ohayi, R. S., Ezugwu, E. C., Chigbu, C. O., Arinze-Onyia, S. U., & Iyoke, C. A. (2015). Prevalence and pattern of rape among girls and women attending Enugu State University Teaching Hospital, southeast Nigeria. *International Journal of Gynecology & Obstetrics, 130,* 10–13.

Pan, G., Zhao, X., Chen, S., Jiang, S., Huang, Y., Zu, Y., & Wang, H. (1992). The ovulation-inducing effect of seminal plasma in the Bactrian camel. In W. R. Allen, A. J. Higgins, I. G. Mayhew, D. H. Snow & J. F. Wade (Eds.), *Proceedings of the First International Camel Conference* (pp. 159–161). Newmarket, England: R & W.

Petralia, S. M., & Gallup, G. G. (2002). Effects of a sexual assault scenario on handgrip strength across the menstrual cycle. *Evolution and Human Behavior, 23,* 3–10.

Poe, N. (2005). Big contradictions in the evolution theory (web discussion log). *Christian Forums.* Retrieved from https://www.christianforums.com

Ramsey, C. N., Abell, T. D., & Baker, L. C. (1986). The relationship between family functioning life events, family structure, and the outcome of pregnancy. *Journal of Family Practice, 22,* 521–527.

Ratto, M. H., Huanca, W., Singh, J., & Adams, G. P. (2006). Comparison of the effect of ovulation-inducing factor (OIF) in the seminal plasma of llamas, alpacas, and bulls. *Theriogenology, 66,* 1102–1106.

Rawls, W. E., Laurel, D., Melnick, J. L., Glicksman, J. M., & Kaufman, R. H. (1968). A search for viruses in smegma, premalignant and early malignant cervical tissues: The isolation of herpes viruses with distinct antigenic properties. *American Journal of Epidemiology, 87,* 647–655.

Reynolds, S. J., Shepherd, M. E., Risbud, A. R., Gangakhedkar, R. R., Brookmeyer, R. S., Divekar, A. D., . . . Bollinger, R. C. (2004). Male circumcision and risk of HIV-1 and other sexually transmitted infections in India. *Lancet, 363*, 1039–40.

Robertson, S. A. (2005). Seminal plasma and male factor signaling in the female reproductive tract. *Cell and Tissue Research, 322*(1), 43–52.

Robertson, S. A., Guerin, L. R., Bromfield, J. J., Branson, K. M., Ahlstrom, A. C., & Care, A. S. (2009). Seminal fluid drives expansion of the CD4+CD25+ T regulatory cell pool and induces tolerance to paternal alloantigens in mice. *Biology of Reproduction, 80*(5), 1036–1045.

Robillard, P. Y., Hulsey, T. C., Dekker, G. A., & Chaouat, G. (2003). Preeclampsia and human reproduction: An essay of long term reflection. *Journal of Reproductive Immunology, 59*, 93–100.

Rogers, A., Iltus, D., & Wooding, S. (2004). Genetic variation at the MC1R locus and the time since loss of human body hair. *Current Anthropology, 45*, 105–108.

Schernhammer, E. S., Kroenke, C. H., Laden, F., & Hankinson, S. E. (2006). Night work and risk of breast cancer. *Epidemiology, 17*, 108–111.

Shapiro, E. (1999). American Academy of Pediatrics policy statement on circumcision and urinary tract infection. *Reviews in Urology, 1*, 154–156.

Sigurdson, K., & Ayas, N. (2007). The public health and safety consequences of sleep disorders. *Canadian Journal of Physiological Pharmacology, 85*, 179–183.

Silva, M. E., Smulders, J. P., Guerra, M., Valderrama, X. P., Letelier, C., Adams, G. P., & Ratto, M. H. (2011). Cetrorelix suppresses the preovulatory LH surge and ovulation induced by ovulation-inducing factor (OIF) present in llama seminal plasma. *Reproductive Biology and Endocrinology, 9*, 1.

Singh, D., & Bronstad, P. M. (2001). Female body odour is a potential cue to ovulation. *Proceedings of the Royal Society of London: B, 268*, 797–801.

Smith, G. N., Walker, M., Tessier, J. L., & Millar, K. G. (1997). Increased incidence of preeclampsia in women conceiving by intrauterine insemination with donor versus sperm for treatment of primary infertility. *American Journal of Obstetrics and Gynecology, 177*, 455–458.

Stear, J. (2005, November). Adaptive function of sleep: Circadiana: The early bird catches the worm [weblog post]. Retrieved from http://circadiana.blogspot.com/2005/11/non-adaptive-function-of-sleep.html

Steube, A. M., Willet, W. C., Xue, F., Michaels, K. B. (2009). Lactation and incidence of premenopausal breast cancer: A longitudinal study. *Archives of Internal Medicine, 169*, 1364–1371.

Stewart, F. H., & Trussell, J. (2000). Prevention of pregnancy resulting from rape: A neglected preventive health measure. *American Journal of Preventive Medicine, 19*, 228–229.

Strasburger, V. C., Jordan, A. B., & Donnerstein, E. (2010). Health effects of media on children and adolescents. *Pediatrics, 125*, 756–767.

Tigreros, N., South, A., Fedina, T., & Lewis, S. (2009). Does fertilization by proxy occur in Tribolium beetles? A replicated study of a novel mechanism of sperm transfer. *Animal Behaviour, 77*, 555–557.

Tobian, A. A., Serwadda, D., Quinn, T. C., Kigozi, G., Gravitt, P. E., Laeyendecker, O., . . . Nowak, R. G. (2009). Male circumcision for the prevention of HSV-2 and HPV infections and syphilis. *New England Journal of Medicine, 360*, 1298–1309.

Trupin, L. S., Simon, L. P., & Eskenzai, B. (1996). Change in paternity: A risk factor for preeclampsia in multipapras. *Epidemiology, 7*, 240–244.

Turner, J. R., Grindstaff, C. F., & Phillips, N. (1990). Social support and outcome in teenage pregnancy. *Journal of Health and Social Behavior, 31*, 43–58.

VanBuskirk, K., Winer, R. L., Hughes, J. P., Feng, Q., Arima, Y., Lee, S. K., . . . Koutsky, L. A. (2011). Circumcision and acquisition of HPV infection in young men. *Sexually Transmitted Diseases, 38*(11), 1074.

Wiechert, U., Halliday, A. N., Lee, D.-C., Snyder, G. A., Taylor, L. A., & Rumble, D. (2001). Oxygen Isotopes and the Moon-Forming Giant Impact. *Science, 294*, 345–348.

Weiss, H. A., Quigley, M. A., & Hayes, R. J. (2000). Male circumcision and risk of HIV infection in sub- Saharan Africa: A systematic review and meta-analysis. *AIDS, 14*, 2361–2370.

Weiss, H. A., Buve, A., Robinson, N. J., Van Dyck, E., Kahindo, M., Anagonou, S., . . . Hayes, R. J. (2001). The epidemiology of HSV-2 infection and its association with HIV infection in four urban African populations. *AIDS, 15*(Suppl 4), S97–S108

Weiss, H. A., Thomas, S. L., Munabi, S. K., & Hayes, R. J. (2006). Male circumcision and risk of syphilis, chancroid, and genital herpes: a systematic review and meta-analysis. *Sexually Transmitted Infections, 82*, 101–110.

Wessel, J., & Buscher, U. (2002). Denial of pregnancy: Population based study. *BMJ, 324*, 458.

Wilcox, A. J., Dunson, D. B., Weinberg, C. R., Trussell, J., & Baird, D. D. (2001). Likelihood of conception with a single act of intercourse: providing benchmark rates for assessment of post-coital contraceptives. *Contraception, 63*, 211–215.

CHAPTER 16 | The "EvoS Effect"

The Influence of Evolutionary Training on Critical Thinking Skills

RICHARD A. KAUFFMAN JR., IAN F. MACDONALD, AND DAVID SLOAN WILSON

> "Before taking this class, I had no idea how relevant evolutionary theory was to my everyday life."
> "This course opened a new way of thinking for me; I now see evolution all around me, from basic human actions to reading history."
> "This class revolutionized my way of thinking."

COMMENTS LIKE THESE ACTUAL quotes from students in Binghamton University's (B.U.) Evolutionary Studies courses (EvoS, pronounced as one word: "Ee-vohs") are common at the end of each semester. Dubbed "the EvoS Effect," students regularly report a transformative experience after engaging in their course material. They report being able to see evolutionary principles at play in all aspects of their lives thereafter and, being "unable to turn it off", this new framework allows the students to better understand their world: particularly when it comes to the human experience. Moreover, many students have suggested that this evolutionary training has lead to improvements in their problem solving, general thinking skills and academic success, even to the extent of increasing their participation and interest in non-EvoS courses.

The EvoS Effect is not restricted to Binghamton University or even to EvoS programs at other universities. We predict that it will occur whenever evolution is taught as a theory that applies to all aspects of humanity in addition to the biological sciences. The study presented in this chapter seeks to understand the impact that the EvoS Effect has on learners, to move beyond anecdotal evidence and to begin quantifying the transformative nature of engaging with an evolutionary framework. Presented through the lens of education science, we assert that incorporating an evolutionary studies program into all levels of educational policy and practice can have profound impacts on students' science literacy and, even more so, on their general thinking skills, e.g. critical thinking and domain knowledge transfer.

Background

Fostering higher-order cognitive skills (HOCS; e.g., critical thinking and knowledge transfer) is an oft-cited goal in education yet is rarely examined as an outcome. Until recently the implicit assumption in education practice has been that, with appropriate instruction, thinking skills naturally take care of themselves. In consequence to HOCS being largely unattended, students do not acquire these skills as much as they could and should. Furthermore, with the emphasis on high-stakes testing, even the most experienced teachers are pressured to rely on rote memorization and the passive application of "learned" algorithms; grades are what count, not learning. In this study, we provide evidence suggesting a science education program rooted in a transdisciplinary evolutionary framework can have a profound impact on fostering HOCS and other habits of mind for life-long learning.

By the very nature of science, science education is well suited for providing a medium through which students can engage in critical thinking. Because scientific thinking is a formalization of critical thinking, it can be used as a model for fostering HOCS at the core of straightforward and easily implemented teaching strategies. If students can develop the skills for science literacy they will know how to think successfully and reach reliable conclusions: abilities valued far beyond scientific and scholarly pursuits.

A critical issue in science education today is the teaching of evolutionary theory. Darwin's theory of evolution is the most powerful explanatory and predictive tool available for the study of living things and the human experience, yet many students receive little or no exposure to this important framework. More so, when it is included in the curriculum, it is primarily taught as a biological topic with little to no application to understanding human affairs.

Three key ideas from science education research informed the study presented in this paper. First, that focusing instruction on <u>nature of science (NOS) skills</u> <u>improves</u> <u>reasoning and critical thinking</u> (CT; Abd-El-Khalick & Lederman, 2000b; Akerson, Abd-El-Khalick, & Lederman, 2000; Duncan & Arthurs, 2012; Lederman, 1992; Sadler, 2004; Schwartz, Lederman, & Crawford, 2004). Second, that <u>improving understanding of evolutionary</u> <u>theory can improve NOS skills</u> (Abd-El-Khalick & Lederman, 2000a; Nickels, Nelson, & Beard, 1996). And third, that <u>using human examples in the teaching of evolution can improve students' understanding of evolution</u> (Nettle, 2010). It follows, then, that an EvoS education—a course emphasizing the general applicability of evolutionary theory, with a particular focus on human-related topics—can be expected to improve students' CT and other HOCS.

This chapter begins with a review of the research on CT instruction in current education practice. The focus then shifts to answering three questions regarding improving CT instruction through evolution education: 1) how science education can be used to increase CT; 2) how evolution education can add to science education; and 3) how EvoS education can add to evolution education. Finally, a study is presented that examines the effects of an evolutionary science course, designed to simultaneously foster an understanding of the NOS, on students' CT skills.

Critical Thinking

> "We don't know what they will need to know. We do know they will need to be skilled in finding out."
>
> PAUL & ELDER, 2007, pg. 2.

Since the age of Socrates, CT has been recognized as both a major goal of educational systems and a challenge in terms of realization. Despite various nuanced differences in definition, all schools of thought generally concur that "[c]ritical thinking is the art of analyzing and evaluating thinking with a view to improving it" (Paul & Elder, 2006). It is inquiry-oriented and encompasses skills of rational, logical, reflective, and consequential evaluative thinking in terms of what to accept or reject and what to believe in, followed by a decision of what to do or not to do about it (Zoller, 1993). As outlined by Paul & Elder (2006), a well cultivated critical thinker:

- raises vital questions and problems, formulating them clearly and precisely;
- gathers and assesses relevant information, using abstract ideas to interpret it effectively;
- comes to well-reasoned conclusions and solutions, testing them against relevant criteria and standards;
- thinks open-mindedly within alternative systems of thought, recognizing and assess, as need be, their assumptions, implications, and practical consequences; and
- communicates effectively with others in figuring out solutions to complex problems.

Furthermore, research suggests that, while simply developing the specific cognitive skills for CT are indeed necessary, learners must be willing, not just able, to think critically. Consistent willingness, motivation, inclination, and a drive to be engaged in critical thinking while reflecting on significant issues, making decisions, and solving problems is the essence of a disposition toward critical thinking (Facione, Facione, & Giancarlo, 1997; Facione, Giancarlo, Facion, & Gainen, 1995).

Critical Thinking and Domain Transfer of Knowledge

Aside from this general understanding of CT ability, there are a variety of specific skills involved with the CT process. One such skill is the transfer of knowledge across disciplines and domains (Zoller, 1999). Domain transfer takes place when an insight or skill learned in one situation is applied in another situation. As a simple example, if someone has just learned how to calculate the per-kilogram price for almonds, they should then be able to calculate the per-kilogram price for mangoes. If they cannot, we would say that the learning has failed to transfer from pricing almonds to pricing mangoes. A more complex, real-world example of knowledge transfer, more immediately relevant to science education, is when a person who learns the principles of wind flow to design a windmill can transfer that knowledge to direct the sail on a sailboat.

A transfer of acquired knowledge and skills certainly does occur to some extent in education, otherwise every educational process would require starting from the ground up. The problem is that it happens much less than one might naïvely expect (Detterman, 1993). Failure to transfer knowledge has been repeatedly shown in both laboratory settings and in the classroom (Barnett & Ceci, 2002; Ennis, 1989; Perkins & Salomon,

1988, 1989). This is largely due to the absence of instructional strategies that emphasize incorporating knowledge learned in other subjects into the different classes.

In current education practice, students learn information in one classroom and then learn a separate compilation of facts in their next class, without being encouraged to link the information. What's learned in one classroom may not transfer outside of that class, and what's learned outside the classroom is rarely encouraged to transfer inside the schoolhouse walls. Studies show that children who do complex mathematical calculations in their street business are unable to do the same math problems when presented formally in a classroom situation (Scribner & Cole, 1973, 1981). This is an important finding because "the failure to transfer means the failure to think" (Haskell, 2000, p. 47).

CT is especially vulnerable to weak transfer skills because CT is "intrinsically general in nature" (Van Gelder, 2005, p. 43). If students fail to transfer learned abilities across domains, then their ability to engage in CT will be severely limited. Nonetheless, transfer is an important component of CT and strategies that emphasize the transfer of learning will benefit general CT skills and vice versa.

Fostering Critical Thinking Skills through Evolution Education

The good news is both the disposition toward CT and CT itself are perceived as being open to educational influence (Brown, 1997), and research suggests that CT can be improved through specific teaching interventions (Abrami et al., 2008; Jensen, 2005/2008; Adey & Shayer, 1990, 1993, 1994; Ernst & Monroe, 2004; Morier & Keeports, 1994; Penningroth, Despain, & Gray, 2007; Wesp & Montgomery, 1998). The research is also clear that this cannot be a matter of implicit expectation; these skills do not develop when direct instructional strategies are absent (Haskell, 2000).

The best instructional approach may be to work simultaneously from both ends: (1) in regular practice of explicit instruction so that skills are exercised, strengthened, and consolidated, and (2) in fostering the understanding and application of the intellectual values ("dispositions") that play a major role when CT skills will be used (Kuhn, 1999). Science education is a natural fit for this two-fold approach to enhancing the development of HOCS—incorporating improved instructional techniques and developing natural dispositions to engage in CT.

Critical Thinking, the Nature of Science, and Science Instruction

Science education has a long list of goals and purposes, nearly all of which center around CT skills (see AAAS, 1990, 1993; Chiappetta & Kobballa, 2009; NRC, 1996, 2012 Zoller, 1993, 1997; Zoller & Tsaparlis, 1997). Scientific thinking *is* critical thinking; when students develop the skills for science literacy, they become naturally engaged in CT practices as they seek to explore and understand the world around them.

Science is not merely a collection of facts, concepts, and useful ideas about nature. Science is a method of investigating nature that discovers and examines reliable knowledge about it. Science education should emphasize the importance of educating students to encourage their curiosity about the world in which they live so that they ask questions about its nature and address those questions through their investigative actions (NRC, 1996, 2012).

One key aspect of science instruction that naturally reinforces HOCS while simultaneously emphasizing content-related curriculum goals is through an increased attention on the Nature of Science (NOS). The NOS consists of those seldom-taught but very important features of working science, e.g., its realm and limits, its levels of uncertainty, its biases, its social aspects, and the reasons for its reliability (AAAS, 1990). When students develop the habits of mind that stress the useful values, attitudes, and skills for understanding the world around them they are developing the fundamental understanding about the NOS.

Although many states include NOS in their science standards, students often have misconceptions about it, and some science teachers share these misconceptions (e.g., Carey & Smith, 1993; Dagher & BouJaoude, 2005; Lederman, 1992; Scharmann & Harris, 1992; Solomon, Scott, & Duncan, 1996). Science is often misrepresented in the media, and classroom teaching can overemphasize what we know rather than how we know it. Consequently, many students see science as a boring enterprise—the tedious accumulation of facts about the world, totally lacking any imagination or creativity. Many students believe that theories can simply be "read off" from the world, that scientific claims can be definitively proved, and that theories are merely speculative guesses without much empirical investigation supporting them and have therefore not yet achieved the privileged status of facts or laws. These misconceptions may be aggravated by teaching what textbooks usually call "*the* scientific method": a linear sequence of steps suggesting scientists follow a single, fixed process to develop laboratory experiments that directly and definitively test hypotheses. Confronting these misconceptions is the first step toward developing a scientifically literate population, one that can think rationally and critically.

Furthermore, an increased understanding of the NOS assists in developing a foundation for transfer skills. As any decent education in science requires that the student come to understand the central role of theory[1] in scientific methodology, students are encouraged to seek an understanding between relationships and to draw on similarities and differences across many fields of knowledge. An emphasis on the role of theory in science allows students to acquire sound conceptual knowledge about their world through the formulation of relationships among ideas (Ausubel, 1963).

When students, and teachers, understand the NOS (AAAS, 1993, Ch. 1) and "Science as a Way of Thinking" (AAAS, 1993, Ch. 12), they are automatically entered into a world of CT and knowledge transfer. Research in the laboratory and classroom confirms that focusing instruction on NOS skills results in improved reasoning and CT (Abd-El-Khalick & Lederman, 2000b; Akerson, Abd-El-Khalick, & Lederman, 2000; Duncan & Arthurs, 2012; Lederman, 1992; Sadler, 2004; Schwartz, Lederman, & Crawford, 2004). In no other curriculum is the demand for CT and other HOCS as important for success as it is in science education.

Nature of Science Instruction, Evolution and Knowledge Transfer

Amongst the chief aims of increasing science literacy, and therefore CT, is developing an increased understanding of evolutionary theory (Rudolph & Stewart, 1998). Generic scientific reasoning existed before Darwin but was insufficient by itself to achieve insights about the living world. By unlocking the theory of evolution by natural selection, Darwin opened a realm of understanding unforeseen by any other scientific accomplishment at the time; causing one scientist to famously remark: "nothing in biology makes sense except in the light of evolution" (Dobzhansky, 1973).

Evolution is recognized as a unifying principle for understanding biological phenomena and provides a coherent framework for organizing biological knowledge and thinking; making an impressive degree of knowledge transfer possible within the biological sciences. What's less understood

---

[1] As a coherent group of general propositions that are used as principles of explanation for a class of phenomena.

about evolution is its transferability to domains beyond biology. Evolution is recognized as a common theme of science[2] that transcends disciplinary boundaries (see AAAS, 1990, Ch. 11; AAAS, 1993, Ch. 11; NRC, 1996, Ch. 6; NRC, 2012, Ch. 4); most importantly, for understanding a panoply of human-related behaviors (Wilson, 2007). By applying the principles of evolutionary science to the thinking about other subjects, students can improve CT by transferring their understanding of evolutionary theory to a diverse range of topics.

Research indicates that NOS instruction can be enhanced through a thorough investigation of evolutionary theory in the curriculum (Abd-El-Khalick & Lederman, 2000a; Nickels, Nelson, & Beard, 1996). Thus, the effects of a course on evolution are two-fold. First, students are able to identify and understand their previously held misconceptions about evolutionary theory. And second, by investigating the major claims of evolutionary science, students must apply the habits of mind for scientific thinking, developing an increased understanding about the NOS while fostering CT. So, it stands to reason that a course on the domain general applicability of evolutionary theory can be expected to have an impact on students' understanding and incorporation of the NOS, and that this course can directly impact CT skills.

The next section provides an overview of the status of evolution education today and then describes how a single course, Evolution for Everyone (as taught through the B.U. EvoS program), can be designed to overcome these challenges. Finally, the actual effects that this course has on students' understanding about evolutionary theory, the NOS, and their CT skills will be examined.

---

[2] "Some important themes pervade science, mathematics, and technology and appear over and over again, whether we are looking at an ancient civilization, the human body, or a comet. They are ideas that transcend disciplinary boundaries and prove fruitful in explanation, in theory, in observation, and in design" (AAAS, 1993, p. 261). The American Association for the Advancement of Science (AAAS, 1990; 1993) identifies four Common Themes of Science: Systems; Models; Constancy & Change [*Evolution*]; and Scale. The National Research Council (NRC, 1996) identified five Unifying Concepts in their 1996 National Science Education Standards: Systems, order, and organization; Evidence, models, and explanation; Change, constancy, and measurement; Evolution and equilibrium; and Form and function. A recent re-envisioning of the NRC Unifying Concepts has resulted in a new framework for the Next Generation Science Standards (NRC, 2012) with seven Crosscutting Concepts: Patterns; Cause and effect; Scale, proportion, and quantity; Systems and system models; Energy and matter: Flows, cycles, and conservation; Structure and Function; Stability and change [*Evolution*]. Though different in quantity and specific titles, each of these views are congruent with the important impact of ideas that transcend disciplinary boundaries for a comprehensive understanding of NOS.

The Current State of Evolution Education

Darwin's theory of evolution is the most powerful explanatory and predictive tool available for the study of living things (Darwin, 1859, 1871; Coyne, 2010; Dawkins, 2009; Miller, 1999; Wilson, 2007). Evolution is accepted and supported by all reputable scientists and new evidence surfaces daily to support its explanatory power. Yet, many students receive little or no exposure to this important concept (Lerner, 2000). Fewer than 50% of Americans accept evolution (Miller, Scott, & Okamoto, 2006). People and groups opposed to the teaching of evolution in the public schools have pressed teachers and administrators to present ideas that conflict with evolution or to teach evolution as a "theory, not a fact." And this cultural movement giving rise to this nonscientific worldview in America has also been gaining strength in other countries (see PACE, 2009 for a summary of such events in Europe).

The effect these movements have had on evolution education is unsettling. These attitudes have contributed to widespread misconceptions about the state of biological understanding specifically, and more generally about what science is and is not. Even worse is a second wall of resistance: amongst those who do accept evolution as a suitable framework for unifying biological knowledge, many don't relate the importance of evolution to the human species. There is a divide within academia regarding the theory's transferability to human affairs and evolution is still taught primarily as a subject in the biological sciences rather than as a theory that can help to unify the human-related disciplines (Ehrenreich & McIntosh, 1997).

Nonetheless, each of these walls of resistance can be surmounted through a course properly designed to increase the acceptance of evolutionary theory through an increased focus on the NOS. Research has demonstrated that understanding the nature of science may be correlated with accepting evolution for both students (Lombrozo et al., 2008; Verhey, 2005; Woods & Scharmann 2001) and high school biology teachers (Rutledge & Mitchell, 2002; Rutledge & Warden, 2000; Scharmann & Harris, 1992; Trani, 2004). And there is some suggestive evidence that learning about NOS can increase acceptance of evolution in teachers (Scharmann & Harris, 1992) and in students (Verhey 2005). In one study, acceptance of evolution in secondary school teachers increased after a workshop that focused on scientific theory and evolution (Scharmann & Harris, 1992). A course developed at Binghamton University is specifically designed to accomplish this goal—to simultaneously foster an understanding of evolutionary theory and NOS through a scientific investigation of the contemporary evolutionary perspective—and stands as the introductory course for the University-wide Evolutionary Studies program.

Evolutionary Studies at Binghamton University

Binghamton University's (B.U.) Evolution for Everyone course is the introductory course for the university's EvoS program, a comprehensive, transdisciplinary effort to teach evolution as part of a scientific approach to all human-related subjects in addition to the biological sciences (O'Brien & Wilson, 2010; O'Brien, Wilson, & Hawley, 2009; Wilson 2005, 2007).

The basic mission of an EvoS program is to expand evolutionary theory beyond the biological sciences, to reach as many students as possible early in their academic careers, and to provide a multicourse curriculum so students can continue their training throughout their academic careers. The courses in B.U.'s EvoS program are designed to not only teach evolutionary theory but to specifically emphasize the transfer of evolutionary concepts across domains to enhance students' understanding of human-related knowledge. Initiated in 2003, it has become the basis of a nationwide consortium of programs funded by the National Science Foundation (http://www.evostudies.org/) that currently includes groups from over 25 institutions.

The cornerstone of an EvoS program is an introductory "Evolution for Everyone" course that immerses students from all backgrounds and academic disciplines in both evolutionary theory and the scientific process. The course has proven successful at influencing the acceptance and understanding of evolutionary theory in students from the sciences and the humanities (O'Brien, Wilson, & Hawley, 2009; Wilson, 2005), and has demonstrated success at reaching students across a multitude of political and religious backgrounds. Here, we continue an investigation of this course by studying the effects of "evolutionary training" on students' CT skills.

Methods

The Course: Evolution for Everyone

B.U.'s version of "Evolution for Everyone" (E4E, hereafter) is a large (approximately 170 students) introductory-level biology course that has no prerequisites and satisfies general education requirements. The course is cross-listed by the biology and anthropology departments, reflecting its emphasis on both biological and human-related topics. Although a 100-level course, it is popular among students from all class levels and a diverse range of majors (see Table 16.1). The basic design of the course includes the following features:

- A large lecture course that provides students from any background with a firm knowledge of the mechanisms of evolution.
- A strong focus on the application of evolutionary theory to human affairs.
- A curriculum that is structured to highlight NOS and the scientific habits of mind. This includes both in-class experiments and a final project that requires the students to propose novel research on topics of their own choice. Basic statistics are also taught through analyses of in-class experiments and evaluating evidence from the literature.
- The course goes beyond lecture mode, incorporating active, in-class demonstrations, discrepant events, and other "minds-on" activities on a regular basis; as well as cooperative discussions and brainstorming tasks.
- A before-and-after assessment (Hawley, Short, McCune, Osman, & Little, 2011) that has demonstrated a positive effect on both the students' knowledge of evolution and their attitudes towards its relevance.
- In addition to bi-weekly lectures, students also attend smaller weekly activity/discussion sessions run by undergraduate teaching assistants.
- Exams are take-home essay format, which allows students to fully digest the course material, emphasizing the application of principles being learned.

The format of the course involves teaching the basic principles of evolution during the first part of the semester, which is then applied to a diverse range of biological and human-related subjects over the rest of the course. This format became the basis of a book by the same name (Wilson, 2007), which is now used as the textbook with supplementary reading from the primary literature.

For additional details about B.U.'s EvoS program, the course curriculum, and examples of instruction style, we direct the reader to the following sources: O'Brien & Wilson, 2010; O'Brien, Wilson, & Hawley, 2009; Wilson, 2005.

TABLE 16.1. Demographic Information of Study Population

DEMOGRAPHIC VARIABLES	NUMBER	PERCENTAGE
Gender		
Male	44	40.7
Female	64	59.3
Class Level		
Freshman	42	38.9
Sophomore	20	18.5
Junior	27	25.0
Senior	18	16.7
Graduate Student	1	0.9
Major		
Accounting	3	2.8
Anthropology[a]	5	4.6
Art	1	0.9
Biochemistry	2	1.9
Bioengineering	2	1.9
Biology	11	10.2
Chemistry	2	1.9
Communications	1	0.9
Computer Science	2	1.9
Education	2	1.9
Engineering	7	6.5
English	2	1.9
Environmental Studies	1	0.9
Geology	1	0.9
Human Development	3	2.8
Integrative Neuroscience	7	6.5
Mathematics[a]	3	2.8
Nursing	1	0.9
Philosophy	2	1.9
Political Science	1	0.9
Psychology	29	26.9
Undeclared	15	13.9
Religion		
Agnostic	19	17.6

TABLE 16.1. Continued

DEMOGRAPHIC VARIABLES	NUMBER	PERCENTAGE
Atheist	11	10.2
Buddhist	2	1.9
Christian	42	38.9
Jewish	15	13.9
Spiritual but not religious	14	13.0
Other/Undeclared	5	4.6
Youth Exposure to Evolution[b]	2.34 (0.66)	

[a]One student was pursuing a dual major and was included in frequency count for both Anthropology and Mathematics.

[b]Demographic variable from Evolutionary Attitudes and Literacy Survey scale. Students evaluated a series of questions related to youth exposure to evolution on a scale from 1 to 7; class mean (standard deviation) is reported.

Participants

Participants were students in B.U.'s Fall 2011 E4E course. Of the 182 students enrolled in the course, only 108 completed both surveys at times one and two (T1 and T2). These discrepancies are in part due to the university's add-drop period, which extends two weeks into the semester

There were 44 males and 64 females, and the population represented a wide variety of majors and academic standings (Freshman, Sophomore, Junior, Senior). As religious affiliation is frequently posited as a major hurdle to evolution education, religious affiliation was also identified. Demographic information is reported in Table 16.1.

Measures

Assessing Course Goals with the Evolutionary Attitudes and Literacy Survey

Since the hypothesis being tested in this paper claims that students who develop a greater understanding of evolutionary theory will have improved CT performance, the overall efficacy of the course at teaching the evolutionary perspective and the NOS were assessed first to determine whether the course was successful in this task. Outcomes were assessed by way of the Evolutionary Attitudes and Literacy Survey (EALS; Hawley, Short, McCune, Osman, & Little, 2011), which was administered at the beginning[3] and end of the course via an anonymous online survey made available to enrolled students. Participation was voluntary and all methods were approved by B.U.'s Institutional Review Board.

The EALS is comprised of 12 scales assessing various belief systems (e.g., religious), attitudes (e.g., about the relevance of evolutionary theory), knowledge of biological topics (i.e., genetics), and science literacy (used as a measure for NOS understanding). The scales were as follows: Attitudes Toward Life, Intelligent Design Fallacies, Young Earth Creationist Beliefs, Moral Objections to Evolution, Social Objections to Evolution, Distrust of the Scientific Enterprise, Relevance of Evolutionary Theory, Genetic Literacy, Evolutionary Knowledge, Evolutionary Misconceptions, Knowledge about the Scientific Enterprise, and Self-exposure to Evolution (items and scale reliabilities reported in Appendix). Responses to each scale item were measured using a 7-point Likert scale from "strongly disagree" to "strongly agree," the center point ("4") representing "neither agree nor disagree." Scores for each of the EALS scales were calculated as a mean of the items in that scale. In addition to assessing beliefs, knowledge, and attitudes, the survey assesses a wide range of background demographic information including political and religious activity and affiliations, intended major, class level, and experience with courses teaching evolution during college and high school.

Correlations and interactions between EALS scales have already been analyzed in previous reports (O'Brien, Wilson, & Hawley, 2009) and relationships between changes in these scales and participants' backgrounds will not be a focus of this study. The EALS is included in this analysis in order to determine whether an "EvoS Effect" can be expected in the study population; in other words, whether the students sufficiently incorporated a deeper understanding of evolutionary theory and the NOS which, according to the hypothesis presented in this paper, should carry over to an increase in CT skills.

Assessing Critical Thinking

The literature identifies two primary modes of analyzing CT: 1) a componential assessment, which employs standardized CT questions (e.g., Facione, Facione, & Sanchez, 1994) that rely largely on formal or informal logic to individually assess specific components of what is generally considered to be CT—e.g., inference, induction, deduction, interpretation,

---

[3] From one week before the first day of classes to the beginning of the second day of class.

analysis, etc.; and 2) a more holistic approach that utilizes rubrics as an assessment strategy, allowing for the analysis of CT in more-general open-ended responses (Paul & Elder, 2007). Because students in E4E regularly participate in surveys and in-class experiments, this study utilized the latter of these two methods of CT assessment to reduce survey fatigue (Porter, Whitcomb, & Weitzer, 2004).

CT was analyzed with a separate pre-/post-course survey, also administered during the first and last weeks of the course. Each survey consisted of six questions that were designed by the researchers to encourage a critical evaluation of a given topic. The survey questions asked students to reason about situations in the biological sciences and situations from human-related topics. This approach was selected to test students' application of CT and the near-transfer of learning to a subject that is typically presented in line with evolutionary theory (i.e., biology), while also testing CT and the far-transfer of evolutionary principles to an area where evolution is less prominent (i.e., the social sciences and human-related affairs).

Eleven questions were designed: five questions related to biological scenarios and six questions concerning situations regarding the human condition (reported in Table 16.2). The questions were randomly assigned to pre- and post-test conditions. Each survey was composed of three questions from the biological category (*Bio*) and three questions from the social sciences (*SS*). One question was repeated on both surveys ("*Carnivores*") to examine differences in responses between survey administrations within a single question. Students were asked to answer each question with three to five sentences and were not otherwise primed to execute CT.

Measuring Critical Thinking with Lexical Content Analysis

CT survey responses were first analyzed with the Linguistic Inquiry and Word Count text analysis program (LIWC, pronounced "Luke"; Pennebaker & Francis, 1996, 1997) in order to identify potential differences in CT between T1 and T2. LIWC analyzes text files and computes 61 language dimensions, each presented as a percentage of total words. Analyses in this study focused on the cognitive mechanism parameters: insight (i.e., *think, consider*), causation (i.e., *because, effect*), discrepancy (i.e., *should, could*), tentativeness (i.e., *maybe, guess*), certainty (i.e., *always, never*), inhibition (i.e., *block, constrain*), inclusion (i.e., *and, with*), and exclusion (i.e., *but, without*). These parameters have been used in prior studies of critical thinking (Carroll, 2007; Pennebaker, Mayne, & Francis, 1997; Pennebaker & Seagal, 1999). Because LIWC identified significant

TABLE 16.2. Critical Thinking Questions and Krippendorf's Alpha Scores of Inter-Rater Reliability

TIME	DOMAIN	NAME	QUESTION	APPROACH	LOGIC	EVIDENCE	CONCLUSION	SUM
Time 1	**Bio**	Carnivores	An ecologist surveying mammal populations in tropical forests observed a greater diversity of carnivores on plots of and with greater plant diversity than those with fewer plant varieties. What factors might explain this correlation?	0.780	0.843	0.799	0.871	0.903
		Vampire Bats	Vampire bats drink blood from large mammals such as livestock. They will starve if they don't have a meal within three days. In addition to feeding directly from mammals, they can also regurgitate food to each other. You are studying the social behavior of vampire bats. What factors do you think will influence their food sharing? Please explain why.	0.879	0.911	0.616	0.954	0.964
		Island Ferns	A habitat conservancy group became alarmed when it performed a survey within a group of islands and found only four species of fern on a small island, compared to 16 species of fern typically found on the larger islands. How might this difference be explained?	0.831	0.894	0.699	0.887	0.945

SS	Cinderella	Versions of the Cinderella story are found in many different cultures all around the world. As is the plight of Cinderella in her story, sociological research has revealed that stepchildren are much more likely to be abused by a guardian than are that parent's own children. Identify some reasons that someone may mistreat their step-children, while at the same time they may be favoring their own children; be sure to explain your answer.	0.779	0.810	0.791	0.813	0.903
	Newspaper	You read a newspaper article reporting that the rate of homicide is lower in regions that have higher life expectancies. How would you explain this pattern?	0.816	0.832	0.205	0.842	0.919
	Obesity	Many health problems, such as heart disease, cancer, and obesity, are caused by the way we eat. Please explain why is it so difficult for us to eat in more healthy ways?	0.766	0.744	0.731	0.442	0.870

(*continued*)

TABLE 16.2. Continued

TIME	DOMAIN	NAME	QUESTION	APPROACH	LOGIC	EVIDENCE	CONCLUSION	SUM
Time 2	**Bio**	Carnivores	An ecologist surveying mammal populations in tropical forests observed a greater diversity of carnivores on plots of land with greater plant diversity than those with fewer plant varieties. What factors might explain this correlation?	0.949	0.877	0.866	0.835	0.936
		Guppies	You are observing guppies in a stream. You notice that guppy school size varies according to the other fish species found in the pool and that school size is largest when an unknown orange-colored fish species is also present. When you research the unknown species, do you expect to identify the orange fish as a predator to the guppy, or not? Why?	0.780	0.843	0.799	0.871	0.903
		Slime Molds	Many organisms, such as aphids, slime molds, and sea anemones, reproduce sexually as well as asexually. Which mode of reproduction do you expect to find occurring during periods of favorable conditions? Please explain why.	0.975	0.963	0.856	0.961	0.986

SS					
Women	Women with higher occupational status and prestige tend to have fewer children than do women with lower occupational status and prestige. Explain why this may variance occur?	0.952	0.953	0.931	0.948	0.983
Cleaning	As a widespread fear of disease and illness sweeps through our world, more and more cleaning supplies are being pushed onto store shelves and sold. Please explain how and why this increase in excessive cleaning patterns may be harming our health?	0.883	0.943	0.888	0.945	0.968
Guns & Snakes	About 10,000 people are killed with guns in the United States annually, while spiders and snakes kill only a handful. Yet, many people tend to fear spiders and snakes about as easily as they do a pointed gun, and more easily than an unpointed gun. How do you explain these responses?	0.954	0.937	0.837	0.937	0.977

differences in the use of cognitive mechanisms between T1 and T2 (as detailed below, also see Table 16.3), a more thorough investigation of CT responses was conducted by employing independent raters to score each response according to a rubric. This rubric and the scoring process are described in the next section.

Measuring Critical Thinking by Independent Raters

DEVELOPING THE CRITICAL THINKING RUBRIC

The rubric used in this study was a modified version of the Association of American Colleges & Universities (AAC&U) Valid Assessment of Learning in Undergraduate Education (VALUE) Critical Thinking

TABLE 16.3. Evolutionary Attitudes and Literacy Survey

SCALE	TIME 1	TIME 2	t-VALUE	COHEN'S d
Evolutionary Knowledge[a]	5.09 (0.82)	5.27 (0.97)	1.930	0.19
Relevance of Evolutionary Theory[b]	5.39 (1.01)	5.70 (1.11)	3.119***	0.30
Evolutionary Misconceptions[a]	3.91 (0.95)	3.66 (0.96)	−2.492*	−0.24
Self Exposure to Evolution[b]	2.17 (0.74)	2.53 (0.82)	5.020***	0.49
Moral Objections[b]	2.04 (0.95)	1.91 (0.92)	−1.515	−0.15
Social Objections[b]	3.07 (1.15)	2.79 (1.18)	−2.627**	−0.25
Intelligent Design Fallacies[a]	2.80 (0.97)	2.59 (1.05)	−3.002***	−0.29
Young Earth Creationist Beliefs[b]	2.42 (1.04)	2.27 (1.15)	−1.672	−0.16
Knowledge of the Scientific Enterprise[a]	5.47 (0.97)	5.52 (1.09)	0.489	0.05
Distrust of Science[b]	2.96 (0.88)	2.62 (0.97)	−4.080***	−0.39
Genetic Literacy[b]	5.12 (0.73)	5.31 (0.93)	2.269*	0.22
Political Activity[b]	2.55 (1.18)	2.59 (1.24)	0.351	0.03
Religious Activity[c]	3.16 (1.68)	3.25 (1.63)	1.236	0.12
Conservative Self-Identity[b]	3.47 (1.40)	3.70 (1.41)	2.327*	0.22
Attitudes Towards Life[b]	3.92 (1.48)	3.84 (1.59)	−0.698	−0.07

NOTES: Table reports Evolutionary Attitudes and Literacy Survey (EALS) scores (and standard deviations) and results of paired samples t-tests between EALS Scores at times 1 and 2, with Cohen's d.

*$p < 0.025$; **$p < 0.01$; ***$p < 0.005$.

[a]df = 105.

[b]df = 106.

[c]df = 107.

rubric (Rhodes & Finley, 2013; available at http://www.aacu.org/VALUE/rubrics/). A modified version of the VALUE rubric was developed because, in its original form, the rubric is designed to be used with text passages that are composed of multiple paragraphs. Since our questions were designed to be answered in only 1-2 paragraphs, the VALUE rubric was modified to be more applicable to the short-answer questions developed for this study. The finalized rubric is provided in Figure 16.1.

RATING RESPONSE SETS

Scoring of the survey responses for CT was performed by three independent raters. All raters were unaffiliated with the course and B.U.'s EvoS program, and remained naïve to the experimental aims. Meetings were conducted in the university library, a neutral location so as to remain unbiased to the departmental "home" for this study. Prior to scoring experimental responses, the raters were given a program of training, which included background reading about measuring CT, a four-week orientation to the rubric, calibration sessions and scoring practice, and discussions of several samples that represented the whole range of possible results.

Inter-rater reliability of CT scores was calculated using Krippendorf's alpha (Hayes & Krippendorf, 2007) for each of the CT components (*Approach, Logic, Evidence,* and *Conclusion*) and the combined sum of component scores for each question (*Sum*). Every component yielded moderate to high measures of inter-rater reliability, except for three *Evidence* scores and one *Conclusion*. All values are reported in Table 16.2.

The low reliability of *Evidence* scores is, in part, the result of very low scores in this category overall. Most responses were scored as "0" for this category, as they were identified to contain no evidence or examples and so a single discrepancy greatly influenced the reliability calculation. For example, of the 166 responses collected for the *Newspaper* question, the *Evidence* scores ($\alpha = .20$) for each rater were: Rater 1–163 responses scored "0" and 3 responses scored "1"; Rater 2–all 166 responses scored "0"; Rater 3–161 scored "0", 3 scored "1", and 1 response scored "2." Given this pattern of ratings and given that the reliability of the *Sums* scores for every question yielded high reliabilities, all scores were retained for further analysis.

Because nearly all scores demonstrated high reliability between raters, mean scores were calculated for each question on each of the individual

	0	Benchmark 1	Milestones 2	Milestones 3	Capstone 4
Approach/Propose Solution	Does not meet benchmark -or- attempt to answer question.	Propose a solution that is vague, simplistic, or cliché.	Two or more solutions that may be vague, simplistic or cliché. -or- One solution that represents a basic understanding.	Proposes one or more solutions from the same point of view that indicate a deep comprehension of the problem. -or- Poorly presents multiple points of view.	Proposes multiple points of view that indicate a deep comprehension
Development of Argument/Logical Support	Does not meet benchmark -or- attempt to answer question.	Provides limited logic that is too general to explain the specific details of the problem.	Has specific support for argument but requires reader to make assumptions.	Argument uses adequate reasoning and support for assertions.	Uses a clear and concise argument using a *systematic* flow of logical support based off the premise.
Evidence/Examples	Does not meet benchmark -or- attempt to answer question.	Example is too vague or general.	Uses one or more clear examples but is limited to the scope of the information given.	Brings in specific, outside, relevant information that is not well-intergrated into response; still leaves something out.	Brings in outside, relevant information to create a scenario that is easily envisioned by the reader, that addresses every proposed solution.
Conclusion	Does not meet benchmark -or- attempt to answer question.	Conclusion is ambiguous or illogical to the premise. (Dosen't make sense.)	States a general conclusion that, bacause it is so general, also applies beyond the scope of the inquiry findings.	States a specific conclusion that is drawn from information used but: does not address all proposed solutions, -or- does not have systematic flow.	Clearly states a conclusion that *systematically* flows from information used. Conclusion will leave no ambiguities to the position; they have addressed every proposed solution.

FIGURE 16.1. Critical thinking/problem solving rubric.

components (*Approach, Logic, Evidence,* and *Conclusion*) and for the total *Sums*. Scores from the pre-test condition were combined into a pre-test (T1) mean score for the biological questions (*Bio*), social sciences questions (*SS*), and a total pre-test score (*Combined*). Similarly, a post-test (T2) mean score was calculated for each of these categories—*Bio, SS,* and *Combined.*

It is important to point out that answers were scored for the inclusion of CT skills and not for correct answers. Independent raters were not provided an "answer key" nor was there a discussion of "correct" answers during the norming sessions. This was done in order to prevent bias in response ratings, as an incorrect answer could still demonstrate good CT skills.

Assessing Transfer of Learning

To test for transfer of learning, each response from the CT survey was evaluated for the incorporation of evolutionary principles by a fourth independent rater. Responses were considered to incorporate an evolutionary perspective if they included any of the following key terms with appropriate application of the principles: adapt, adaptation, drift, evolution, evolve, evolutionary, fitness, selection, trade-off, mismatch, byproduct, and variation. For this reason, this fourth independent rater was selected from B.U.'s EvoS population, as knowledge of evolutionary principles was required for accurate assessment.

Responses incorporating evolutionary reasoning into their answers were scored as "1," while answers which did not directly employ an evolutionary perspective were scored as "0." Responses containing key terms but lacking an obvious evolutionary rationale were not considered cases of knowledge transfer. Total pre- and post-test transfer scores were calculated for each participant as a sum of each score in the individual categories—*Bio* and *SS*, and for the *Combined* scores.

Results

Analysis of Evolutionary Attitudes and Literacy Survey

A paired samples *t*-test of scores between T1 and T2 of the EALS revealed a greater understanding of evolutionary theory and its general applicability beyond the confines of biology. Scores also reflect a better understanding of the NOS. Mean scores for T1 and T2 are reported in Table 16.3.

Students reported a significant increase in understanding the *Relevance of Evolutionary Theory* ($t(106) = 3.119$, $p < .005$), and a decrease in *Evolutionary Misconceptions* ($t(105) = -2.492$, $p < .015$). Students also reported an increase in their *Self-exposure to Evolution* ($t(106) = 5.020$, $p < .001$), suggesting their willingness to further increasing their understanding of evolution outside of any class requirements. Students also reported a marginally significant increase in *Evolutionary Knowledge* ($t(105) = 1.931$, $p = .056$).

This demonstrated increase in students' knowledge of evolutionary theory was accompanied by a decrease in beliefs that typically contend with evolutionary thinking. Students reported a decrease in beliefs regarding *Social Objections to Evolutionary Theory* ($t(106) = -2.627$, $p = .010$), and a decrease in *Intelligent Design Fallacies* ($t(105) = -3.002$, $p < .005$); accompanied by reported, however non-significant, decreases in the *Moral Objections to Evolutionary Theory* ($t(106) = -1.515$, $p = .133$), and decreased *Young Earth Creationist Beliefs* ($t(106) = -1.672$, $p = .098$). Though differences between T1 and T2 in *Moral Objections* and *Young Earth Creationist Beliefs* were not significant, they are still important to include as an analysis of the overall trend of differences from the beginning to end of the semester.

Regarding NOS measures, while *Knowledge of the Scientific Enterprise* did not increase ($t(105) = 0.489$, $p = .626$), students reported a decrease in their *Distrust of Science* ($t(106) = -4.080$, $p < .001$). Future studies would benefit with the inclusion of a direct measure of knowledge about the NOS but these current scales serve the function of identifying an improvement in student's general understanding about the NOS.

Religious and political attitudes were generally unaffected, however students reported a significant increase in *Conservative Self-Identity* ($t(106) = 2.327$, $p < .025$). Further implications regarding political and religious affiliations will be discussed elsewhere, as they are outside the scope of this study.

The pattern of results presented here demonstrates that students sufficiently incorporated a deeper understanding of evolutionary theory and the NOS through participation in the course. Therefore, an analysis of CT skills was conducted to examine if this "EvoS Effect" carries over to an increase in CT skills.

TABLE 16.4. Linguistic Inquiry and Word Count

LEXICAL DIMENSION	TIME 1	TIME 2	t-VALUE	COHEN'S d
Cognitive Mechanisms	16.84 (6.57)	17.51 (8.62)	1.541	0.09
Insight	2.03 (2.86)	1.39 (2.49)	−4.184***	−.24
Causation	3.04 (3.00)	4.63 (4.87)	6.888***	0.39
Discrepancy	2.79 (3.25)	2.38 (3.85)	−2.007*	−0.11
Tentative	4.06 (3.68)	2.88 (3.61)	−5.655***	−0.32
Certainty	0.70 (1.34)	0.46 (1.53)	−2.833**	−0.16
Inhibition	0.40 (1.11)	0.78 (2.42)	3.513***	0.20
Inclusive	3.49 (3.20)	4.30 (4.65)	3.550	0.20
Exclusive	2.40 (2.98)	2.19 (3.13)	−1.180	−0.07

NOTES: Table reports means (standard deviations) and results of independent samples t-test, with Cohen's d. df = 1,223.

*$p < 0.05$; **$p = 0.005$; ***$p < 0.001$.

Analysis of Critical Thinking

Analysis of Critical Thinking with Lexical Content Analysis

Table 16.4 presents the results of a paired samples t-test between LIWC scores from T1 and T2 for each of the cognitive mechanisms (*Insight, Discrepancy, Tentative, Certainty, Inhibition, Inclusive, Exclusive*). Results revealed significant differences between scores in all categories of cognitive mechanisms except *Exclusive*. At T2, students were more likely to include words expressing *Causation, Inhibition*, and *Inclusion*. They were less likely to include words expressing *Insight, Discrepancy, Tentativeness*, and *Certainty*. Implications of these results will be interpreted in the discussion section.

While LIWC can identify differences in the use of cognitive terms, these results do not lend themselves to an interpretation for assessing any increase or decrease in CT—they merely suggest that a difference could exist based on word use. An analysis of response scores by the independent raters was conducted to examine the quality of CT in the survey responses.

Analysis of Critical Thinking Scores Assigned by Independent Raters

Paired samples t-tests between CT scores from T1 and T2 were run for each CT component and the sums of CT scores for the *Bio* questions, *SS* questions, and the *Combined* questions. Paired samples t-tests were also performed

TABLE 16.5. Critical Thinking Questions

QUESTION SET	CT COMPONENT	TIME 1	TIME 2	t-VALUE	COHEN'S d
Carnivores[a]	Approach	2.20 (0.41)	2.49 (0.61)	4.753***	0.48
	Logic	2.14 (0.52)	2.20 (0.57)	0.954	0.10
	Evidence	0.02 (0.17)	0.16 (0.45)	2.762**	0.28
	Conclusion	1.98 (0.58)	2.06 (0.51)	1.356	0.14
	Sum	6.33 (1.36)	6.90 (1.84)	3.220**	0.33
Bio[b]	Approach	2.08 (0.38)	2.21 (0.26)	3.233**	0.32
	Logic	2.01 (0.43)	2.05 (0.36)	0.967	0.09
	Evidence	0.04 (0.14)	0.06 (0.16)	1.324	0.13
	Conclusion	1.91 (0.36)	1.82 (0.35)	−2.018*	−0.20
	Sum	6.03 (1.12)	6.14 (0.96)	0.993	0.10
SS[c]	Approach	2.09 (0.37)	2.33 (0.41)	5.589***	0.54
	Logic	1.94 (0.38)	2.08 (0.43)	3.074**	0.30
	Evidence	0.26 (0.42)	0.26 (0.40)	0.107	0.01
	Conclusion	1.70 (0.30)	2.07 (0.40)	10.098***	0.97
	Sum	5.99 (1.20)	6.74 (1.46)	5.432***	0.52
Combined[c]	Approach	2.05 (0.29)	2.12 (0.35)	1.909	0.18
	Logic	1.95 (0.35)	2.05 (0.42)	2.409*	0.23
	Evidence	0.07 (0.26)	0.10 (0.30)	0.831	0.08
	Conclusion	1.85 (0.38)	1.95 (0.35)	2.589**	0.25
	Sum	5.94 (1.09)	6.34 (1.17)	3.672***	0.35

NOTES: Table reports scores (standard deviations) and results of paired samples t-tests for Critical Thinking questions at Times 1 and 2. Questions were analyzed by type (single question measure: Carnivores, Bio, SS) and for the combined values. Cohen's d effect sizes were calculated for each.

*$p < 0.05$, **$p < 0.015$, ***$p < 0.001$.

[a]$df = 96$.

[b]$df = 105$.

[c]$df = 107$.

between the pre- and post-test scores on the individual *Carnivores* question, so as to measure the differences on a single item between T1 and T2. Full results are presented in Table 16.5; Figure 16.2 displays the scores of CT pre and post *Sums* for each question set (i.e., *Carnivores*, *Bio*, *SS*, and *Combined*).

Beginning with the repeated-item *Carnivores* question, results of a paired samples *t*-test indicate that students scored higher on overall CT at

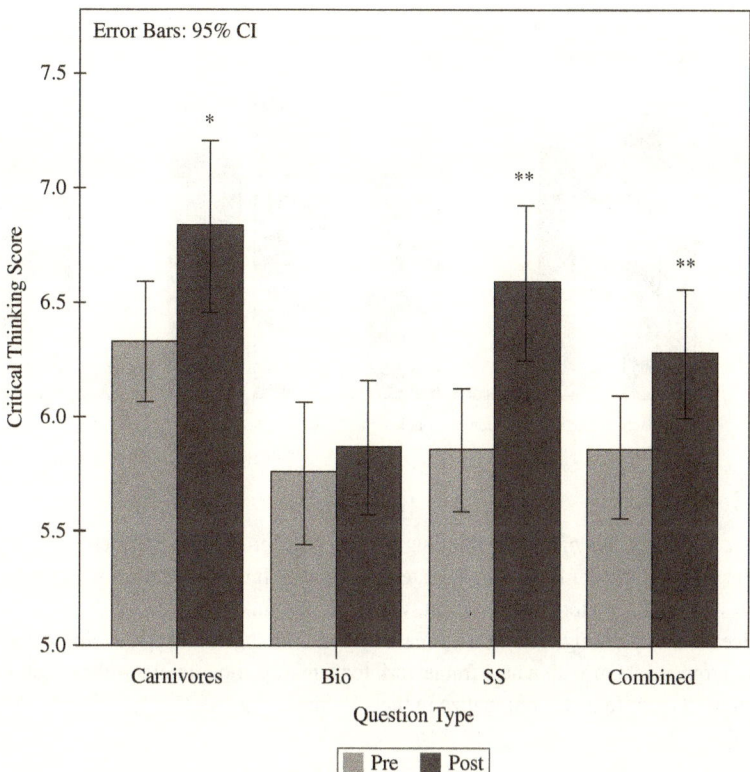

FIGURE 16.2. Critical thinking scores (*Sums*) at times 1 and 2 for each question type and for the combined set. *$p < 0.015$; **$p < 0.001$.

T2 than at T1 (*Sum: t*(96) = 3.200; $p < .005$). Analysis of individual CT components attributes this improvement in overall CT to a better *Approach* to answering the question ($t(96) = 4.753, p < .001$), and by including better *Evidence* to justify their claims ($t(96) = 2.762, p < .010$).

Results for *Bio* questions, *SS* questions, and the *Combined* questions also demonstrated increased CT scores at T2. Students demonstrated the largest increases in CT on the *SS* responses, and improved on every dimension except for *Evidence* ($t(107) = 0.107, p = .915$). Interestingly, the least impressive of these results was in the *Bio* questions, where only *Approach* resulted in a significant increase from T1 to T2 ($t(105) = 3.233, p < .005$), and students actually wrote worse conclusions to their responses at T2 for the *Bio* questions ($t(105) = -2.018, p < .050$). Figure 16.3 displays the scores of CT components in the *Bio* and *SS* responses.

In the *Combined* question sets, students demonstrated an overall significant improvement on all components of CT except for *Approach*

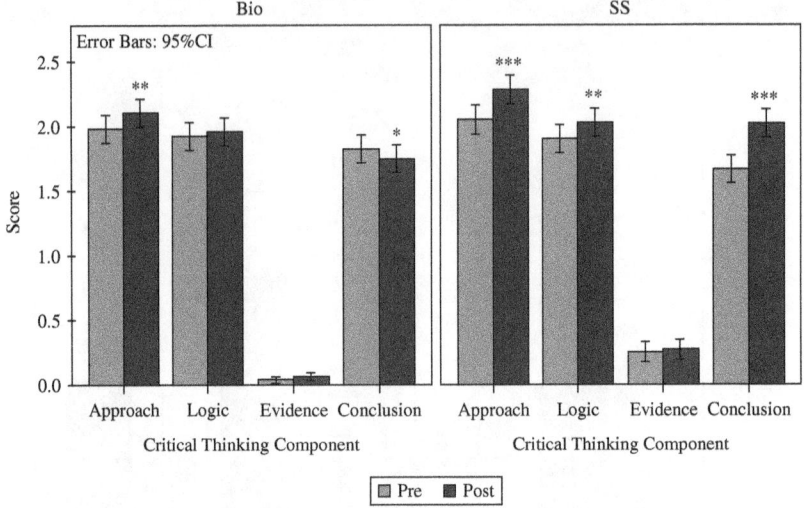

FIGURE 16.3. Scores of critical thinking components for Bio and SS questions. Results indicate that students demonstrated the greatest improvements of CT in the SS questions. This pattern suggests that, while students may have already been in the practice of incorporating evolutionary reasoning into biological topics, evolutionary theory provided them with a new framework for thinking about human-related subjects and allowed them to better rationalize in these domains. $*p < 0.05$; $**p < 0.015$; $***p < 0.001$.

($t(107) = 1.909$, $p = .059$; marginal significance) and the inclusion of *Evidence* ($t(107) = -0.408$; $p = .408$).

Analysis of Knowledge Transfer

The transfer of evolutionary principles was analyzed for the repeated-item *Carnivores* question, and for the combined *Bio* questions, *SS* questions, and the *Total* response set. Prior to a comparison of the scores between the pre- and post-test conditions, the scores of the *Bio* and the *SS* questions were compared with each other to identify the likelihood of students incorporating evolutionary principles into responses from each of these domains. Results of a paired samples *t*-test revealed that at T1 students were more likely to apply evolutionary principles in their responses to biological topics ($M = 0.30$ ($SD = 0.52$)) than in the social sciences ($M = 0.13$ ($SD = .39$); $t(107) = -2.732$, $p < .010$). This bias was no longer present at T2, where students were equally as likely to apply evolutionary reasoning into their responses for the *SS* questions ($M = 0.43$ ($SD = 0.69$)) as they were for the biological questions ($M = 0.51$ ($SD = 0.63$); $t(107) = -1.174$,

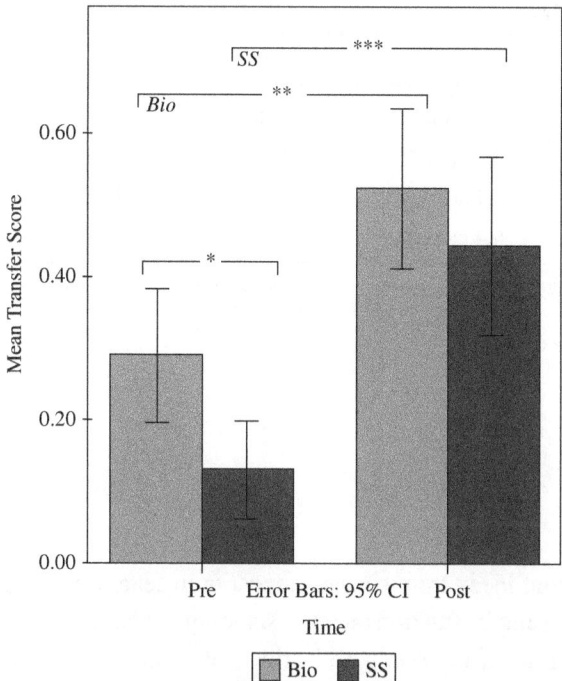

FIGURE 16.4. Transfer scores at times 1 and 2 in *Bio* and *SS* responses. Results indicate that students were more likely to incorporate evolutionary reasoning at Time 1 into responses for biological questions than for the human-related topics. This bias was removed at Time 2, where students were equally as likely to incorporate evolution into their responses from either domain (*Bio* or *SS*). For both domains, students were more likely to incorporate evolutionary thinking at Time 2. $*p < 0.010$; $**p = 0.005$; $***p < 0.001$.

$p = .243$). Figure 16.4 displays these differences between *Bio* and *SS* transfer scores at T1 and T2.

Following this preliminary analysis, scores between T1 and T2 were compared. Analysis of the single-item *Carnivores* question did not reveal a significant difference between the inclusion of evolutionary principles in students' responses between T1 and T2 ($t(96) = -0.241$, $p = .810$). However, when comparing the full sample of responses, results revealed that students were more likely to incorporate an evolutionary explanation at the end of the semester than they were at the start ($t(107) = 4.783$, $p < .001$). This result was demonstrated in both the biological questions ($t(107) = 2.859$, $p = .005$) and the human-related questions ($t(107) = 4.573$, $p < .001$). Results are reported in Table 16.6.

TABLE 16.6. Domain Transfer of Knowledge

QUESTION SET	TIME 1	TIME 2	t-VALUE	COHEN'S d
Carnivores[a]	0.08 (0.28)	0.07 (0.26)	−0.241	−0.02
Bio[b]	0.30 (0.52)	0.51 (0.63)	2.859*	0.28
SS[b]	0.13 (0.39)	0.43 (0.69)	4.573**	0.44
Combined[b]	0.43 (0.66)	0.94 (1.10)	4.783**	0.46

NOTES: Table reports scores (and standard deviations), and results of paired samples t-tests for Domain Transfer of Knowledge, with Cohen's d effect sizes.

*$p = 0.005$; **$p < 0.001$.

[a]$df = 96$.

[b]$df = 107$.

Discussion

Three important ideas from science education research informed this study. First, that focusing instruction on understanding the Nature of Science (NOS) results in improved reasoning and critical thinking (CT). Second, that an improved understanding of evolutionary theory can improve NOS skills. And third, that understanding evolution can be improved by focusing instruction on humans. Thus, this study sought to test the hypothesis that a course emphasizing the general applicability of evolutionary theory should be expected to improve CT, particularly by way of increased knowledge transfer.

Discussion of Evolutionary Attitudes and Literacy Survey

Evolution for Everyone (E4E), B.U.'s introductory EvoS course, is a course that is designed explicitly around these themes. Results of the EALS confirmed that participation in the course decreased students' misconceptions about evolution and NOS, and reduced students' reports of holding conflicting beliefs against evolution. With the EALS serving as a confirmation that these stated aims of the course—improving evolutionary understanding—had been achieved, we then proceeded to examine how receiving this "evolutionary training" impacted students' CT skills, and if this "EvoS Effect" transfers across subject domains.

Discussion of LIWC Results

Through a lexical content analysis of survey results, responses were found to differ significantly between T1 and T2 in the use of words related to cognitive mechanisms, the categories most directly related to CT. Of the cognitive mechanisms, the most direct connection to CT is in the *causation* (i.e., because, hence) category. After participation in E4E, students used higher frequencies of words related to *causation* when answering the CT survey questions, indicating more thoughtful cause-and-effect responses.

Interpretations of the other cognitive mechanisms categories are slightly less straightforward. After the course, students were less *tentative* (i.e., maybe, perhaps) in their responses which would typically indicate that they had higher certainty in their answers but LIWC results also indicate that students were less-likely to use words indicating *certainty* (i.e., always, never). This pattern of results could indicate that while students were more confident in their answers they were also less-likely to speak in absolute terms, allowing more flexibility and open-mindedness (a characteristic associated with CT, and NOS education). The decrease in *insight* (i.e., think, consider) scores may indicate that students had more knowledge-informed responses and did not need to speculate as much in their explanations after participation in the class as they did at the beginning.

While LIWC is able to identify differences in the lexical content of writing samples, the program is unable to assess the contextual quality of the responses. Therefore, any assessment of CT from LIWC results is only speculative. The purpose of incorporating LIWC analysis here serves as a preliminary assessment as to whether differences in CT could be expected with a more thorough investigation. Because LIWC results did identify significant differences in the use of cognitive terms between T1 and T2, the researchers proceeded to investigate the quality of responses for differences in CT.

Discussion of Critical Thinking Assessment

A comparison of the scores assigned by naïve raters indicated a significant improvement in CT (*Sums*) at the end of the semester. Of the CT categories scored by the raters, the largest improvements were observed in students' Development of Argument (*Logic*) and the formulation of *Conclusions*.

Overall, students showed very little improvement in the incorporation of evidence/examples into their responses. This is not surprising since students were not permitted to look up information when answering the questions and the instructors intentionally avoided any conversations about these specific topics during the semester. Therefore, students were not necessarily equipped with sufficient background knowledge to incorporate specific examples for each question. Nonetheless, even without knowing any exact details of the situations presented in each question, students were better able to rationalize a meaningful response at T2, demonstrating the impact of presenting evolutionary theory as a general explanatory framework that can be integrated with knowledge across subject domains.

When the questions were analyzed by topic: biological or human-related questions, students demonstrated the largest improvements in their responses to the human-related questions. This suggests that, while a course on evolutionary theory improved overall CT on all questions, it provided a valuable new framework for organizing thinking about human affairs—an area where students are not typically encouraged to incorporate evolutionary thinking in their other courses.

Evolutionary Training and the Transfer of Learning

Beyond the assessment of general CT skills, this study specifically sought to examine the effect that evolutionary training has on domain transfer of learning. The transfer of learned information from one context to another is an overarching goal in education, and is a basic requirement for CT (Perkins & Salomon, 1988; Zoller, 1999). However, much research has demonstrated that students often fail to apply knowledge and skills learned in one context to other situations (Detterman, 1993; Ennis, 1989; Perkins and Salomon, 1989). One of the key explanations behind this disconnect is that the current emphasis on content-specific instruction and assessment leaves little room in curriculum-instruction-assessment practices for incorporating cross-cutting concepts. Content-specific instruction is not optimal for teaching students to transfer more abstract critical thinking skills across domains (Halpern, 1998). Evolutionary theory is distinctive in this regard, wherein a course that emphasizes increasing an understanding of evolutionary content can simultaneously extend the principles across examples from a wide variety of domain contexts in order to reinforce the general applicability of the theory.

Our results show that prior to participation in the course students were far more likely to apply evolutionary reasoning to questions about biological topics than they were to apply an evolutionary framework to human-related situations. This bias was no longer present after participating in E4E, and students were equally likely to incorporate evolutionary reasoning for the social science questions as they were for questions about biology.

A comparison between T1 and T2 responses revealed an overall increase in the application of evolutionary principles after participation in the course. And, like the results described for CT, the largest increase in the application of evolutionary theory was observed in the questions related to human affairs. This pattern of results demonstrates that students were not only more inclined to explain biological situations from an evolutionary perspective after taking an evolutionary science course but also that students were inclined to transfer evolutionary principles across disciplines, deepening their understanding of human-related issues.

Limitations

The study presented here provides evidence that a single course on evolutionary theory which is designed to simultaneously emphasize a deeper understanding of the NOS can have considerable impact on students' CT skills. More work must be done to further investigate these effects. At this time, the design of this study cannot discriminate whether the observed improvements can be attributed to either the learning of a domain-general evolutionary perspective or to the increased understanding of the NOS, or whether they operate in unison. Also, the assessment did not include a comparison with another course. This is in part because it is difficult to know what an appropriate control group would be. Nonetheless, this study would benefit from further investigation of these results.

A second limitation of this study that arises from the absence of a comparison group is in considering the general increase of CT during the course of routine college participation. Since most disciplines and general education classes stress critical thinking, it is reasonable to suggest that these measures could improve over any semester in the absence of evolutionary training. However, where improving CT is a widely-reported aim for most college and university programs, research has revealed that this goal often remains unrealized (Arum & Roska, 2011; Roksa & Arum, 2011). A study that followed 2,322 traditional-age students from the fall of 2005 to the spring of 2009 and examined testing data and student surveys at a broad range of 24 U.S. colleges and universities revealed that 45% of students made no significant improvement in their CT, reasoning or writing skills during the first two years of college. After four years, 36% showed no significant gains in higher-order thinking skills.

While more work is needed with appropriate control groups to fully investigate the general trajectory of CT development of students with and without evolutionary training, we further analyzed the results to speak to some of the implications associated with this concern: If most classes are designed to enhance CT skills, one might suspect that this increase would be greater for first semester freshman when compared to upperclassmen. A comparison of CT results revealed no significant differences between class standing (i.e., freshmen, sophomores, juniors and seniors) in both the pre-assessment ($F(3,100) = .866$, $p = .461$) and post-assessment scores ($F(3,100) = .140$, $p = .936$). Seniors scored the lowest in CT at the start of the semester ($M = 5.65$, $SD = 0.86$) and showed the greatest overall improvement at T2 ($M = 6.24$, $SD = 0.90$), followed by freshmen (T1: $M = 5.92$, $SD =\_1.16$; T2: $M = 6.33$, $SD = 1.16$), sophomores (T1: $M = 6.05$, $SD = 0.62$; T2: $M = 6.21$, $SD = 0.98$) and juniors (T1: $M = 6.15$, $SD = 1.20$; $M = 6.33$, $SD = 1.30$). This pattern of results reveals that CT skills improved as much for upper-classmen as they did for freshmen, providing further support for a specific EvoS Effect in this sample.

Perhaps the most suggestive results from this study are those that demonstrate that students increase the transfer of evolutionary principles to domains beyond the biological sciences—a practice that is uncommon with traditional evolution instruction. This could also be improved with further research that tests a population of past students (and a comparison group) in a manner which does not have any direct connection to an evolutionary science course. In the current methodology, students may have been primed to answer with evolutionary thinking since the survey was disseminated in connection to the class. However, in support of the conclusions drawn from this study, both the pre- and post-tests were provided as voluntary research participation through the course and any priming for evolutionary thinking would be present at both times. With these priming effects being equal, the observed results can be considered a reliable demonstration of improvement in the transfer of evolutionary principles across domains.

Conclusion

The theory of evolution is impressive, productive and important. To become a serious biologist, one needs to have a grasp of what a domain-general evolutionary theory means. To be an informed citizen, one should have a general understanding of not only what the theory claims but of how those claims are substantiated through scientific methodology. If students master an understanding of evolutionary thinking they will know how to think successfully and reach reliable conclusions; such ability will prove valuable in any human endeavor. The results presented here demonstrate that a single course on evolutionary theory that stresses evolution across domains increases students' acceptance and understanding of evolution, the NOS, and, crucially, CT performance.

It is striking, and worth emphasizing, that a traditional evolution education does not produce the EvoS Effect. Traditionally, evolution has been treated as a short, stand-alone unit in the biology curriculum, rather than a unifying theme that can improve understanding across the entire discipline. Sometimes, this topic is skipped entirely (Berkman & Plutzer, 2011). Occasionally, psychology courses will include a short "evolutionary psychology" unit, but, again, this inclusion is largely reduced to one or two classes primarily focused on evolved sex differences and "the mating mind." With this limited exposure, students are not taught to understand evolutionary theory as a toolkit that can help them understand diverse concepts and develop well-reasoned conclusions in their biology classes, let alone the more-broad applications characteristic of the EvoS Effect.

And, even when courses do attempt to feature evolution, human-related applications are limited, if not absent, in traditional evolution education. In this study, we have demonstrated that the greatest benefit from the EvoS Effect comes from improving students' thinking about humans. Without intentional instruction, this is severely limited in the traditional methods of teaching evolution. Students tend not to transfer the knowledge to the human-related subjects on their own; it must be part of their education.

In its current form at Binghamton University, Evolution for Everyone represents a successful experiment in evolution education, one that attempts to engage students of all academic and personal backgrounds in evolutionary studies. Focusing on human affairs and an interactive approach, the material is accessible and intriguing to students of all academic and personal backgrounds. We hope to export the course in an effort to replicate not so much its curriculum but its ideals: to enhance the teaching of science education, particularly evolution education, and to give all undergraduates the skills necessary to ask questions about their world and answer them effectively.

While increasing the emphasis on developing higher-order cognitive skills in all subjects is a widely recognized goal in education, this study focused on science education because changing the way we do science education can drive reform across all of education. Science education is inquiry-based; is hands-on; is engaging; and it changes the way students and teachers interact. It changes how we think about education.

Placing evolutionary science at the center of this reform can provide a means for not only improving student performance but also for improving the performance of teachers, administrators, and policy-makers. Evolutionary science can provide a general theoretical framework for understanding the complex interactions within the educational theatre and can inform the design of curriculum-instruction-assessment strategies, educational environments, and educational policies to best promote learning.

Acknowledgments

We would like to offer our sincere gratitude to a few people who were instrumental in the fruition of this study: Daniel Talamo, Chelsea Laber, William Lauffer, Matt Raskin, Nia Pellone, and Michael Sussin for the many hours of work they put into the development of the critical thinking rubric; Samantha Kuchlik for assisting with the development of early drafts of the critical thinking questions; and to the independent raters for the unwavering dedication to your task. To Ken Kurtz for consultation and assistance with early phases of experimental design and implementation. Also to Patricia Hawley, Texas Tech, and Steve Short, Kansas University, for providing us with the Evolutionary Attitudes and Literacy Survey. And, most of all, to the Fall 2011 Evolution for Everyone students.

Appendix
Questions and Cronbach's Alphas for All Scales in the Evolution Literacy Survey for Times 1 and 2, administered at the Beginning and End of the Course

Questions	Number of Items	Time 1	Time 2
Evolutionary Knowledge	N = 8	0.670	0.813
In most populations, more offspring are born than can survive.			
Individuals don't evolve, species do.			
Mutations can be passed down to the next generation.			

Questions	Number of Items	Time 1	Time 2
Increased genetic variability makes a population more resistant to extinction.			
The more recently species share a common ancestor, the more closely related they are.			
Natural selection is the only cause of evolution.[a]			
Mutations occur all the time.			
Individuals better suited to their environment are more likely to survive and reproduce.			
Relevance of Evolutionary Theory	N = 9	0.896	0.928
The theory of evolution helps us understand plants.			
Evolutionary theory is highly relevant for biology.			
The theory of evolution helps us understand animals.			
The theory of evolution helps us understand human origins.			
For explaining human behavior, evolutionary theory is irrelevant.[a]			
Evolutionary theory is highly relevant for the social sciences (e.g., anthropology, psychology, sociology).			
Evolutionary theory is highly relevant for the humanities (e.g., history, literature, philosophy).			
Evolutionary theory is relevant to our everyday lives.			
The theory of evolution helps explain the world as it is in the present.			
Evolutionary Misconceptions	N = 6	0.627	0.587
Natural selection is a random process.			
Natural selection is synonymous (means the same) as evolution.			
Characteristics acquired during the lifetime of an organism are passed down to that individual's offspring.			
Species evolve to be perfectly adapted to their environments.			
Evolution means progression towards perfection.			

Questions	Number of Items	Time 1	Time 2
Evolution is a linear progression from primitive to advanced species.			
Self Exposure to Evolution	N = 5	0.822	0.805
I've visited evolution related web sites (e.g., Science Daily, Pharyngula, Edge.org).			
I've watched evolution related videos on the web (e.g., Ted.com, YouTube).			
I read science magazines featuring evolution (e.g., *Discover, National Geographic, Nature*).			
I've watched nature shows that discussed evolution (e.g., PBS/Nova, Discovery, National Geographic).			
I've read evolution related books (e.g., by Richard Dawkins, E. O. Wilson, Steven Pinker).			
Moral Objections	N = 5	0.594	0.525
People who accept evolution do not believe in God.			
People who accept evolution as fact are immoral.			
If you accept evolution, you really can't believe in God.			
Darwinism strips meaning from our lives.			
People can be moral and believe in evolution at the same time.[a]			
Social Objections	N = 6	0.833	0.812
The theory of evolution has contributed to racism.			
Applying the theory of evolution to human affairs implies we are not fully in control of our behavior.			
The theory of evolution has contributed to sexism.			
The theory of evolution has contributed to an increase in abortion.			
The theory of evolution has contributed to genocide (the deliberate killing of a group based on nationality, race, politics, or culture).			

Questions	Number of Items	Time 1	Time 2
The theory of evolution has contributed to an increase in euthanasia (the act of killing someone painlessly or allowing to die to stop the suffering; also called mercy killing).			
Intelligent Design Fallacies	N = 11	0.809	0.809
There is scientific evidence that humans were created by a supreme being or intelligent designer.			
There is no evidence that humans evolved from other animals.			
The theory of evolution is a matter of faith and belief, just like religion.			
Humans were specially designed.			
There are no transitional fossils (remains of life forms that illustrate an evolutionary transition).			
It is statistically impossible that life arose by chance.			
The theory of evolution does not explain similarities or differences between chimps and humans.			
Complex biological systems cannot come about by slight successive modifications (i.e., they are irreducibly complex).			
Evolution is a theory in crisis.			
Evolution violates the 2nd law of thermodynamics (that systems move toward disorder, not order).[b]			
Natural selection cannot create complex structures; it is like a tornado blowing through a junkyard and creating a 747.			
Young Earth Creationist Beliefs	N = 9	0.868	0.898
I read the bible literally.			
God created humans in their present form.			
Humans never could have been related to apes.			
The Earth isn't old enough for evolution to have taken place.			
There was a time when humans and dinosaurs lived on earth together.			

Questions	Number of Items	Time 1	Time 2
Present animal diversity can be explained by the Great Flood.			
Young Earth Creationist Beliefs (Continued)			
A majority of present-day geological features are the result of the Great Flood.			
Adam and Eve of Genesis are our universal ancestors of the entire human race.			
All modern species of land vertebrates are descended from those original animals on the ark.			
Knowledge about the Scientific Enterprise	N = 6	0.753	0.802
Good theories can be proven by a single experiment.[a]			
For scientific evidence to be deemed adequate, it must be reproducible by others.			
Scientific ideas can be tested and supported by feelings and beliefs.[a]			
Scientific explanations can be supernatural.[a]			
Theories requiring more untested assumptions are generally better than theories with fewer assumptions.[a]			
Good theories give rise to testable predictions.			
Distrust of the Scientific Enterprise	N = 7	0.522	0.570
Contemporary methods of determining the age of fossils and rocks are untrustworthy.[b]			
The data used to support evolution is untrustworthy.			
The theory of evolution is capable of explaining the diversity of life.[a]			
Evolutionary theorists believe that if something is natural then it is good or right.			
Evolutionary theorists believe that inevitable inequality is morally acceptable.[c]			
Evolutionary theorists believe that because the strongest survive, it's a mistake to help the weak.[c]			
The available data are ambiguous as to whether evolution actually occurs.[d]			

Questions	Number of Items	Time 1	Time 2
Genetic Literacy	N = 9	0.642	0.765
Humans share a majority of their genes with chimpanzees.[e]			
Humans share more than half of their genes with mice.[e]			
Ordinary tomatoes do not have genes, whereas genetically modified tomatoes do.[a,e]			
Today it is not possible to transfer genes from one species of animal to another.[e]			
All plants and animals have DNA.[e]			
Humans have somewhat less than half of the DNA in common with chimpanzees.[a,e]			
You can see traces of our evolutionary past in human embryos.			
Humans developed from earlier life forms.			
Mutations are never beneficial.[a]			
Political Activity	N = 6	0.900	0.934
To what degree are you political?			
To what degree are you politically active?			
To what degree are you politically aware/up-to-date?			
To what degree do your political views influence your daily life?			
To what degree do your political views influence your decisions?			
To what degree do your political views influence courses you enroll in?			
Religious Activity	N = 6	0.948	0.935
To what degree are you religious?			
To what degree does religion impact your daily life?			
To what degree does your religion influence your decisions?			
To what degree do you participate in religious activities?			
How much do you believe in God?			
Religion is especially important to me because it answers many of my questions about the meaning of life.[f]			

Questions	Number of Items	Time 1	Time 2
Conservative Self-Identity	N = 5	0.790	0.786
To what degree are you conservative?			
In general, how do you self-identify politically?[g]			
In general how liberal/conservative are you on social issues (abortion, same-sex marriage, flag burning, etc.)?[h]			
In general how liberal/conservative are you on economic issues (welfare, taxation, free market policies, etc.)?[h]			
In general how liberal/conservative are you on foreign policy and defense issues (defense spending, combating terrorism, pre-emptive war)?[h]			
Attitudes toward Life[e]	N = 3	0.693	0.773
Life begins at conception.			
After conception, a developing human is only a cluster of cells, and it makes no sense to discuss its moral condition.[a]			
All stages of human life—embryo, fetus, child, adult—should have the same legal protections.			

NOTES: Cronbach's alphas are a measurement of the inter-correlation of all items in a scale (measured on a 0–1 scale).

[a]Items are reverse coded (i.e., 1→7, 2→6, etc.) in order to maintain positive correlations between all scale items.

[b]See also Ingram and Nelson (2006).

[c]Item was drawn from R. Deaner (personal communication with P. Hawley, January 20, 2009, as reported in Hawley, Short, McCune, Osman, & Little, 2011).

[d]See also Rutledge and Sadler (2007).

[e]From Miller, Scott, and Okamoto (2006).

[f]See also Dudley and Cruise (1990).

[g]From ANES (2009).

[h]From Carney, Jost, Gosling, and Potter (2008).

References

Abd-El-Khalick, F., & Lederman, N. G. (2000a). Improving science teachers' conceptions of nature of science: A critical review of the literature. *International Journal of Science Education, 22,* 665–701.

Abd-El-Khalick, F., & Lederman, N. G. (2000b). The influence of history of science courses on students' views of nature of science. *Journal of Research in Science Teaching, 37,* 1057–1095.

Abrami, P. C., Bernard, R. M., Borokhovski, E., Wade, A., Surkes, M. A., Tamim, R., & Zhang, D. (2008). Instructional interventions affecting critical thinking skills and dispositions: A stage 1 meta-analysis. *Review of Educational Research, 78,* 1102–1134.

Adey, P. S., & Shayer, M. (1990). Accelerating the development of formal thinking in middle and high school students. *Journal of Research in Science Teaching, 27,* 267–285.

Adey, P. S., & Shayer, M. (1993). An exploration of long-term far-transfer effects following an extended intervention program in the high school science curriculum. *Cognition and Instruction, 11,* 1–29.

Adey, P. S., & Shayer, M. (1994). *Really raising standards: Cognitive intervention and academic achievement.* London, England: Routledge.

Akerson, V. L., Abd-El-Khalick, F., & Lederman, N. G. (2000). Influence of a reflective explicit activity-based approach on elementary teachers' conceptions of nature of science. *Journal of Research in Science Teaching, 37,* 295–317.

American Association for the Advancement of Science (AAAS). (1990). *Science for all Americans.* New York, NY: Oxford University Press.

American Association for the Advancement of Science (AAAS). (1993). *Benchmarks for science literacy.* New York, NY: Oxford University Press.

Arum, R., & Roksa, J. (2011). *Academically adrift: Limited learning on college campuses.* Chicago, IL: University of Chicago Press.

Ausubel, D. (1963). *The psychology of meaningful verbal learning.* New York, NY: Grune & Straton.

Barnett, S. M., & Ceci, S. J. (2002). When and where do we apply what we learn? A taxonomy for far transfer. *Psychological Bulletin, 128,* 612–637.

Berkman, M. B., & Plutzer, E. (2011). Defeating creationism in the courtroom, but not in the classroom. *Science, 331,* 404–405.

Brown, A. (1997). Transforming schools into communities of thinking and learning about serious matters. *American Psychologist, 52,* 399–413.

Carey, S., & Smith, C. (1993). On understanding the nature of scientific knowledge. *Educational Psychology, 28,* 235–251.

Carroll, D. W. (2007). Patterns of student writing in a critical thinking course: A quantitative analysis. *Assessing Writing, 12,* 213–227.

Chiappetta, E. L., & Koballa, T. R. (2009). *Science instruction in the middle and secondary schools: Developing fundamental knowledge and skills* (7th ed.). Upper Saddle River, NJ: Merrill.

Coyne, J. A. (2010). *Why evolution is true.* Oxford, England: Oxford University Press.

Dagher, Z. R., & BouJaoude, S. (2005). Students' perceptions of the nature of evolutionary theory. *Science Education, 89,* 378–391.

Darwin, C. R. (1859). *On the origin of species by means of natural selection, or the preservation of favoured races in the struggle for life.* London, England: Murray.

Darwin, C. R. (1871). *The descent of man, and selection in relation to sex.* London, England: Murray.

Dawkins, R. (2009). *The greatest show on earth: The evidence for evolution.* New York, NY: Free Press.

Detterman, D. K. (1993). The case for the prosecution: Transfer as an epiphenomenon. In D. K. Detterman & R. J. Sternberg (Eds.), *Transfer on trial: Intelligence, cognition and instruction* (pp. 1–24). Norwood, NJ: Ablex.

Dobzhansky, T. (1973). Nothing in biology makes sense except in the light of evolution. *American Biology Teacher, 35*, 125–129.

Duncan, D. K., & Arthurs, L. (2012). Improving student attitudes about learning science and student scientific reasoning skills. *Astronomy Education Review, 11*(1), 010102.

Ehrenreich, B., & McIntosh, J. (1997, June 9). The new creationism: Biology under attack. *The Nation,* 11–16.

Ennis, R. H. (1989). Critical thinking and subject specificity: Clarification and needed research. *Educational Researcher, 18*(3), 4–10.

Ernst, J., & Monroe, M. (2004). The effects of environment-based education on students' critical thinking skills and disposition toward critical thinking. *Environmental Education Research, 10*, 507–522.

Facione, N. C., Facione, P. A., & Sanchez, C. A. (1994). Critical thinking disposition as a measure of competent clinical judgment: The development of the California Critical Thinking Disposition Inventory. *The Journal of Nursing Education, 33*, 345–350.

Facione, P. A., Facione, N. C, & Giancarlo, C. A. (1997). The motivation to think in working and learning. In E. Jones (Ed.), *Preparing competent college graduates: Setting new and higher expectations for student learning* (pp. 67–79). San Francisco, CA: Jossey-Bass.

Facione, P. A., Giancarlo, C. A., Facione, N. C., & Gainen, K. J. (1995). The disposition toward critical thinking. *Journal of General Education, 44*, 1–25.

Halpern, D. F. (1998). Teaching critical thinking for transfer across domains: Disposition, skills, structure training, and metacognitive monitoring. *American Psychologist, 53*(4), 449–455.

Haskell, R. E. (2000). *Transfer of learning: Cognition and instruction.* Washington, DC: Academic Press.

Hawley, P. H., Short, S. D., McCune, L. A., Osman, M. R., & Little, T. D. (2011). What's the matter with Kansas? The development and confirmation of the Evolutionary Attitudes and Literacy Survey (EALS). *Evolution: Education and Outreach, 4*, 117–132.

Hayes, A. F., & Krippendorff, K. (2007). Answering the call for a standard reliability measure for coding data. *Communication Methods and Measures, 1*, 77–89.

Jensen, E. (2005/2008). *Teaching with the brain in mind.* Alexandria, VA: Association for Supervision and Curriculum Development.

Kuhn, D. (1999). A developmental model of critical thinking. *Educational Researcher, 28*(2), 16–46.

Lederman, N. G. (1992). Students' and teachers' conceptions of the nature of science: A review of the research. *Journal of Research in Science Teaching, 29*, 331–359.

Lerner, L. S. (2000). *Good science, bad science: Teaching evolution in the states.* Washington, DC: Thomas B. Fordham Foundation.

Lombrozo, T., Thanukos, A., & Weisberg, M. (2008). The importance of understanding the nature of science for accepting evolution. *Evolution: Education and Outreach, 1*, 290–298.

Miller, J. D., Scott, E. C., & Okamoto, S. (2006). Public acceptance of evolution. *Science, 313*, 765–766.

Miller, K. R. (1999). *Finding Darwin's God: A scientist's search for common ground between God and evolution*. New York, NY: Cliff Street.

Morier, D., & Keeports, D. (1994). Normal science and the paranormal: The effect of a scientific method course on students' beliefs in the paranormal. *Research in Higher Education, 35*, 443–453.

National Research Council (NRC). (1996) *National science education standards*. Washington, DC: National Academies Press.

National Research Council (NRC). (2012). *A framework for K–12 science education: Practices, crosscutting concepts, and core ideas*. Washington, DC: National Academies Press.

Nickels, M. K., Nelson, C. E., & Beard, J. (1996). Better biology teaching by emphasizing evolution & the nature of science. *American Biology Teacher, 59*, 332–336.

Nettle, D. (2010). Understanding of evolution may be improved by thinking about people. *Evolutionary Psychology, 8*, 205–228.

O'Brien, D. T., & Wilson, D. S. (2010). Using "Evolution for Everyone" as a guide for new general education courses in evolution. *EvoS, 2*, 1–11.

O'Brien, D. T., Wilson, D. S., & Hawley, P. H. (2009). "Evolution for Everyone": A course that expands evolutionary theory beyond the biological sciences. *Evolution: Education and Outreach, 2*, 445–457.

Parliamentary Assembly of the Council of Europe (PACE). (2009). *The dangers of creationism in education*. Retrieved from http://www.assembly.coe.int/Main.asp?link=/Documents/WorkingDocs/Doc07/EDOC11297.htm

Paul, R., & Elder, L. (2006). *The miniature guide to critical thinking: Concepts & tools*. Dillon Beach, CA: Foundation for Critical Thinking.

Paul, R., & Elder, L. (2007). Consequential validity: using assessment to drive instruction. *Foundation for Critical Thinking*. Retrieved from http://www.criticalthinking.org/files/White%20PaperAssessmentSept2007.pdf

Pennebaker, J. W., & Francis, M. E. (1996). Cognitive, emotional, and language processes in disclosure. *Cognition and Emotion, 10*, 601–626.

Pennebaker, J. W., & Francis, M. E. (1997). *Linguistic inquiry and word count, the second version: LIWC*. Austin, TX: University of Texas, Department of Psychology.

Pennebaker, J. W., & Seagal, J. (1999). Forming a story: The health benefits of narrative. *Journal of Clinical Psychology, 55*, 1243–1254.

Pennebaker, J. W., Mayne, T. J., & Francis, M. E. (1997). Linguistic predictors of adaptive bereavement. *Journal of Personality and Social Psychology, 72*, 863–871.

Penningroth, S., Despain, L., & Gray, M. (2007). A course designed to improve psychological critical thinking. *Teaching of Psychology, 34*, 153–157.

Perkins, D. N., & Salomon, G. (1988). Teaching for transfer. *Educational Leadership, 46*, 22–32.

Perkins, D. N., & Salomon, G. (1989). Are cognitive skills context-bound? *Educational Researcher, 18*(1), 16–25.

Porter, S. R., Whitcomb, M. E., & Weitzer, W. H. (2004). Multiple surveys of students and survey fatigue. *New Directions for Institutional Research, 2004*(121), 63–73.

Rhodes, T. L., & Finley, A. P. (2013). *Using the VALUE rubrics for improvement of learning and authentic assessment*. Washington, DC: Association of American Colleges & Universities.

Roksa, J., & Arum, R. (2011). The state of undergraduate learning. *Change: The Magazine of Higher Learning, 43*(2), 35–38.

Rudolph, J. L., & Stewart, J. (1998). Evolution and the nature of science: On the historical discord and its implications for education. *Journal of Research in Science Teaching, 35*, 1069–1089.

Rutledge, M., & Mitchell, M. (2002). High school biology teachers' knowledge structure, acceptance and teaching of evolution. *Teaching Evolution, 64*, 21–28.

Rutledge, M., & Warden, M. (2000). Evolutionary theory, the nature of science, and high school biology teachers: Critical relationships. *American Biology Teacher, 62*, 123–131.

Sadler, T. D. (2004). Informal reasoning regarding socioscientific issues: A critical review of research. *Journal of Research in Science Teaching, 41*, 513–536.

Scharmann, L. C., & Harris, W. M. (1992). Teaching evolution: Understanding and applying the nature of science. *Journal of Research in Science Teaching, 29*, 375–388.

Schwartz, R. S., Lederman, N. G., & Crawford, B. A. (2004). Developing views of nature of science in an authentic context: An explicit approach to bridging the gap between nature of science and scientific inquiry. *Science Education, 88*, 610–645.

Scribner, S., & Cole, M. (1973). Cognitive consequences of formal and informal education. *Science, 182*, 553–559.

Scribner, S., & Cole, M. (1981). *The psychology of literacy*. Cambridge, MA: Harvard University Press.

Solomon, J., Scott, L., Duncan, J. (1996). Large-scale exploration of pupils' understanding of the nature of science. *Science Education, 80*, 493–508.

Trani, R. (2004). I won't teach evolution; It's against my religion. And now for the rest of the story . . . *American Biology Teacher, 66*, 419–427.

Van Gelder, T. (2005). Teaching critical thinking: Some lessons from cognitive science. *College Teaching, 53*, 41–48.

Verhey, S. D. (2005). The effect of engaging prior learning on student attitudes toward creationism and evolution. *Bioscience, 55*, 996–1003.

Wesp, R., & Montgomery, K. (1998). Developing critical thinking through the study of paranormal phenomena. *Teaching of Psychology, 25*, 275–278.

Wilson, D. S. (2005). Evolution for everyone: How to increase acceptance of, interest in, and knowledge about evolution. *PLoS Biology, 3*(12), e364.

Wilson, D. S. (2007). *Evolution for everyone: How Darwin's theory can change the way we think about our lives*. New York, NY: Bantam Dell.

Woods, C. S., & Scharmann, L. C. (2001). High school students' perceptions of evolutionary theory. *Electronic Journal of Science Education, 6*(2), 1–21.

Zoller, U. (1993). Lecture and learning: Are they compatible? Maybe for LOCS: Unlikely for HOCS. *Journal of Chemical Education, 70*, 195–197.

Zoller, U. (1997). The traditional-to-innovative switch in college science teaching: An illustrative case study on the reform trail. In M. W. Carpio (Ed.), *From traditional approaches toward innovation* (pp. 1–10). Arlington, VA: NSTA.

Zoller, U. (1999). Teaching college science towards the next millennium: Are we getting it right? *Journal of College Science Teaching, 29*, 409–414.

Zoller, U., & Tsaparlis, G. (1997). Higher and lower-order cognitive skills: The case of chemistry. *Research in Science Education, 27*, 117–130.

CHAPTER 17 | Our Evolutionary Underpinnings
The Past, Present, and Future of Evolution Education in the United States

AMANDA L. GLAZE

CHANGE IS A CONCEPT simple enough in nature and one with which we are intimately familiar. Surroundings change, times change, and people change greatly over a lifetime, both physically and intellectually. Change is so much a part of our mantra that we have an adage that puts it in perspective:

The Only Constant Is That Everything Changes

Over generations we observe organisms around us, including ourselves, as we come and go, changing along the way. We have no issue saying that life on Earth is not the same as it was a generation, or a host of generations, ago regardless of our beliefs and culture. So why is the idea of all life changing, of human evolution, one that continues to be surrounded by controversy and conflict? The place where this is most visible is in classrooms around the United States. It is here that the lessons of our past and the conflict in our present intersect to frame a national debate between the scientific community and the general public that will inform and decide our future.

Evolution can be defined most simply as descent with modification and more completely as the collection of theories that explain how descent with modification occurs in all living things. It is the underpinning collection of theories in the life sciences, a unifying concept by which all other concepts make sense (Dobzhansky, 1973). Not only does it connect the various realms of biology, from the cellular to the organismal level, it also compliments a wide range of other scientific fields of study such as biochemistry, anthropology, and geology (Plutzer & Berkman, 2008; Wiles, 2010).

In the nearly 200 years since the voyage of the *HMS Beagle*, much has been done to study, explain, and define the many mechanisms of evolution as well as to present this evidence openly for public consumption (Dawkins, 2009; Smithsonian's National Museum of Natural History, 2010a; University of California Museum of Paleontology, 2016). The evidence supporting biological evolution is so robust that the scientific community widely, if not unanimously, accepts that evolution is indeed occurring and that it does so in all living things (Wiles, 2010). As such, evolution is viewed as one of the most important theoretical understandings in biological studies (Plutzer & Berkman, 2008), and you would be hard-pressed to find professors in higher education that would disagree with the teaching of evolution as a mandatory part of a program in the sciences (Branch, Scott, & Rosenau, 2010). However, the case is much the opposite in primary and secondary education in the United States, where the conflict over the teaching and learning of evolution has deep historical roots and goes back almost as far as the original publication of Darwin's (1859) *On the Origin of Species by Means of Natural Selection* (Moore, 2004; Moore & McComas, 2016). It is this centuries-old conflict, with the scientific community on one path and a large portion of the public sector on the other, that frames the state of evolution education in the modern era where the controversy continues to be highly publicized (Plutzer & Berkman, 2008; Pobiner, 2016).

One key area of research exploring perceptions surrounding evolution in both the public and academic settings is the measure of levels of acceptance of evolution (Glaze & Goldston, 2015; Pobiner, 2016). Acceptance of evolution looks beyond what can be learned in the content and examines instead the level at which concepts and their implications are internalized and valued by individuals. There are a wide range of influential factors identified that impact acceptance of evolution among science teachers, preservice teachers, and students across levels and across oceans (Akyol, Tekkaya, & Sungur, 2011; Glaze, Goldston & Dantzler, 2015; Ha, Haury, Nehm, 2012; Herman, 2016; Rutledge & Sadler, 2011). These influences, both internal and external, interact, diverge, and intersect with our existing world view, the lens of our beliefs and experiences that we use to measure all other experiences (Cobern, 1994; Herman 13) in ways that are complex (Glaze et al., 2015; Nelson, 2012; Ranney & Thanukos, 2011).

Statistically, the most influential of these include religiosity or how beliefs impact decision-making (Deniz, Cetin, & Yilmaz, 2011; Glaze et al., 2015; Herman, 2016; Rissler, Duncan & Caruso, 2014; Wiles, 2008) and understanding of the nature of science—scientific ways of knowing, process skills, and practices (Akyol, Tekkaya, & Sungur, 2011; Glaze et al., 2015; Lederman, Antinck, & Bartos, 2014; Rutledge & Warden, 2000).

In stark contrast to findings of research into knowledge and acceptance of other scientific content, knowledge of evolution content has demonstrated little to no statistically significant impact on acceptance of evolution in studies (Baker, 2013; Glaze et al., 2015; Matthew, 2001; Sinatra, Southerland, McConaughy, & Demastes, 2003). Sadly, we only have adequate opportunities to address understandings and knowledge in classroom teaching and learning, making it easy for students to learn what they must to pass a class and promptly return to their prior ways of thinking (Verhey, 2005). Although the evolution controversy is primarily viewed as an issue among the nonscience public only, research suggests that misconceptions are pervasive not only in the public sphere but among students, including science majors, nonmajors, and future science teachers (Glaze et al., 2015; Nadelson & Sinatra, 2009; Nehm, Kim, & Sheppard, 2009; Rissler et al., 2014) as well as university faculty (Rice, Clough, Olson, Adams, & Colbert, 2016).

Evolution Education in Historical Context

Throughout history, there are examples of controversy when it comes to science and religious beliefs (Hellman, 1998). What is important to point out is that it is not per se a matter of religion versus science so much as it is a matter of concern when new knowledge and understanding being generated in the sciences presents a challenge to long-held religious doctrine or personal belief (Kahan, 2017). There is as much history of reconciliation between science and religion as there is conflict, as our understanding of our universe has grown and these understandings challenge older ways of knowing (Colburn & Henriquez, 2006; Murphy, Hickey, & Beggs, 2010). However, in the case of evolution, there are special circumstances that frame an ongoing conflict between the personal beliefs closely held by many in the United States and scientific explanations of the world around us (Bowman, 2008). World view and contributing beliefs, whether religious or otherwise, are formed early in life (Cavallo & McCall, 2008) and crafted over the course of a lifetime at a level impossible to parallel in the classroom (Verhey, 2005). This is as much an issue in the 21st century as it has been since the beginnings of evolutionary understandings in the 1800s.

When we talk about the history of evolutionary thinking, teaching, and learning in the United States, there is a rich history of conflict. As a largely agrarian society in its early generations, the education system of the United States was, and in many places still is, community based and guided by the moral compass of those served. Even in the 21st century, we see pushback, mostly across the Midwest, Southeast, and other rural pockets across the country, where state and local groups continue to fight the presence of national standards and other nationalized aspects of education by creating their own standards and dictates for education within the state or region (Matzke, 2016). The education system itself has changed dramatically in the last century, going from community-focused optional training to being highly regulated, accredited, and standardized at a national level (Urban & Wagoner, 2004). Topics have changed from health and hygiene to the more classical curriculum to the accelerated college and career readiness programs that drive education in a global economy. Students are now surrounded by technology in the classroom that is light years beyond the technology that landed man on the moon (Urban & Wagoner, 2004). Yet throughout education reform and curriculum development, conflict between knowledge and beliefs persists, specifically when it comes to teaching that impacts, contradicts, or is perceived to undermine community or cultural underpinnings (Nadelson & Nadelson, 2010). This is precisely where the evolution controversy is situated, among the intersections of science and society, where deeply rooted beliefs and novel understandings about the natural world diverge (Baker, 2013; Kahan, 2017; Moore, Brooks, & Cotner, 2011).

Almost immediately following the publication and public arguments surrounding Charles Darwin's (1873) *The Descent of Man and Selection in Relation to Sex*—the follow up to *Origin of Species* that addressed the nature of human evolution—there were movements against the theory of evolution in Europe. In the United States, however, the conflict was slower to catch on, and it was not until the turn of the century that the debate moved into mainstream conversation by way of Christian writers and speakers who challenged evolution as upsetting to Christian beliefs (Masci, 2009). Consequently, public discourse on evolution coincided with the division of the Protestant church in the United States and was, in fact, one of the prime issues where the evangelical (creationist) and liberal modernists clashed prior to the formal division of Protestant belief systems (Masci, 2009).

Challenges to evolution from the pulpit led to calls to legally fight the spread of what had by then been called "Darwinism." As a result, laws were drafted to prevent the teaching of evolution and protect young minds. One such law, the Butler Act of 1925, was put in place in Tennessee and actively banned the teaching of evolution at all levels of public education as well as any other perspective contradictory to special creation of mankind. Perhaps the most publicized battle in the United States was the infamous Scopes "Monkey" trial (otherwise known as *The State of Tennessee vs. John Thomas Scopes*) that took place in Dayton, Tennessee in 1925 (Linder, 1995; Moore & McComas, 2016). For many, *Scopes* was the early defining moment in the battle over evolution education in the United States.

The *Scopes* trial certainly represents the spirit of widely publicized and dramatized clashes over what should and should not be taught in public schools, characterizing the evolution discussion as a battlefield for souls, pitting science and evolution on one side against salvation and belief on the other. It was, in fact, only the beginning in a long line of conflicts that took place in public education over the teaching of evolution in the classroom. Even before Butler was officially on the books, state boards of education in California, North Carolina, Texas, and Louisiana had come forward with statements or actions to prevent or discourage the teaching of evolution and other states including Oklahoma, Kentucky, and Florida had antievolution laws or bills on the books (Elsberry, 2001). With the outset of two world wars and other conflicts that tied up popular conversation, evolution teaching and learning was placed on a back burner until the 1960s when changing mindsets following the launch of *Sputnik* shifted the focus of education, and the nation, to science (Urban & Wagoner, 2004).

The middle of the 20th century is remarkable for a great number of reasons, scientifically, culturally, and socially. From the perspective of evolutionary science, the years between 1945 and 1975 mark the turning point where scientific ways of thinking became more valued in the popular mindset than in any time in United States history. During this time, laws on the books since the 1920s came under the scrutiny of the court system and rulings were made that would frame what would and would not be accepted in classrooms around the United States.

A great deal of time could be spent discussing the cases at length; their historical background, proceedings, and subsequent rulings represent a rich history of judicial interpretation relative to science education. However, the handful of cases summarized here provide a strong framework to capture in snapshot where we have been and how we arrived at

TABLE 17.1. Key Legal Cases Impacting Evolution Teaching and Learning

CASE	SUMMARY OF RULING
Epperson v. Arkansas (1968)	Governments cannot prohibit evolution teaching to tailor to religious groups.
McLean v. Arkansas Board of Education (1981)	Fair and balanced treatment of evolution and creationism is a violation of the Establishment Clause; creation science is not science and has no place in science curriculum.
Edwards v. Aguillard (1987)	The Louisiana Creationism Act endorses a religious belief system in violation of Establishment Clause; it is unconstitutional for a government entity to require teaching of creationism.
Webster v. New Lenox School District (1990)	Districts can protect students by prohibiting teachers from teaching creationism without violating the teachers' right to religious freedom.
Peloza v. Capistrano School District (1994)	Requiring science teachers to teach science (evolution) does not violate their religious freedom rights under the constitution.
Freiler v. Tangipahoa Parish Board of Education (1997)	Teachers cannot be required by their district/state to read a disclaimer about evolution and creationism in the classroom prior to teaching.
Selman v. Cobb County School District (2005)	Disclaimers denouncing evolution in ways incongruous with scientific fact cannot be required in textbooks.
Kitzmiller v. Dover (2005)	Intelligent design is not scientific in nature and cannot be separated from its creationist foundations, therefore teachers cannot be required to teach it in science class.

the point we are today in regard to the legal issues of evolution education. Table 17.1 provides a brief overview of some of the key cases in US judiciary history that have impacted evolution teaching and learning in public schools and establish a framework for the teaching of evolution at present. Justices have, on multiple occasions, addressed the role of science in public schools as well as nonscientific alternative conceptions that are commonly put forward as scientific alternatives to that science. Additional rulings have supported their findings that states and other government bodies cannot require the teaching of nonscientific topics such as creationism and intelligent design in public school classrooms as doing so would represent establishment of a religious preference in violation of the Establishment Clause. The courts have also provided legal protection for teachers and districts when they take actions to protect students from being taught non-scientific alternatives in science classrooms. Early attempts to circumvent rulings on evolution have also been struck down, including the use of nonscientific disclaimers, both verbal and written.

The Reality of Evolution Education in the Present

In the nearly 100 years since *Scopes,* there has been little change in the overall public perception of the teaching and learning of evolution in the United States, even as the popularity and need for science has increased. Masci (2007) lamented that 20 years after the ruling in *Kitzmiller,* which many thought would bring an end to attempts at antiscience legislation and board decisions, people in the United States continue to fight over the teaching of evolution in public schools. Sadly, another decade has passed, and we find instead of rectification a resurgence of attempts to prevent the teaching of evolution and allow nonscientific alternatives into science classrooms via "academic freedom" bills that have been submitted in many states and have passed into law in a surprising number of places. Using language carefully crafted to avoid raising the red flags of the court system, academic freedom laws purport to protect teachers who are "teaching scientific subjects in an objective manner" but in reality are written to target topics such as evolution and climate change by allowing teachers to avoid the topic or introduce alternative concepts under the guise of "critical analysis" (National Center for Science Education, 2016).

Louisiana and Tennessee have successfully passed such bills, leading to a revival of attempts in Alabama, Oklahoma, Arizona, Missouri, Montana, Texas, Florida, New Mexico, South Carolina, Iowa, Michigan, Maryland, Kentucky, Colorado, Virginia, and South Dakota between 2004 and 2015 (National Center for Science Education, 2016). In most places, these bills have been quashed in subcommittee, but a surprising number have made it to the floor for consideration only to be tabled for later discussion and failing to be considered before closure. Wesley Elsberry (2001) takes this further by looking at the attempts at antievolution legislation, court cases, and political action prior to 2001. Comparison of the National Center for Science Education report and Elsberry's collection from the 20th century, there is a pattern of increased pressure from members of state and federal Senates and Houses of Representatives to interfere with the teaching of evolution in public education (Elsberry, 2001; National Center for Science Education, 2016).

When attempts at legislation to prevent the teaching of evolution failed, the focus shifted to removing the burden of requirement to teach evolution. In 2000, Lawrence Lerner released the revision of the 1998 report card scores for state standards as they related to evolution. In this revision, only 31 states, four years out from the release of the National Science Education Standards (1996), had evolution standards that were considered passing when evaluated for proper coverage, application, and terminology surrounding the teaching of evolution (Lerner, 2000a). Further notation was applied to states where textbook disclaimers were applied, such as the one that prominently featured in Alabama's science textbooks statewide; for states who included creationist language in their standards; and those where human evolution and, in some cases, even the word "evolution" itself, were purposely excluded from consideration (Lerner, 2000b).

In the *Next Generation Science Standards* (NGSS Lead States, 2013), evolution receives its most comprehensive coverage to date, embedded as early as primary school and including a wide range of topics that flow throughout the course of life science study. Sadly, the *Next Generation Science Standards* are optional for consideration by states, many of whom elect instead to write their own standards at the state level, washed of the so-called controversial topics that make some uncomfortable on grounds of beliefs and distrust of science (Baker, 2013; Fisk & Dupree, 2014; Moore & Kraemer, 2005). For those of us who advocate for science education, it is an ongoing battle to ensure that the science education standards remain rigorous, that teachers have support to teach science accurately and inclusively, and that the protections afforded to teachers and students regarding the teaching of science are observed (Mead & Mates, 2009). It is relevant to note that standards alone are not enough to ensure accurate and adequate teaching of evolution; however, the existence of standards that directly address evolution represent forward progress in the long history of evolution education (Moore, 2002).

Curriculum and Support for 21st Century Learners and Beyond

The future of evolution education, both in the United States and abroad, lies in building stronger relationships between scientists and the people. To reach a point where evolution is embraced by society, much work is needed and the journey will be long and arduous (Glaze & Goldston, 2015). One of the biggest hurdles to science literacy is the distrust that many in the public have for the sciences and for scientists as the proprietors of these practices (Pobiner, 2016). Although they are respected for the levels of thinking and academic study they represent, scientists are often seen by the public as cold and untrustworthy (Fisk & Dupree, 2014). Science denialism is on the rise, a new challenge to science where research, logic, and data are characterized as falsehoods and debates are generated where there is no debate among scientists (Diethelm & McKee, 2009). When antiscience and science denialism are internalized by communities, they become a part of the world view shared by those who are a part of that community. As a result, denialism and distrust bleed across topics, coloring views and conversations about evolution, climate change, stem cell research, and other issues that impact the scientific literacy of our society (Nadelson & Hardy, 2015).

In response to the challenges of science denialism in the 21st century, it is up to scientists, science educators, and other science enthusiasts to find ways to bridge the gaps between science and the public, to use tools like social media and other informal science communication to break down barriers of distrust and be the face of science. Changing the focus of our communication in the sciences is the first step to changing how the public views science, especially evolution. There exists an invisible division of the masses, a proverbial "us" or "them"—scientist or layperson—where scientists share their science with their peers through publication and presentation and only that which catches the attention of the media, whether positive or negative, is what reaches the ears and eyes of the public. So much of what divides the two is based on misunderstanding and misrepresentation with very little of the discord based on legitimate scientific conflict. Consequently, leaders of many of major Christian religious groups in the United States and abroad are in agreement that evolution is not conflicting with Christian beliefs, with the noted exception of biblical literalist traditions (Meadows, Doster, & Jackson, 2000; Murphy et al., 2010), and efforts are underway to bridge the gaps in communication between religion and science (Smithsonian's National Museum of Natural History, 2010a).

Another key mechanism of change is to reevaluate how we approach the training of experts, whether they be future scientists or future teachers (Glaze & Goldston, 2015; Pobiner, 2016). So much of what we have done historically has been based in traditional ways of teaching, of direct instruction and content. What has long been missing is a return to the practice of science with a strong parallel to the nature, history, and philosophy of science as a way of knowing and exploring the world (Pickstone, 2000).

When scientists face the same struggles with understandings of the nature of science as students and even members of the public, there is a strong need to connect everyone more deeply with scientific practice as well as understanding to remedy the misconceptions that people hold (Lederman et al., 2014; McComas, 1996; Rice et al., 2016; Yates & Marek, 2014).

In addition to seeking catalysts of conceptual change and addressing misconceptions in our preparation programs, it is imperative that our reform efforts extend to support and advocacy for teachers, the teaching and learning of evolution, and of science in general. Evolution, while a part of national and many state standards, is something that for teachers is often met with criticism by their students, community, and administration (Hermann, 2013). For others, it is a personal conflict between their world view—a culmination of experiences, beliefs, and culture—and the topics they are asked to teach in the classroom (Glaze & Goldston, 2015; Glaze et al., 2015). Although some have extensive content training, others, especially at the middle and elementary grade levels, have little experience with the content and even less with the methods, misconceptions, and potential approaches to teaching that would enable them to adequately teach evolution in their classrooms in light of perceived pressures (Athanasiou & Papadopolou, 2011; Campos, 2013). Therefore, it is important for scientists and lovers of science alike to come together and form partnerships to mentor and support teachers, to provide content support and training, to give time to be in classrooms and take action for science literacy and evolution education. To meet a controversy head on, it is imperative that we all take steps to be a part of the change we want to see in the system and in others.

Thoughts in Closing

A great deal of change has come about in education since the days of Darwin and the classical curriculum, yet when it comes to teaching and learning evolution there is still a great deal of ground that must be covered to meet, and someday exceed, expectations for a scientifically literate society. In the 21st century, we have seen strides made to change how teachers and students address evolution: specifically including evolution in the Next Generation Science Standards, building on earlier standards by shifting the focus on evolutionary concepts from higher grades down to primary grades to lay a foundation for deeper understanding and growth, and court cases that have provided support for teaching evolution and preventing nonscientific alternatives from being presented in the classroom. Despite these strides, teaching and learning evolution is still a subject of controversy in the public sphere (Plutzer & Berkman, 2008; Pobiner, 2016). Through ongoing outreach from scientists and other professionals in the field, it is possible to have a lasting positive impact on how we talk about evolution, in the classroom and with the public (Pobiner, 2016).

The focus of science education should remain on science literacy, on ensuring that all people—regardless of their beliefs, background, or creed—understand how science works and have the ability to logically evaluate the world around them (AAAS, 1990). A part of this is ensuring that students across all levels are taught not only scientific content but also scientific ways of knowing, our scientific history, and the philosophy that underlies the practice of science as a field. It is in understanding that science and beliefs are different perspectives used to explain, that one does not have to choose between being a scientist and embracing his or her culture, religion, or community, and that they are just different perspectives of the same world, we can have more meaningful conversations about the places where science and society intersect and resolve conflict.

Evolution is not controversial in the scientific community at large, nor are the theories that are accepted in explanation of how this descent with modification is taking place. The controversy surrounding evolution in the public sphere is a product of misunderstanding of how scientific ways of knowing differ from those of philosophy, religion, and emotion. Evolution itself is not a matter of scientists pushing their own form of beliefs but is a widely accepted and well-evidenced collection of scientific theories that provide for us the best explanation, at present, of the changes taking place in living things and the interrelatedness of all life on Earth.

References

Akyol, G., Tekkaya, C., & Sungur, S. (2010). The contribution of understanding of evolutionary theory and nature of science to pre-service science teachers' acceptance of evolutionary theory. *Procedia Social & Behavioral Sciences, 9*, 1889–1893.

American Association for the Advancement of Science (AAAS). (1990). *Project 2061: Science for all Americans.* New York, NY: Oxford University Press.

Athanasiou, K., & Papadopolou, P. (2011). Conceptual ecology of the evolution acceptance among Greek education students: Knowledge, religious practices and social influences. *International Journal of Science Education, 34*, 903–924.

Baker, J. O. (2013). Acceptance of evolution and support for teaching creationism in public schools: The conditional impact of educational attainment. *Journal for the Scientific Study of Religion, 52*, 216–228.

Bowman, K. L. (2008). The evolution battles in high school science classes: Who is teaching what? *Frontiers in Ecology and the Environment, 6*(2), 69–74.

Branch, G., Scott, E., & Rosenau, J. (2010). Dispatches from the evolution wars: Shifting tactics and expanding battlefields. *Annual Review of Genomics and Human Genetics, 11*, 317–338.

Campos, R. (2013). Early evolution of evolutionary thinking: Teaching biological evolution in elementary schools. *Evolution Education and Outreach, 6*, 25.

Cavallo, A. M. L., & McCall, D. (2008). Seeing may not mean believing: examining students' understandings and beliefs in evolution. *The American Biology Teacher, 70*, 522–530.

Cobern, W. W. (1994). Belief, understanding, and the teaching of evolution. *Journal of Research in Science Teaching, 31*, 583–590.

Colburn, A., & Henriques, L. (2006). Clergy views of evolution, creationism, science & religion. *Journal of Research in Science Teaching, 43*, 419–442.

Darwin, C. (1859). *On the origin of species by means of natural selection, or, the preservation of favoured races in the struggle for life.* London, England: J. Murray.

Dawkins, R. (2009). *The greatest show on earth: the evidence for evolution.* New York, NY: Free Press.

Deniz, H., Cetin, F., & Yilmaz, I. (2011). Examining the relationships among acceptance of evolution, religiosity, and teaching preference for evolution in Turkish preservice biology teachers. *Reports of the National Center for Science Education, 31*(4), 1.1–1.9.

Diethelm, P., & McKee, M. (2009). Denialism: What is it and how should scientists respond. *European Journal of Public Health, 19*, 2–4.

Dobzhansky, T. (1973). Nothing in biology makes sense except in the light of evolution. *The American Biology Teacher, 35*, 125–129.

Edwards v. Aguillard, 482 U.S. 578 (1987).

Elsberry, W. R. (2001). *Anti-evolution and the law.* Retrieved from http://www.antievolution.org/topics/law/

Epperson v. Arkansas, 393 U.S. 97, 37 U.S. Law Week 4017, 89 S. Ct. 266, 21 L. Ed 228 (1968).

Fiske, S. T., & Dupree, C. (2014). Gaining trust as well as respect in communicating to motivated audiences about science topics. *Proceedings of the National Academy of Sciences, 111*(Supp 4), 13593–13597.

Freiler v. Tangipahoa Board of Education, No. 94-3577 (E.D. La. Aug. 8, 1997).

Glaze, A. L., & Goldston, M. J. (2015). Evolution and science teaching and learning in the United States: A critical review of literature 2000–2013. *Science Education, 99*, 500–518.

Glaze, A. L., Goldston, M. J., & Dantzler, J. (2014). Evolution in the southeastern United States: Factors influencing acceptance and rejection in pre-service science teachers. *International Journal of Science and Mathematics Education, 13*, 1189–1209.

Ha, M., Haury, D. L., & Nehm, R. H. (2012). Feeling of certainty: uncovering a missing link between knowledge and acceptance of evolution. *Journal of Research in Science Teaching, 49*, 95–121.

Hellman, H. (1998). *Great feuds in science: Ten of the liveliest disputes ever.* New York, NY: Wiley.

Hermann, R. S. (2013). On the legal issues of teaching evolution in public schools. *The American Biology Teacher*, *75*, 539–543.

Hermann, R. S. (2016). Elementary education majors' views on evolution: a comparison of undergraduate majors understanding of natural selection and acceptance of evolution. *Electronic Journal of Science Education*, *20*(6), 21–44.

Kahan, D. M. (2017). Ordinary science intelligence: A science comprehension measure for use in the study of risk perception and science communication. *Journal of Risk Research*, *20*, 995–1016.

Kitzmiller v. Dover Area School District, 400 F. Supp. 2d 707, Docket No. 4cv2688 (2005).

Lederman, N. G., Antinck, A., & Bartos, S. (2014). Nature of science, scientific inquiry, and socio-scientific issues arising from genetics: A pathway to developing a scientifically literate citizenry. *Science & Education*, *23*, 285–302.

Lerner, L. S. (2000a). Good and bad science in US schools. *Nature*, *407*, 287–290.

Lerner, L. S. (2000b). *The state of state standards*. Washington, DC: Thomas B. Fordham Foundation.

Linder, D. O. (1995). *State v. John Scopes*: A Final Word. *Famous Trials*. Retrieved from http://law2.umkc.edu/faculty/projects/ftrials/scopes/finalword.html

Masci, D. (2007). Twenty years after a landmark Supreme Court decision, Americans are still fighting about evolution. *Pew Research Center*. Retrieved from http://www.pewforum.org/2007/06/13/twenty-years-after-a-landmark-supreme-court-decision-americans-are-still-fighting-about-evolution/

Masci, D. (2009). Overview: The conflict between religion and evolution. *Pew Research Center*. Retrieved from http://www.pewforum.org/2009/02/04/overview-the-conflict-between-religion-and-evolution

Matthews, D. (2001). Effect of a curriculum containing creation stories on attitudes about evolution. *The American Biology Teacher*, *63*, 404–409.

Matzke, N. J. (2016). The evolution of antievolution policies after *Kitzmiller vs. Dover*. *Science*, *351*, 28–30.

McComas, W. F. (1996). Ten myths of science: Reexamining what we think we know about the nature of science. *School Science and Mathematics*, *96*, 10–16.

McLean v. Arkansas Board of Education, 529 F. Supp. 1255, 1258-1264 (ED Ark. 1982).

Mead, L. S., & Mates, A. (2009). Why science standards are important to a strong science curriculum and how states measure up. *Evolution: Education and Outreach*, *2*, 359–371.

Meadows, L., Doster, E., & Jackson, D. F. (2000). Managing the conflict between evolution & religion. *The American Biology Teacher*, *62*, 102–107.

Moore, R. (2002). Teaching evolution: do state standards matter? *BioScience*, *52*, 378–381.

Moore, R. (2004). How well do biology teachers understand the legal issues associated with the teaching of evolution? *Bioscience*, *54*, 860–865.

Moore, R., Brooks, D. C., & Cotner, S. (2011). The relation of high school biology courses and students' religious beliefs to college students' knowledge of evolution. *American Biology Teacher*, *73*, 222–226.

Moore, R., & Kraemer, K. (2005). The teaching of evolution and creationism in Minnesota. *The American Biology Teacher*, *67*, 457–466.

Moore, R., & McComas, W. F. (2016). *The Scopes monkey trial*. Charleston, SC: Arcadia.

Murphy, C., Hickey, I., & Beggs, J. (2010). All Christians? Experiences of science educators in Northern Ireland. *Cultural Studies of Science Education, 5,* 79–89.

Nadelson, L. S., & Hardy, K. K. (2015). Trust in science and scientists and the acceptance of evolution. *Evolution: Education and Outreach, 8,* art. 9.

Nadelson, L. S., & Nadelson, S. (2010). K–8 educators' perspectives and preparedness for teaching evolution topics. *Journal of Science Teacher Education, 21,* 843–858.

Nadelson, L. S., & Sinatra, G. M. (2009). Educational professionals' knowledge and acceptance of evolution. *Evolutionary Psychology, 7,* 490–516.

National Academy of Sciences. (1996). *National science education standards.* Washington, DC: National Academy Press.

National Center for Science Education. (2016, November 23). *Chronology of academic freedom bills.* Retrieved from https://NationalCenterforScienceEducation.com/library-resource/chronology-academic-freedom-bills

Nehm, R. H., Kim, S. Y., & Sheppard, K. (2009). Academic preparation in biology and advocacy for teaching evolution: biology versus non-biology teachers. *Science Education, 93,* 1122–1146.

Nelson C. E. (2012). Why don't undergraduates really "get" evolution? What can faculty do? In K. S. Rosengren, S. K. Brem, M. E. Evans, & G. M. Sinatra (Eds.), *Evolution challenges: Integrating research and practice in teaching and learning about evolution* (pp. 311–347). New York, NY: Oxford University Press.

NGSS Lead States. 2013. *Next generation science standards: For states, by states.* Washington, DC: National Academies Press.

Peloza v. Capistrano Unified School District, 37 F. 3rd 517 (1994).

Pickstone, J. V. (2000). *Ways of knowing: A new history of science, technology and medicine.* Chicago, IL: University of Chicago Press.

Plutzer, E., & Berkman, M. (2008). Trends evolution, creationism, and the teaching of human origins in schools. *Public Opinion Quarterly, 72,* 540–553.

Pobiner, B. (2016). Accepting, understanding, teaching and learning (human) evolution: Obstacles and opportunities. *American Journal of Physical Anthropology, 159,* 232–274.

Ranney M. A., & Thanukos, A. (2011). Accepting evolution or creation in people, critters, plants, and classrooms: the maelstrom of American cognition about biological change. In R. Taylor & M. Ferrari (Eds.), *Epistemology and science education: Understanding the evolution vs. intelligent design controversy* (pp. 143–172). Oxford, England: Routledge.

Rice, J. W., Clough, M. P., Olson, J. K., Adams, D. C., & Colbert, J. T. (2016). University faculty and their knowledge and acceptance of biological evolution. *Evolution: Education and Outreach, 8,* art. 8.

Rissler, L., Duncan, S. I., & Caruso, N. M. (2014). The relative importance of religion and education on university students' views of evolution in the Deep South and state science standards across the United States. *Evolution: Education & Outreach, 7,* art. 24.

Rutledge, M. L., & Sadler, K. C. (2011). University students' acceptance of biological theories—Is evolution really different? *Journal of College Science Teaching, 41*(2), 1–43.

Rutledge, M. L., & Warden, M. A. (2000). Evolutionary theory, the nature of science and high school biology teachers: Critical relationships. *The American Biology Teacher, 62*, 23–31.

Selman v. Cobb County School District, 449 F.3d 1320 (11th Cir. 2006).

Sinatra, G. M., Southerland, S. A., McConaughy, F., & Demastes, J. W. (2003). Intentions and beliefs in students' understanding and acceptance of biological evolution. *Journal of Research in Science Teaching, 40*, 510–528.

Smithsonian's National Museum of Natural History, Broader Social Impacts Committee. (2010a). What does it mean to be human? The challenges posed by evolution. *The Smithsonian Institution's Human Origins Program.* Retrieved from http://humanorigins.si.edu/about/broader-social-impacts-committee

University of California Museum of Paleontology. (2016, November 28). *Understanding evolution.* Retrieved from evolution.berkeley.edu/evolibrary/search/topics.php?topic_id=14

Urban, W. J., & Wagoner, J. L. (2004). *American education: A history.* Boston, MA: McGraw-Hill.

Verhey, S. D. (2005). The effect of engaging prior learning on student attitudes toward creationism and evolution. *Bioscience, 55*, 996–1003.

Webster v. New Lenox School District #122, 917 F. 2d 1004 (1990).

Wiles, J. R. (2008). Factors potentially influencing student acceptance of biological evolution. *Dissertation Abstracts International, 71–12, Section A.*

Wiles, J. R. (2010). Overwhelming scientific confidence in evolution and its centrality in science education—and the public disconnect. *Science Education Review, 9*, 18–27.

Yates, T. B., & Marek, E. A. (2014). Teachers teaching misconceptions: A study of factors contributing to high school biology students' acquisition of biological evolution-related misconceptions. *Evolution: Education and Outreach, 7*, art. 7.

CHAPTER 18 | Reconciling Evolution with a Christian Identity

A Professional Development Workshop to Reduce Anxiety and Enhance Self-Efficacy for Science Teachers

PATRICIA H. HAWLEY, RACHAEL K. PHILLIPS, AND MATTHEW S. OLSON

Of course the elephant in the room is religious opposition to teaching evolution. Sometimes it's standing there quietly, with everyone politely ignoring it, and sometimes it's trumpeting loudly, stomping its feet and throwing the furniture around.

(SCOTT, 2010, p. 241)

You won't find any opposition to the idea of evolution among sophisticated, educated theologians. It comes from an exceedingly retarded, primitive version of religion, which unfortunately is at present undergoing an epidemic in the United States.

(DAWKINS, 2005)

THESE QUOTES SET UP what we believe to be an undeniable tension. Especially in Texas, faith is an important part of teachers' identities, which has documentable impact on the degree to which teachers teach evolution according to state standards (Shankar & Skoog, 1993); their loyalty to their faith can sometimes be at odds with their need to fulfill pedagogical responsibilities. At the same time, the Dawkins's (2005) quote in the epigraph illustrates how this aspect of teachers' identities is under consistent and fervent attack by educational outsiders where religion is repeatedly and publicly denigrated. Those who favor both science and religion as important epistemologies find themselves in sometimes deep epistemological anxiety. Can we help teachers convey good science and help them resolve some of their anxieties?

This chapter describes a professional development workshop designed for inservice and preservice teachers that both implicitly and explicitly attempts to clarify key misconceptions about the nature of science, evolution, and natural selection in ways that are informed by the epistemological anxieties experienced by teachers and in terms provided by the source that accelerates those anxieties, namely, the intelligent design movement.

The Nature of the Problem

Understanding biological evolution and the nature of science are essential to achieving scientific literacy (National Academy of Science, 1998). And yet, understanding, acceptance (Miller, Scott, & Okomoto, 2006) and willingness to teach evolution without reservations (Berkman, Pacheco, & Plutzer, 2008; Berkman & Plutzer, 2010; Osif, 1997) have remained critically low in the United States for decades. Reception to evolutionary theory is undermined by a number of factors, including personal political ideologies and religious beliefs (Hawley, Short, McCune, Osman, & Little, 2011; Mead, Clarke, Forcino, & Graves, 2015; Miller & Toth, 2014; Rissler, Duncan, & Caruso, 2014; Rosengren, Evans, Brem, & Sinatra, 2012; Scott, 1997, 2005; Sinatra, Southerland, McConaughy, & Demastes, 2003). Indeed, evolution is an emotional topic (Bland & Morrison, 2015), and teaching is an emotional profession (Griffith & Brem, 2004). The theory of evolution—especially as it relates to humans—historically has been met with hostility like no other scientific topic (Branch, Scott, & Rosenau, 2010), and science classrooms have been transformed into political and religious battlefields (Berkman & Plutzer, 2010). This state of affairs leaves teachers questioning their ability to effectively teach science (i.e., self-efficacy; Akyol et al., 2012).

The domain where belief interfaces with acceptance and understanding is complex (Ha, Haury, & Nehm, 2012; Southerland, Sinatra, & Matthews, 2001). We maintain that anxieties stemming from epistemological conflict and one's place in nature (we refer them as *epistemological and existential anxieties*; see also Wilensky, 1997) translate not only into reduced self-efficacy at best, but also avoidance of evolution altogether. Consequently, teachers downplay evolution's importance or integrate "both sides" of the "evolution controversy" (Pobiner, 2016). For some, especially those who adhere to more traditional and conservative belief systems, the solution may be to protect and defend one's beliefs by refusing to teach the standards, discounting evidentiary support for evolution, minimizing or skipping evolutionary concepts that are perceived as impediments to faith (e.g., origin of humans, macroevolution, descent with modification), or activating counterstrategies consistent with their religious identities (e.g., teaching the Creation Story in Genesis). As a consequence, K–12 students and their teachers either avoid or lack exposure to the theory and accordingly lack content knowledge (Clores & Limjap, 2006; Lombrozo, Thanukos, & Weisberg, 2008).

The State of the Field

For the purposes of organizing how evolution training for teachers and others has been addressed historically, we have divided the approaches into three categories (see also Hawley & Sinatra, 2018).

Level 1: Content-Based Approaches

Teaching and learning about evolution is more difficult than other science-related topics for several reasons. The first reason for difficulty is related to content. All too often biology teachers feel underprepared to confidently address the details of the theory requisite for student understanding (e.g., Nadelson & Nadelson, 2010). A number of problem concepts have been documented (Mead & Scott, 2010a, 2010b). Accordingly, content-based workshops and more targeted efforts can help clarify troublesome central organizing concepts (e.g., Mead, 2009), increase knowledge about the nature of science, and decrease reliance on faulty conceptions (e.g., Nehm & Schonfeld, 2007). But at the same time, simply providing biological content, as in a university class, only takes us so far. Namely, some university courses on evolution do not enhance attitudes toward the theory or mitigate the sticky misconceptions about it (Short & Hawley, 2014). Moreover, even the more targeted reflective approaches do not appear to reliably shift preferences toward actually teaching evolution (Nehm, Kim, & Sheppard, 2009; Nehm & Schonfeld, 2007). Indeed, teachers with backgrounds in biology may not differ in attitudes toward evolution from teachers without such backgrounds (Losh & Nzekwe, 2011; Nadelson & Sinatra, 2009).

Content-based remediations tend to ignore "the elephant in the room"; that is, such workshops are seldom designed or administered with explicit sensitivity to the existential and epistemological anxieties of the participants. Indeed, delivering additional content to those who are ambivalent about evolution for existential reasons may lead to additional retreat (emotional and motivational withdrawal, or *reactance*; Brehm, 1966) if those participants perceive that they are being confronted with antireligious material. Without acknowledgement of alternate approaches born of faith, students are left on their own to draw difficult epistemological comparisons (Meadows, 2009; Southerland, Golden, & Enderle, 2012; Southerland & Scharmann, 2013). In the end, "biology only" or "science only" approaches, from a psychological perspective, may have the iatrogenic effects of increasing antievolution attitudes rather than mitigating them.

Level 2: Cognitive Change Approaches

In addition to reduced teaching self-efficacy due to gaps in content knowledge, there are cognitive barriers to proficient teaching. Common, predictable, and persistent misconceptions plague learners of all levels (e.g., heritability of acquired traits, essentialism, and evolving out of "need"; Cunningham & Wescott, 2009; Heddy & Nadelson, 2013; Shtulman & Calabi, 2012). Importantly, common misconceptions about evolution reveal deficiencies in learners' scientific reasoning processes (Lawson & Worsnop, 1992). For example, biological essentialism has been shown to be a major stumbling block for both understanding and acceptance of evolution (Evans, 2008; Kelemen, 2004). Especially important for our present purposes, developing and veteran teachers are prone to fallacious reasoning; they, in turn, pass these misconceptions on to their students (Yates & Marek, 2013, 2014). Business-as-usual content-based remediations are not enough to undo these cognitive obstacles. To this end, specialized instructional practices have been developed and tested that are designed to promote cognitive shifts regarding sticky topics (e.g., climate change; Sinatra, Kardash, Taasoobshirazi, & Lombardi, 2012) by incorporating elements and practices that enhance cognitive, affective, and behavioral engagement (e.g., Teaching for Transformative Experience in Science; Heddy & Nadelson, 2013; Pugh & Girod, 2007). Even as these treatments have been shown to be effective, they have not been developed expressly for the purposes of teacher training as of this writing (but see Heddy & Sinatra, 2013; Meadows, 2009).

Level 3: Epistemological Anxiety, Motivation, and Identity

As previously discussed, there is an elephant in the room that is suboptimally yet routinely ignored (i.e., religiosity) when it comes to evolution education generally and teacher support specifically. Namely, neither content-focused nor cognitive change-focused remediations address head-on the epistemological and existential anxieties that evolution engenders, especially in those teachers with backgrounds in conservative, Christian denominations (Berkman & Plutzer, 2010). When religiosity is part of one's core identity (as is the norm in West Texas), these identity features can be triggered in the presence of evolutionary content that confronts beliefs associated with them (e.g., special creation is incongruous with human descent with modification; Southerland & Scharmann, 2013). This dissonance (a disconcerting disconnect between the two epistemologies, faith and science) may result in existential anxiety and thus ambivalence toward the theory, or even flat out denial of its relevance (Griffith, & Brem, 2004). At the same time, teachers are bombarded with messages from the popular media that claim that their faith and scientific knowledge cannot be reconciled (e.g., Ham, 2015; Dawkins, 1997, 2005; cf. Miller, 2007). Even if teachers' beliefs were not under threat, they may nonetheless experience anxiety because they are ill-prepared (due to gaps in teacher training programs) to discuss students' questions that are driven by the students' own epistemological anxieties and existential concerns. Even in universities where preservice teachers are trained, such discussions are avoided altogether because most university science instructors—who may be atheistic or agnostic themselves—are unaware of teachers' needs in this domain and therefore cannot well advise them about how to deal with issues outside of philosophical materialism (Barnes & Brownell, 2016).

For those teaching evolution or remediating deficiencies in reasoning relevant to the nature of science and evolution (especially descent with modification), ignoring these identity features (e.g., "difference blindness"; Southerland, Smith, Sowell, & Kittelson, 2007) puts our curricular interventions at risk for failure. This risk is especially urgent in light of the fact that students in Colleges of Education tend to be more religious than other majors (Paz-y-mino & Espinosa 2010; Pobiner, 2016). As a result, and important for our present purposes, learners are vulnerable to retreating back into heuristics, many of which are readily supplied by the intelligent design movement, the slogans of which quickly assuage existential anxieties in a direction antithetical to sound science education ("I didn't evolve from monkeys," "The second law of thermodynamics makes evolution impossible," etc.).

The Present Work

The present workshop was carefully designed to address all three targeted levels found in the field thus far (Hawley & Sinatra, 2018). As such, it stands as a unique integration. Though biological content (Level 1) is neither ignored nor downplayed, content was carefully chosen in light of its utility for the intelligent design movement, and, accordingly, presented in a targeted way to catalyze cognitive, affective, and behavioral engagement consistant with cognitive change models (Level 2). Moreover, unique to the present workshop, Levels 1 and 2 were shaped explicitly to serve Level 3 goals, namely, to reduce existential anxieties stemming from perceived epistemological conflict.

Declawing the Dinosaurs

The intelligent design rhetoric functions as an accelerant to teachers' slow burn of worry. By casting doubt on physical evidence while explicitly aligning evolution with irreligiosity and "evil," the message is clear; you shouldn't believe it, but if you do, your salvation and eternal existence are in peril. In the end, the movement has effectively created a culture of doubt and distrust in the scientific enterprise and in the science teachers themselves, thus promulgating misinformation and resistance.

Scientific accounts of said rhetoric have focused on its veracity and, accordingly, have countered its talking points with stern explications of "the facts" (Dawkins, 1997, Ham & Nye, 2014). The presentation of these "facts" by the scientific community can backfire; rather than "clarify misinformation" and thus win hearts and minds, these campaigns may in fact inadvertently widen the gap between two competing cultural ideologies (Hawley & Sinatra, 2018; Kahan, 2015).

What hasn't been done heretofore is to utilize the intelligent design rhetoric directly to address misconceptions from a content perspective (Level 1), while at the same time inspire proscience attitude change consistent with cognitive change models (Level 2) in individuals who embrace a Christian identity (Level 3). That is, rather than considering the talking points a source of neutral misinformation, the present authors thought of them more as a playbook. As a playbook, the rhetoric (e.g., eloquently revealed in the 1995 State of Alabama School Board biology textbook disclaimer) begs to be deconstructed in terms of end goals, motivation, and process (see Hawley & Phillips, 2017, for complete explication). We reasoned, since the talking points function to amplify existential anxiety and fear, that revealing the rhetoric *as a motivated and pointed strategy* would reduce said anxiety and fears, because teachers would be prepared to address ambiguities for themselves and others. Furthermore, if such careful deconstruction were to be presented to science teachers in West Texas in a way respectful to their cultural identities (cf. the Dawkins, 2005, as previously quoted), not only would their negative affect surrounding the theory and teaching it decrease, but also their self-efficacy for managing such material and student questions would increase.

Rhetoric of Intelligent Design Movement

The rhetoric of the intelligent design movement fuels teachers' anxieties about teaching evolution. Rather than helping resolve epistemological anxieties in ways that foster devotion to both faith and science, the rhetoric exploits these anxieties preferentially for faith and to the detriment of science. It is our view that—rather than unknowingly promulgating "misconceptions"—the movement has developed effective, highly focused talking points, reproduced in many websites, publications, and videos, that enjoy consensus from the stakeholders of the movement (Hawley & Phillips, 2017; e.g., evolution is random, complexity cannot be explained, gaps in the fossil record, second law of thermodynamics). Indeed, some websites are dedicated to teaching young people to ask questions that are designed to throw their teachers off balance and undermine their credibility (Dembski, 2011). Moreover, this movement implicitly, if not explicitly, casts doubt on teachers' intentions, undermining teachers' confidence yet further.

Our view is that these talking points are rather laser focused on the Achilles' heels of an untrained or modestly trained audience, such as the public at large and teachers who have taken a limited number of biology classes. That is, how does one address the accusation that evolution violates physical laws or that there are no Precambrian fossils? Both of these claims implied by these questions are patently false and yet are unlikely to have been addressed in biology classes, which are typically not designed to address the talking points of the intelligent design movement (or students' anxieties about the boundedness of science and faith as epistemologies). In this sense, these "talking points" function as a Trojan horse; a subversive technique disguised as a helpful clarification that in reality lures the target (here, teachers) to open the door to creationism in the science classroom for the sake of "fairness," "open mindedness," or "critical thinking" (Hawley & Phillips, 2017). As such, the rhetorical strategies played a key and central role in the organization of our curricular materials and consumables.

Conducting the Workshop

Pilot Group

Our first administration of the workshop was conducted in a local regional Education Service Center that serves West Texas. Fifty K–12 science teachers were participants in a service center grant and, as such, were required to attend a weeklong series of professional development workshops focused on science teaching and content. The present workshop occurred on the final day. Men comprised 26% of the sample, the participants' average age was 37.7 years, and the average time of service was 9.23 years. Christians of various denominations comprised 74% of the sample (36% of which identified as Baptist), 10% identified as atheist or agnostic, and 10% identified as "spiritual but not religious."

Setting the Tone

To maintain an atmosphere consistent with "an equitable approach to science" (Southerland & Scharmann, 2013, p. 64) and to communicate our conciliatory stance regarding faith (i.e., respect for their religious identities), we opened by explicitly stating our intention to inform and arm the teachers about the "debates" *without impacting their faith*. To drive this point home, throughout the workshop we had contributions from our second author and workshop co-presenter (a theistic evolutionist and science teacher) about how she dealt with her own anxieties with teaching evolution. Nothing more was said about the spiritual stance of the first or third authors, the faculty running the workshop.

Open Discussion of Fears

We opened the day by asking our teachers "What scares you about evolution?" and "What scares you about *teaching* evolution?" To encourage cooperative discourse, we used the well-known Kagan strategy of "placemats," which are simply a pieces of paper divided into quarters with space for each group member to write their thoughts. These mats encourage cooperative thinking and discussion while keeping participants engaged and on topic (see Kagan & Kagan, 2009), expectations we wished to carry through the entire day. Their "fears" were then publically placed for the group to discuss. The fears included community backlash, ostracism from religious and political groups, pressure from family, retribution from administration, internal epistemological conflict, and long-held misconceptions about the theory itself. Notably, only one of the listed fears related to content (i.e., misconceptions). To us, this spoke volumes about the need to directly address anxieties even before content can be effectively delivered.

Common Understanding of Evolution

Before addressing the primary goal of the workshop (reconciliation; cf. Meadows, 2009), the third author (a biology professor) presented briefly on what is meant by evolution, adaptation, natural selection, descent with modification, teleology, and common ancestry.

The Handbook

Our handbook was constructed with teachers in mind and designed (by the first author) as a practitioner tool. Even though in some respects the final product looked like colorful workbook that teachers would use in their own classrooms, its content was scholarly, theory driven, and sophisticated from a content perspective, and yet written in a language common to practitioners. Divided into four modules, this handbook carves a path from dissonance to reconciliation while considering the religious and professional identities of those reading it. Throughout the handbook are call-outs referred to as "Rachael's Revelations." These are the real-life reactions and insights of the second author, a practicing Christian and former high school science teacher (i.e., a cultural insider). The purpose of these revelations was not only to reveal one teacher's journey to epistemological reconciliation, but also to facilitate connection and assure our participants that we were not there to judge their faith. Also throughout the handbook, we brought participants' focus to hot-button words and phrases (we labeled them "triggers") which we also summarized in the "Words to Watch For" appendix. We closed the handbook by walking teachers through a deconstruction of creationist rhetoric in the Alabama State School Board of Education disclaimers (Hawley & Phillips, 2017) and Texas state standards. These exercises highlight the effectiveness of intelligent design tactics and politicization of K–12 education.

Module 1: What You Know, How You Know It, and What It Means (Semantics)

To begin, we laid the foundation for an advanced lesson on epistemology (Module 2), including differentiating methodological and philosophical materialism. Our purpose for Module 1 was to give our participants a language with which to speak about the differing kinds of knowledge they hold by helping them understand the differences between truths, facts, and beliefs, and to help them define their own without creating a hierarchy. Additionally, we reintroduced the concepts of hypothesis, theory, and law to help clarify their scientific meanings, which are easily lost in the mire of "the debates." Semantics, the branch of linguistics dealing with meaning, is a driving force behind much of the debate over evolution that encourages divisiveness rather than reconciliation.

Truths, Beliefs, and Facts

What we believe is unique to the present authors' approach was a discussion of the terms *truth, belief,* and *fact,* which, we are as often conflated by scientists as they are by creationists (e.g., "If you want to assert a truth, first make sure it's not just an opinion that you desperately want to be true"; deGrasse Tyson, 2013). Our basis for this distinction was the *Oxford English Dictionary*'s early representation of "truth" (with old Germanic and Icelandic roots) *not as factual accuracy*, but rather related to faith or loyalty as pledged in a promise or solemn engagement. The initial use of the term to us conveys why "truth" is in the Bible as a matter of course (over 150 times), but "fact" is not (Hawley & Phillips, 2016). In this sense, despite not requiring evidence, truths are both highly unshakeable and yet distinctive to the individual. This use of the term as a rule is familiar to churchgoers. In contrast, "belief" is merely a neutral "mental conviction" not requiring validation. As such, it is less weighty than a "truth" to the faithful. A "fact" is an observation that has been confirmed and verified. Our perspective is that "truth" is an important concept and trigger word for Christians of certain denominations.

After defining and exemplifying the terms through our own lenses, we asked the teachers to engage in a "stop and think" exercise with their peers; namely, we asked the participants to identify the sources of their own truths, beliefs, and facts, while emphasizing that they are likely to differ from one another. Thus, we circumscribed a language with weighty epistemological implications at the outset. We then applied this language throughout the workshop while at the same time assuring the participants that we were not the authority from which to receive truths, all the while pointing out "Science does not necessarily give us meaning. . . . it gives us a description of the natural world" (Long, 2012; p. 126).

Interestingly, religious audience members (of both the workshop and academic audiences to whom the workshop has been disseminated) identified with and thus greatly appreciated the distinctions between truths, beliefs, and facts. From our perspective, giving the teachers a taxonomy for "all the things they know" at the outset was a critical and highly effective first step to epistemological reconciliation without even saying the word *epistemology*.

Not all feedback has been positive, however. Self-identified atheists (a vocal minority) have tended to react negatively to this content; namely, one angrily asserted that "facts" are true, and thus facts are "truths" (which could be argued to be the hallmark of an atheistic position[1]). On the other extreme, a biblical literalist asserted that the "truths" of the Bible are scientific "facts."[2] To us, this discussion illustrates exactly why the "two camps" fail to communicate effectively and create an environment where cognitive change is nearly impossible.

Truth Violations (Christian Stumbling Blocks to Evolution)

The next section challenged audience members to think about some "facts" of evolution that may also be "truth violations" to some denominations of Christianity. At this point, workshop attendees worked out for themselves that understanding evolution is not the same as accepting evolution and that where "facts" conflict with "truths" is a domain that provokes intense existential anxiety in the faithful. Again, we never evaluated their truths for "fact value." If we were to do so, we would have violated our own lesson plan and subverted our own learning goals. Demanding evidence for a truth would be doing exactly what we told them not to do.

---

[1] "Gravity is not a version of the truth. It is the truth. Anybody who doubts it is invited to jump out of a tenth-floor window" (Dawkins, 2009).

[2] "No apparent, perceived, or claimed evidence in any field, including history and chronology, can be valid if it contradicts the Scriptural record" (Ham, 2015).

Theory, Law, and Hypothesis

The semantics of science have been turned into rhetorical weapons by the intelligent design movement, which intentionally arms the public with such slogans as, "It's *just* a theory," "If it's so strongly supported, why is it not a law?" and "If evolution is a theory (like creationism or the Bible), why then is it taught as fact?" (Hawley & Phillips, 2017; Stoptera, 2014). This motivated rhetoric informed our treatment of the basic language of science.

The module begins with the basic definition of theory, law, and hypothesis. Directly following, the workshop participants were asked to respond to the previously illustrated common claims creationist claims. Through a "Rachael's Revelation," we pointed out that (in contrast to "truth," "belief," and "fact"), scientific concepts *do* have a hierarchy, where "theory" resides in the top position of value. Theories organize and explain laws, for example. Theories do not become laws with more evidence. As explanatory frameworks, theories and theory development enjoy high status in science. At this point, the workshop participants were directly deconstructing the rhetoric of the intelligent design movement and gaining a deeper understanding of semantics and their important implications.

Module 2: Science and Faith as Epistemologies

Ways of Knowing and Explaining the World

Inspired by the work of Eugenie Scott (e.g., Scott, 2005), we described the strengths and boundaries of authority, revelation, logic, intuition, and science and differentiated their respective roles for religion and science (religion utilizes authority, revelation, and intuition whereas science relies on authority, logic, and empirical evidence; Scott, 2000). A "Rachael's Revelation" shares her personal insight that her particular questions can have more than one answer depending on the lens through which she is looking. To her, as a theistic evolutionist, science tells her that humans originated from an extinct ancestor, and faith tells her that human origins were God's plan. Special cases were made for authority and intuition as they seem to be especially manipulated by the intelligent design movement. Those holding the view of biblical inerrancy (e.g., Gish, 2012) claim that there are eternal costs to doubting "God's word as fact," and intuition suggests that complex structures cannot evolve such that evolution consequently "doesn't make sense."

Authority

We described religious authority as often best not questioned and documents and icons not desecrated as they are "sacred" (i.e., deserving veneration and reverence). In contrast, authorities are not unquestioned in science, nor are they or their documents worthy of reverence. Special note was added that this point applies to Charles Darwin himself and that his "word" is not "final" but instead serves as a starting point for contemporary evolutionary work. Here we advise the teachers to be wary of intelligent design rhetoric that claims that Darwin has some exalted, revered position in scientific circles and that his writings represent some form of "sacred scientific belief."

Intuition

Intuition was described as having a modest role in science but a huge role in our daily lives, and this is why we are so susceptible to it. We used a Paley quote to illustrate that "science" was intuitive in the early 19th century and that his view of the world (one evident of design) forms the foundation for the modern belief in an intelligent designer ("the only force capable for creating complexity"). We pointed out that "complexity" is a trigger word for Christians and of high utility value to the intelligent design movement as it directs one straight to Behe's treatises on "irreducible complexity" (e.g., Behe, 1996).

Module 3: Does Accepting Evolution Mean You Are an Atheist?

Acceptance of Evolution and Belief Preferences

One of the key themes of our workshop is the point that extreme voices are the ones that are presently heard (e.g., Ham, Dawkins, Nye, Krauss). These voices tend to occupy positions of either young earth creationism (e.g., biblical literalism and inerrancy; Ham, Morris, Gish) or moderate to strident atheism (Nye, Dawkins, Krauss). Teachers, in contrast, tend to occupy any one of many belief preferences between these two extremes. The *raison d'etre* of the workshop was to speak to "the cautious 60%" (Berkman & Plutzer, 2010). Borrowing again from Scott (2000), we replicated and expanded her degree of acceptance continuum. An interactive activity prompted the workshop participants to place themselves and others around them on the continuum allowing them to explore and articulate the implications of differing types of Christianity in their personal circles. Upon learning that several categories on the continuum expressly involve theistic evolutionism, a good proportion of the participants found a category within which they comfortably reside and a place which neither conflicts with being an effective science teacher nor a good Christian.

Materialism

We closed the module with a segment differentiating methodological materialism (explores cause in the material/natural world without denial of existence of the supernatural) and philosophical materialism (denies existence of supernatural world altogether) with the intention of giving the teachers tools for personal reconciliation. Discussion points highlighted the place in the science classroom for methodological materialism without necessitating philosophical materialism. That is, teachers worked out that they can comfortably convey the standards of science all the while believing in (though not testing or discussing) God. This context invited a deep discussion of the boundaries of science and faith.

Module 4: Setting the Record Straight

Many of the claims of the intelligent design movement are motivated by the desire to frighten Christians away from science in general ("materialism") and evolution in particular (e.g., Center for the Renewal of Science and Culture, 1998; Hawley & Phillips, 2016). Moreover, websites and media presentations appear to be encouraging and empowering young students to undermine their teachers with these false claims and rhetorical questions. At the same time, training programs are not systematically preparing teachers to quickly and succinctly debunk these tactics (e.g., "Evolution is like a tornado blowing through a junkyard and building a 747" and "Why are there still monkeys?"). When students go beyond scientific hesitation and leap into existential skepticism, they are not just questioning evolutionary theory but also their teacher's authority to educate them. In contrast to approaches that are content focused and are designed to teach physical laws and tree thinking in detail, our own strategy was to reveal the source and motivation behind the tactics. Our aim was to arm teachers with quick explanations for students (cf. http://www.talkorigins.org), rather than push teachers into long unwinnable arguments. At the same time, we delivered content (very pointed to the tactics) about tree thinking, the fossil record, the Cambrian explosion, the second law of thermodynamics, and the role of randomness in natural selection. The format of this module was a practitioner-friendly "facts and fibs" design.

Mini Lesson by a Biologist

The third author, a biologist and instructor of evolution, further debunked a number of myths associated with evolution, such as the fallacy of irreducible complexity and how genomic and organismal complexity do not undermine evolution by natural selection. This lesson additionally emphasized the logical structure of intelligent design arguments and focused specifically on false premises and nonsequiturs. Finally, he closed by addressing the false creationist claim that evolution is a "theory in crisis," by explaining that it is resolutely backed by the scientific community and to claim otherwise would be tantamount to throwing fuel on a nonfire.

Mini Lesson by a Texas Science Teacher

Very recently, the Texas science standards have undergone heated revisions and are much improved from their previous editions. Yet, as has been pointed out, enhancing policy and standards to bolster support for sound evolution education will have little effect if teacher ambivalence or antipathy stand in the way of following the standards (Ha, Baldwin, & Nehm, 2015).

The second author began by reminding teachers that their contracts with their local school boards bind them to the Texas Education Code, which itself is part of the Texas Constitution. Therefore, each standard is a piece of legislation carrying legal weight. The aim here was not to scare them into teaching according to the standards and best practices, but rather to remind them that the consequences for not doing so could be far-reaching and quite damaging. This tactic invited a fruitful opportunity to articulate as a group the difficult position teachers are in to follow the law, all the while dealing with doubts about evolution from themselves, their school administrators, and their communities. The lesson ended with the teachers identifying creationist language that still exists in the current Texas standards and interpreting this language in terms of the previous lessons (e.g., "theory, not fact", "random forces", "lack of transitional forms").

Additional Resources

All participating teachers were provided two HHMI videos (Carroll, Kingsley, Wiseman, & Ruse, 2005; Miller, 2006), several Web-based resources from National Academy of Science (http://www.nas.edu/evolution/), and several book recommendations from the National Academy

Press (National Academy of Sciences, 1998, 2014; National Academy of Sciences and Institute of Medicine, 2008).

Participant Reception and Outcomes

Teachers must balance their sense of duty to their profession, the demands and concerns of their community, their response to the larger political climate, and their own personal beliefs. (Griffith & Brem, 2004, p. 792)

Teachers are at the forefront of the culture wars where they are asked to do the near impossible, as reflected in the above quote. At the same time, there is a lack of educational support for the anxieties they experience when two different epistemologies collide.

These anxieties were made plain even before the workshop during discussions with their science curriculum specialist. Just the thought of a workshop on evolution was enough to catalyze dissonance. For this reason, we took special care to assure the teachers at the outset that we were not only not going to undermine their faith but rather support it by distinguishing their "truths" (untouchable and timeless) from "facts" (ever-changing in science). Having a veteran science teacher as a co-instructor, who is also a theistic evolutionist, was useful for our approach.

In the end, the response from the teachers was overwhelmingly positive. Forty-six evaluations from the participants yielded an average 6.27 on a 7-point scale from 1 (poor) to 7 (excellent) on clarity, informativeness, usefulness of the workshop, and quality of the handbook. Even a year later, a number of teachers informed us (via a Qualtrics survey) that they still refer to the handbook for answering questions for themselves and getting ideas on how to direct students. That is not to say that all were pleased or felt well served: One of the attendees was highly agitated throughout the workshop, scored the workshop as "poor" overall, and added the comment, "Content was NOT helpful and attempted to confuse the teachers and push athesism [sic]." On the other hand, a minister/science teacher was very pleased: "I've served my church as an ordained minister for more than ten years, and I'm passionate about the influence of the gospel of Jesus Christ. Nonetheless, the workshop influenced my passion in both camps: creationism and evolution." Indeed, one might expect young earth creationists as well as strident atheists to be less well served than the majority of the teachers who find themselves between the extremes on faith.

TABLE 18.1. Number of Times Workbook Was Consulted One Year Later

Think about your use of the workbook. How many times have you looked at it since the workshop..."	Average number of times workbook was consulted
... for your own thinking	3.62
... to answer student questions, enhance student understanding, or navigate student discussions.	2.88
... to navigate discussions with your family	3.00
... to navigate discussions with your co-workers.	2.88
... when reading (internet, Nye's book) or seeing the news?	4.43

Note: Based on 21 responses on a zero- to 20-point scale.

Additionally, although the explicit outcomes to the data collection protocol are reported elsewhere (Hawley & Sinatra, 2018), the workshop significantly reduced negative emotions about (a) evolution, (b) teaching it, and (c) addressing student questions concerning it and, at the same time, increased self-reported self-efficacy in teaching it. Our results are especially powerful because our audience was mostly veteran teachers who are typically less receptive to new information or willing to shift their world views. Moreover, as a group, the teachers identified as politically conservative and devotedly Christian, scoring themselves on average above a 5 on a 7-point scale on each dimension.

Finally, the utility of our efforts can be evidenced in teachers' reports about how often they consulted with our material even after the workshop had ended. One year later, we had the opportunity to query the teachers again. With the stem, "Think about your use of the workbook. How many times have you looked at it since the workshop for . . . ," we tally the average number of times the teachers self-reported having consulted the handbook in the numerous contexts, as reported in Table 18.1.

Conclusion

American education majors are the most religious of all college majors and become more so while in college (Kimball et al., 2009). When people are presented with information that contradicts deeply held cultural values (e.g., descent with modification) and threaten "truths" that carry the existential weight of eternity (e.g., salvation), most do not easily incorporate such knowledge and, instead, retrench further within the certainty of the system of knowledge they hold (Kahan et al., 2007; Long, 2012; Nyhan & Reifler, 2010). With this in mind, it becomes easier to understand why those with a strong Christian belief system readily accept spiritual answers (e.g., the Creation Story) and hesitate to accept scientific explanations (e.g., the fossil record). The faithful, many of them teachers, become easy targets for those who wish to undermine any attempt to reconcile faith and science (as Duane Gish does explicitly; Gish, 2012). Under the guise of providing comfort (reducing anxiety), the antievolution movement suggests that the unknowable answers can be found in long standing authority and text (God and the Bible).

We agree with Southerland and Scharmann (2013): The false dichotomy pitting faith and science against each other is a broken pedagogical paradigm that propagates the assumption that students and teachers must abandon part of their identity (religious/spiritual beliefs) to practice good science. In response, we propose a shift to teaching epistemological reconciliation rather than disassociation when covering topics such as evolution.

References

Ham, K.A., & Nye, W. S. (2014).*Is Creation a viable model of origins?* [Video file]. Retrieved from https://www.youtube.com/watch?v=z6kgvhG3AkI.

Akyol, G., Tekkaya, C., Sungur, S., & Traynor, A. (2012). Modeling the interrelationships among pre-service science teachers' understanding and acceptance of evolution, their views on nature of science and self-efficacy beliefs regarding teaching evolution. *Journal of Science Teacher Education, 23*, 937–957.

Barnes, M. E., & Brownell, S. E. (2016). Practices and perspectives of college instructors on addressing religious beliefs when teaching evolution. *CBE Life Sciences Education, 15*, art. 18.

Behe, M. J. (1996). *Darwin's black box: The biochemical challenge to evolution.* New York, NY: Simon and Schuster.

Berkman, M. B., Pacheco, J. S., & Plutzer, E. (2008). Evolution and creationism in America's classrooms: A national portrait. *PLoS biology, 6*(5), e124.

Berkman, M., & Plutzer, E. (2010). *Evolution, creationism, and the battle to control America's classrooms.* Cambridge, England: Cambridge University Press.

Bland, M. W., & Morrison, E. (2015). The experimental detection of an emotional response to the idea of evolution. *The American Biology Teacher, 77*, 413–420.

Branch, G., Scott, E. C., & Rosenau, J. (2010). Dispatches from the evolution wars: Shifting tactics and expanding battlefields. *Annual Review of Genomics and Human Genetics, 11*, 317–338.

Brehm, J. W. (1966). *A theory of psychological reactance.* Oxford, England: Academic Press.

Carroll, S. B., Kingsley, D. M., Wiseman, J., & Ruse, M. (2005). Discussion: Reconciling religion and science [Video file]. *HHMI.* Retrieved from https://www.hhmi.org/biointeractive/reconciling-religion-and-science

Center for the Renewal of Science and Culture. (1998). *The wedge document.* Seattle, WA: Discovery Institute. Retrieved from https://ncse.com/files/pub/creationism/The_Wedge_Strategy.pdf

Clores, M. A., & Limjap, A. A. (2006). Diversity of students' beliefs about biological evolution. *Asia Pacific Journal of Education, 26,* 65–77.

Cunningham, D. L., & Wescott, D. J. (2009). Still more "fancy" and "myth" than "fact" in students' conceptions of evolution. *Evolution: Education and Outreach, 2,* 505–517.

Dawkins, R. (1997). The "Alabama insert": A study in ignorance and dishonesty. *Journal of the Alabama Academy of Science, 68,* 1–16.

Dawkins, R. (2005, April 30). The atheist. *Salon.* Retrieved from http://www.salon.com/2005/04/30/dawkins/.

The genius of Charles Darwin, series producer and director, Russell Barnes; IWC Media. (2009). Spring, MD: Acorn Media Group: Athena.

deGrasse Tyson, N. (2013). If you want to assert a truth, first make sure it's not just an opinion that you desperately want to be true [Tweet]. Retrieved from https://twitter.com/neiltyson/status/350753146445893633

Dembski, W. A. (2011). Ten questions to ask your biology teacher about design. Retrieved from http://www.arn.org/docs/dembski/TenQuesDes.pdf

Evans, E. (2008). Conceptual change and evolutionary biology: A developmental analysis. In S. Vosniadou (Ed.), *International handbook of research on conceptual change* (pp. 263–294). New York, NY: Routledge.

Gish, D.T. (2012). *Letters to a theistic evolutionist.* Noble, OK: Icon.

Griffith, J. A., & Brem, S. K. (2004). Teaching evolutionary biology: Pressures, stress, and coping. *Journal of Research in Science Teaching, 41,* 791–809.

Ha, M., Baldwin, B. C., & Nehm, R. H. (2015). The long-term impacts of short-term professional development: Science teachers and evolution. *Evolution: Education and Outreach, 8,* 1–23.

Ha, M., Haury, D. L., & Nehm, R. H. (2012). Feeling of certainty: Uncovering a missing link between knowledge and acceptance of evolution. *Journal of Research in Science Teaching, 49,* 95–121.

Ham, K. (2015, August 10). Statement of faith. *Answers in Genesis.* Retrieved from https://answersingenesis.org/about/faith/.

Hawley, P. H., & Phillips, R. K. (2016). *Declawing the dinosaurs: Reducing teachers' anxiety about evolution in the science classroom.* Lubbock, TX: Texas Tech University.

Hawley, P. H., & Sinatra, G. M. (2018). Declawing the dinosaurs in the science classroom: Reducing Christian teachers' anxiety and increasing their efficacy for teaching evolution. *Journal of Research in Science Teaching,* 1–27. https://doi.org/10.1002/tea.21479

Hawley, P. H., Short, S. D., McCune, L. A., Osman, M. R., & Little, T. D. (2011). What's the matter with Kansas?: The development and confirmation of the Evolutionary Attitudes and Literacy Survey (EALS). *Evolution: Education and Outreach, 4,* 117–132.

Heddy, B. C., & Nadelson, L. S. (2013). The variables related to public acceptance of evolution in the United States. *Evolution: Education and Outreach*, *6*, 1–14.

Heddy, B. C., & Sinatra, G. M. (2013). Transforming misconceptions: Using transformative experience to promote positive affect and conceptual change in students learning about biological evolution. *Science Education*, *97*, 723–744.

Kagan, S., & Kagan, M. (2009). *Kagan cooperative learning*. San Clemente, CA: Kagan.

Kahan, D. M. (2015). Climate-science communication and the measurement problem. *Political Psychology*, *36*, 1–43.

Kahan, D. M., Braman, D., Gastil, J., Slovic, P., & Mertz, C. K. (2007). Culture and identity-ss protective cognition: Explaining the white-male effect in risk perception. *Journal of Empirical Legal Studies*, *4*, 465–505.

Kelemen, D. (2004). Are children "intuitive theists"? Reasoning about purpose and design in nature. *Psychological Science*, *15*, 295–301.

Kimball, M.S., Mitchell, C. M., Young-Demarco, L. C. (2009). *Empirics on the origins of preferences: The case of college major and religiosity*. (No. w15182). Cambridge, MA: National Bureau of Economic Research.

Lawson, A. E., & Worsnop, W. A. (1992). Learning about evolution and rejecting a belief in special creation: Effects of reflective reasoning skill, prior knowledge, prior belief and religious commitment. *Journal of Research in Science Teaching*, *2*, 143–166.

Lombrozo, T., Thanukos, A., & Weisberg, M. (2008). The importance of understanding the nature of science for accepting evolution. *Evolution: Education and Outreach*, *1*, 290–298.

Long, D. E. (2012). The politics of teaching evolution, science education standards, and being a creationist. *Journal of Research in Science Teaching*, *49*, 122–139.

Losh, S. C., & Nzekwe, B. (2011). Creatures in the classroom: Preservice teacher beliefs about fantastic beasts, magic, extraterrestrials, evolution and creationism. *Science & Education*, *20*, 473–489.

Mead, L. S. (2009). Transforming our thinking about transitional forms. *Evolution: Education and Outreach*, *2*, 310–314.

Mead, L. S., Clarke, J. B., Forcino, F., & Graves, J. L., Jr. (2015). Factors influencing minority student decisions to consider a career in evolutionary biology. *Evolution: Education and Outreach*, *8*, 6.

Mead, L. S., & Scott, E. C. (2010a). Problem concepts in evolution part I: Purpose and design. *Evolution: Education and Outreach*, *3*, 78–81.

Mead, L. S., & Scott, E. C. (2010b). Problem concepts in evolution part II: Cause and chance. *Evolution: Education and Outreach*, *3*, 261–264.

Meadows, L. (2009). *The missing link: An inquiry approach for teaching all students about evolution*. Portsmouth, NH: Heinemann.

Miller, J. D., Scott, E. C., & Okamoto, S. (2006). Public acceptance of evolution. *Science*, *313*, 765–766.

Miller, J. S., & Toth, R. (2014). The process of scientific inquiry as it relates to the creation/evolution controversy. *The American Biology Teacher*, *76*, 238–241.

Miller, K. R. *Evolution: Fossils, genes, and mousetraps* [Video file]. *HHMI*. Retrieved from https://www.hhmi.org/biointeractive/evolution-fossils-genes-and-mousetraps

Miller, K. R. (2007). *Finding Darwin's God: A scientist's search for common ground between God and evolution*. New York, NY: Perennial.

Miller, K. R. (n.d.) *Deconstructing the Alabama disclaimer.* Retrieved from http://www.millerandlevine.com/km/evol/disclaimer.html

Nadelson, L. S., & Nadelson, S. (2010). K–8 educators perceptions and preparedness for teaching evolution topics. *Journal of Science Teacher Education, 21,* 843–858.

Nadelson, L. S., & Sinatra, G. M. (2009). Educational professionals' knowledge and acceptance of evolution. *Evolutionary Psychology, 7,* 490–516.

National Academy of Sciences. (1998). *Teaching about evolution and the nature of science.* Washington, DC: National Academy Press.

National Academy of Sciences. (2014). *Thinking evolutionarily: Evolution education across the life sciences* (S. Olsen, Ed.). Washington, DC: National Academy Press.

National Academy of Sciences and Institute of Medicine. (2008). *Science, evolution, and creationism.* Washington, DC: National Academy Press.

Nehm, R. H., & Schonfeld, I. S. (2007). Does increasing biology teacher knowledge of evolution and the nature of science lead to greater preference for the teaching of evolution in schools? *Journal of Science Teacher Education, 18,* 699–723.

Nehm, R. H., Kim, S. Y., & Sheppard, K. (2009). Academic preparation in biology and advocacy for teaching evolution: Biology versus non-biology teachers. *Science Education, 93,* 1122–1146.

Non Serviam. (2009). *Richard Dawkins and postmodernism* [Video file]. Retrieved from https://www.youtube.com/watch?v=rGOrBCkQ6WY.

Nyhan, B., & Reifler, J. (2010). When corrections fail: The persistence of political misperceptions. *Political Behavior, 32,* 303–330.

Osif, B. A. (1997). Evolution and religious beliefs: A survey of Pennsylvania high school teachers. *The American Biology Teacher,* 552–556.

Paz-Y-Mino, C., & Espinosa, A. (2010). Integrating horizontal gene transfer and common descent to depict evolution and contrast it with "common design." *Journal of Eukaryotic Microbiology, 57,* 11–18.

Pobiner, B. (2016). Accepting, understanding, teaching, and learning (human) evolution: Obstacles and opportunities. *American Journal of Physical Anthropology, 159,* 232–274.

Pugh, K. J., & Girod, M. (2007). Science, art, and experience: Constructing a science pedagogy from Dewey's aesthetics. *Journal of Science Teacher Education, 18,* 9–27.

Rissler, L. J., Duncan, S. I., & Caruso, N. M. (2014). The relative importance of religion and education on university students' views of evolution in the Deep South and state science standards across the United States. *Evolution: Education and Outreach, 7,* 24.

Rosengren, K. S., Evans, E. M., Brem, S., & Sinatra, G. M. (Eds.). (2012). *Evolution Challenges: Integrating research and practice in teaching and learning about evolution.* Oxford, England: Oxford University Press.

Scott, E. C. (1997). Antievolution and creationism in the United States. *Annual Review of Anthropology, 26,* 263–289.

Scott, E. C. (2000). Not (just) in Kansas anymore. *Science, 288,* 813–815.

Scott, E. C. (2005). *Evolution vs. creationism: An introduction.* Berkeley, CA: University of California Press.

Scott, E. C. (2010). Listening to teachers. *Evolution: Education and Outreach, 3,* 241–244.

Shankar, G., & Skoog, G. D. (1993). Emphasis given evolution and creationism by Texas high school biology teachers. *Science Education, 77,* 221–233.

Short, S. D., & Hawley, P. H. (2014). The effects of evolution education: Examining attitudes toward and knowledge of evolution in college courses. *Evolutionary Psychology, 13,* 67–88.

Sinatra, G. M., Kardash, C. M., Taasoobshirazi, G., & Lombardi, D. (2012). Promoting attitude change and expressed willingness to take action toward climate change in college students. *Instructional Science, 40,* 1–17.

Sinatra, G. M., Southerland, S. A., McConaughy, F., & Demastes, J. W. (2003). Intentions and beliefs in students' understanding and acceptance of biological evolution. *Journal of Research in Science Teaching, 40,* 510–528.

Southerland, S. A., Golden, B., & Enderle, P. (2012). The bounded nature of science: An effective tool in an equitable approach to the teaching of science. In *Advances in nature of science research* (pp. 75–96). Dordrecht, The Netherlands: Springer.

Southerland, S. A., & Scharmann, L. C. (2013). Acknowledging the religious beliefs students bring into the science classroom: Using the bounded nature of science. *Theory into Practice, 52,* 59–65.

Southerland, S. A., Sinatra, G. M., & Matthews, M. R. (2001). Belief, knowledge, and science education. *Educational Psychology Review, 13,* 325–351.

Southerland, S. A., Smith, L. K., Sowell, S. P., & Kittleson, J. M. (2007). Resisting unlearning: Understanding science education's response to the United States' national accountability movement. *Review of Research in Education, 31,* 45–77.

Stoptera, M. (2014). 22 Messages from creationists to people who believe in evolution. *Buzzfeed.* Retrieved from http://www.buzzfeed.com/mjs538/messages-from-creationists-to-people-who-believe-in-evolutio#.bnmEpoK0w.

Wilensky, U. (1997). What is normal anyway? Therapy for epistemological anxiety. *Educational Studies in Mathematics, 33,* 171–202.

Yates, T. B., & Marek, E. A. (2013). Is Oklahoma really OK? A regional study of the prevalence of biological evolution-related misconceptions held by introductory biology teachers. *Evolution: Education & Outreach, 6,* 1–20.

Yates, T. B., & Marek, E. A. (2014). Teachers teaching misconceptions: A study of factors contributing to high school biology students' acquisition of biological evolution-related misconceptions. *Evolution: Education and Outreach, 7,* 7.

CHAPTER 19 | **The Evolutionary Studies Summer Institute at New Paltz**
A High-Impact, Condensed, Interdisciplinary Educational Experience for Teachers

GLENN GEHER, AILEEN TOBACK, AND NICOLE WEDBERG

FOR OVER A DECADE, a number of colleges and universities around the world have been offering students academic programs in the growing interdisciplinary field of Evolutionary Studies (EvoS). Initially designed by David Sloan Wilson and his colleagues at Binghamton University (see Wilson, Geher, & Waldo, 2009; O'Brien, Wilson, & Hawley, 2009; Wilson, 2007), the idea of an interdisciplinary EvoS program is rooted in the interconnected suppositions that (a) Darwin's ideas regarding evolutionary processes have profound implications for understanding phenomena across nearly all academic areas and (b) most students in modern university settings receive no background in evolution whatsoever (Glass, Wilson, & Geher, 2012).

When Wilson took the initial steps to spearhead the first EvoS program at Binghamton in about 2003, he did so after surveying a broad subsection of students at Binghamton University, one of the most highly selective universities in the Northeast. What he found (O'Brien et al., 2009) was very revealing. These researchers obtained several enlightening results, including a finding suggesting that that biology students at Binghamton generally demonstrated a strong factual understanding of evolutionary principles relative to students from other backgrounds. Regardless, the bottom line of that research clearly was that taking a broad-reaching Evolution for Everyone course had the effect of advancing an understanding and appreciation of evolution for students across a variety of disciplines. Importantly, note that these ideas are summarized and elaborated on in David Sloan Wilson's (2007) book by the same name, *Evolution for Everyone*.

Solving America's Evolution Problem

So just past the turn of the 21st century at a top American university, O'Brien et al. (2009) found that Darwin's major ideas, and all of their implications for being human, were all but fully lost on students in higher education who have not taken a basic course such as Evolution for Everyone. When people think of America's evolution problem, this is not typically where the mind goes. This said, it's clear that America has an evolution problem that is well-beyond the better-publicized finding that a large proportion of American's "don't believe in evolution." Perhaps the even more dramatic problem, documented by a variety of scholars (Wilson, Geher, & Waldo, 2009), is this: Colleges and universities in the United States are generally failures when it comes to exposing bright young minds to the basic ideas of evolution.

The first-ever EvoS program in the United States, developed by David Wilson and his colleagues at Binghamton, was designed to address this problem. The program, since replicated at SUNY New Paltz as well as at several other universities around the globe, exposes students from any academic major to the basic ideas of evolution (in a lower-level course titled "Evolution for Everyone") as well as to a set of classes across various academic disciplines (e.g., anthropology, economics, geology, psychology, etc.) that provide applications of evolutionary principles. Further, this program includes a capstone course in the form of a class that surrounds an external speaker series related to evolution that includes speakers from a broad array of disciplines, addressing different classes of phenomena from an evolutionary perspective.

Between 2008 and 2011, the EvoS programs at Binghamton and New Paltz were awarded a total of $500,000 from the National Science Foundation to help expand the EvoS model to other colleges and universities. As documented in several of our publications (Chang, Geher, Wilson, & Waldo, 2011), this grant was highly successful, leading to many tangible outcomes such as the peer-reviewed *EvoS Journal* and the highly accessed website, http://www.evostudies.org, which boasts serving as host to the world's largest database of free and streaming videos connected to evolutionary topics.

As documented in several other publications (see, e.g., Geher, 2014), the interdisciplinary EvoS model of evolution education is highly effective, leading to such intellectual outcomes as (a) the ability for students to understand the principles of evolution in a deep manner, (b) the ability for students to apply evolution in a wide area of areas, (c) developing students' capacity to see evolutionary principles as applicable to many important issues of humanity (e.g., education, mental health, physical health, etc.), and (d) the ability for students to use evolutionary ideas to create bridges between the sciences and other academic areas, such as the humanities.

An Intensive Evolutionary Studies Institute Designed for Teachers

In 2015, based largely on the success of the EvoS model of undergraduate education, the EvoS faculty at SUNY New Paltz developed a week-long, intensive version of the EvoS curriculum designed to provide participants essentially with a condensed and intensive version of the EvoS curriculum. The week-long, 45-hour Evolutionary Studies Summer Institute was designed largely with teachers in mind—with the goal of helping teachers from various fields develop the kind of deep and broad evolution education that results from the EvoS curriculum.

The Institute was designed to include five days of intensive coursework, with each day include two academic sessions (except for the fifth day, which included an evolution-focused hiking field trip into our local Shawangunk Mountains in partnership with the Mohonk Preserve). The instructors of the sessions were all New Paltz EvoS faculty, representing a broad array of disciplines, including biology, anthropology, psychology, and education.

Sessions were 3.5 hours each, and they addressed the following topics:

- Basics of biological evolution;
- Genetics;
- Human evolution;
- Evolution and behavior;
- Evolution and the human condition;
- Controversies surrounding evolution education; and
- Evolution education and the Next Generation Science Standards.

Further, for each of the first four days of the Institute, in between sessions, a selected video from the http://www.evostudies.org database was shown to participants over lunch and our assistant, a graduate student with a focus on evolutionary studies (third author of this chapter, N.) led discussions designed to get the participants to think about the implications of the work being presented.

On the final day of the Institute, participants joined Institute Director Glenn Geher (first author) for an extensive hike into the mountains that included interpretation of geological features as well as adaptations of various flora and fauna. Specific topics discussed during the hike included a presentation on evidence for undersea life in the rocks from the Devonian epoch, adaptations to extreme conditions of the pitch pines and mountain blueberry plants, and various adaptations of animals encountered along the way, including a skink, a red-tailed hawk, and turkey vultures. The ability to integrate the topic of evolution into the hike was clearly a feather in the cap of the educational experience.

At the end of weeklong Institute, eight participants, all secondary teachers from the mid-Hudson region of New York (with half being part of New York's Master Teacher Program), received certificates of completion and attended a reception, which allowed for further discussion of the material with one another along with faculty of the Institute.

Note that the teacher-participants all received continuing-education credit for their participation. As teachers in all states need to regularly engage in continuing-education coursework each year, the Institute had a built-in clientele in local teachers. Participants paid $450 for the 45-hour version of the Institute (or $400 for the 34-hour version that did not include the field trip on the last day). The Institute was held on the campus at SUNY New Paltz and internal processes via the University, including the assistance from the Office of Continuing Education, were employed to assist with logistics. Also, lunches were provided and financial support was provided by several offices on campus (including the College of Liberal Arts and Sciences, the School of Education, the EvoS Program, and the Honors Program).

Survey of Institute Participants

Informal feedback from participants was positive along the way, suggesting that participants perceived the Institute as a useful and provocative educational experience. At the end of the final session, participants were asked to complete an anonymous survey—with a set of open- and closed-ended questions tapping their perceptions of the Institute in a systematic manner.

Goals of the Current Analysis

The goals of the current analysis are to assess outcomes associated with participation in the EvoS Institute, to identify success in line with the Institute's mission, and to determine implications for future occurrences. The current analysis evaluates the Institute's effectiveness in helping teachers master the content and curriculum for future education, and it identifies areas for improvement or future modification.

Method

Participants and Procedure

Participants were eight regional secondary school teachers, four of whom were from the New York State Master Teacher program. The Master Teacher Program is a special program of the governor designed to incentivize extended learning among the state's most successful teachers. Most of the teachers were science teachers, and several teach biology at the high school level.

At the end of the classroom-experience portion of the Institute, the program assistant gave out a survey to all of the participants to be completed anonymously. Upon completion of the survey, participants were thanked and debriefed before continuing for the last part of the Institute.

Measures

The survey included 14 items to be rated on a 1 (*strongly disagree*) to 5 (*strongly agree*) Likert scale, and each item had room for participants to provide open-ended comments related to that item. The survey also asked participants for overall open-ended comments at the end.

Results

Descriptive statistics for the closed-ended items regarding the Institute are presented in Table 19.1 along with all open-ended responses.

TABLE 19.1. Descriptive Statistics for Closed-Ended Items

ITEM	M	SD	COMMENTS
1. The curriculum for the Institute was well-organized.	4.75	0.46	Very well organized; very well thought out.
			Broad range of different educational disciplines; deepened my understanding of evolution.
			decent amount of overlap; could be eliminated with pre-Institute meeting.
2. The curriculum for the Institute was effective in terms of its breadth of content.	4.63	0.52	Really opened my mind to numerous approaches to evolution.
			Overlap hindered optimization of time/material!!
			Very diverse; interesting to have academics in all areas addressing central unifying topic.
3. The curriculum for the Institute was effective in terms of the depth of the content presented.	4.75	0.46	Lectures start with familiar content and then build; providing excellent dept.
			Depth was detailed; however could have been more topic specific
			Great material; good level.
			Didn't feel anything dumbed down for benefit of secondary ed teachers. . . .this was appreciated!!
4. Overall, the faculty provided expert-level presentations of the material.	4.75	0.46	Presenters were great; providing content in a way that is digestible.
			Absolutely; loved Spencer's talk; Glenn is excellent at presenting Evo-Psych in a way that is balanced; the best talks provided discussion-like forum.
			PowerPoints and discussions solid; however more lab time needed.

TABLE 19.1. Continued

ITEM	M	SD	COMMENTS
5. Overall, the educational part of the Institute was effective.	4.36	0.75	Provided more than adequate information allowing me to reconstruct my evolution unit.
			Speakers were excellent; more lab experiences could be included; more accent on applicability to classroom could be useful.
			A balance of how to teach to younger children yet a clearer understanding of key idea for "us" needed more focusing on.
			Would have liked to see connection between info & the secondary classroom; maybe involve a secondary teacher in presentation?
6. The lunch/video sessions were educational.	3.25	0.89	Appreciated videos, but would have enjoyed some time outside and break from lectures.
			Didn't find last video informative; nor first as I am familiar with Darwin's history; Gordon Gallup was not targeting our audience; his claims were sensational and not supported well; Glenn did excellent job of damage control after video.
			Breaks were needed more.
			While I enjoyed videos, perhaps not during lunch for some down time.
			Lunch/video sessions were too much; a true break is needed.
			Liked having videos during lunch; think some would have preferred break from content, but I liked the continuation of material via video.

(continued)

TABLE 19.1. Continued

ITEM	M	SD	COMMENTS
			Interesting, but hard to sit through after sitting all morning; could have watched on our own and possibly spent lunch discussing.
7. The lunch/video sessions were well-run.	4.25	1.16	Videos were well timed and Nicole did a fine job managing discussion.
			No issues; next time maybe better lunches? Salad? Coffee in morning?
			Smaller video segments would be more powerful to target key issues.
			Well run but a true break needed.
			Not sure really much to run; but anything that occurred ran smoothly.
			??? We just watched the video.
8. Overall, the Institute was well-organized.	4.88	0.35	Good job on organizing!
			This has great potential; could be a star attraction for teachers!
9. Logistical issues were addressed effectively by the administration of the Institute.	4.88	0.35	Everything was smooth.
10. The days for the Institute were an appropriate length.	4.42	0.66	Days were long, but I expected this; could benefit from breaking the day up better; maybe walking discussions, more labs; have to go to different rooms; glad to be able to obtain hours in one week though.
			Targeting fewer specific elements for each perspective in psychology, anthropology, and health would be more beneficial & applicable to younger audiences (middle and high school).

TABLE 19.1. Continued

ITEM	M	SD	COMMENTS
			Would be okay with a break at lunch.
			Tough to stay with it for 8 hours.
			45 hrs. great opportunity to get 3 credits; true lunch break would not subtract from hours counted.
			A bit long; but prefer long days over four days as opposed to short periods over many days.
			They were long; would have liked more time for short breaks; more lab work; more interactive aspects rather than just lecture.
11. The Institute was reasonably priced.	4.50	0.76	Would appreciate a slightly lower price.
			Solid bargain for teachers.
			NYSMT program covered costs, but if it hadn't, price reasonable.
			Well run but a true break needed.
			Didn't pay the fee, but feel reasonable for teachers looking for opportunities for additional course work.
12. I would recommend the Institute to friends in the future.	4.88	0.35	Already have!
			Great examples helped highlight concepts along with discussions clarifying misconceptions around evolution.
			Absolutely.
			This should have been sold out!
			Yes, but only certain friends due to "academic intensity."
13. I believe that my understanding of basic principles of evolution has improved as a result of my being in the Institute.	4.25	1.16	The last course I took on Evo. was 15 yrs. ago, so definitely has enhanced my understanding [of] misconceptions around evolution.

(*continued*)

TABLE 19.1. Continued

ITEM	M	SD	COMMENTS
			I had a fairly strong understanding of basic principles to begin with; giving a pre-seminar poll might be helpful to assess our baseline understanding. This could inform speakers' presentations.
			Great examples helped highlight concepts along with discussions clarifying
			My rating of 3 rather than 5 has to do with personal educational background.
14. I believe that my appreciation of the breadth of topics connected to evolution has increased as a result of the Institute.	4.50	0.93	Sadly to say much self-learning has been my pathway, and participating with colleagues has been beneficial to strengthen my confidence in education of this topic.
			The Institute needs to be advertised/marketed to all teachers, not just science faculty.
			Loved seeing evolution in a biological sense carried over to other disciplines in a meaningful, comprehensive manner.
Additional comments			Really enjoyed!!! The depth & wealth of information was truly outstanding.
			The sessions run as discussions were more enjoyable and I think more valuable than the sessions that were more instructor dominated.
			Great stuff; Need more hands-on lab experiences, 9 1/2 hrs. of talk is a lot!; super potential here.
			This was great! I seriously left every day with a smile and positive, exciting ideas floating round in my head.
			I think the Institute was very informative but am interested in how we can use this to educate our student—more so than just being more informed. How can we use this resource to enhance our instruction & how can we help each other to accurately teach evolution?

These data reflect relatively high ratings and positive opinions about the Institute overall. Participants had positive perceptions of the breadth and depth of the material, the caliber of the faculty, and the organization of the overall experience. Representative comments (which are represented in full in Table 19.1) include the following:

"Loved seeing evolution in a biological sense carried over to other disciplines in a meaningful, comprehensive manner."

"Really enjoyed!!! The depth & wealth of information was truly outstanding."

"This was great! I seriously left every day with a smile and positive, exciting ideas floating round in my head."

"Provided more than adequate information allowing me to reconstruct my evolution unit."

"Broad range of different educational disciplines; deepened my understanding of evolution."

"Really opened my mind to numerous approaches to evolution."

"Very diverse; interesting to have academics in all areas addressing central unifying topic."

"Didn't feel anything dumbed down for benefit of secondary ed teachers . . . this was appreciated!!"

A Secondary Teacher's Perspective on the Institute

Nothing in biology makes sense except in the light of evolution.
—Theodosius Dobzhansky, 1973

As a secondary science teacher with a strong passion for evolution and an alumna of the EvoS Sumer Institute, I (the second author of this chapter, AT) am uniquely positioned to comment on the benefits of the Institute from the inside. This section offers my perspective in this particular context.

With the advent of the introduction of the Next Generation Science Standards, the fundamental theoretical construct of biological evolution has been incorporated as a learning progression, being introduced as early as kindergarten. Identified as a disciplinary core idea, Biological Evolution: Unity and Diversity, has met at least two of the following criteria and, ideally, all four:

- Have broad importance across multiple sciences or engineering disciplines or be a key organizing concept of a single discipline;
- Provide a key tool for understanding or investigating more complex ideas and solving problems;
- Relate to the interests and life experiences of students or be connected to societal or personal concerns that require scientific or technological knowledge;
- Be teachable and learnable over multiple grades at increasing levels of depth and sophistication (NGSS Lead States, 2013).

Often as educators, we find ourselves under the time and content constraints of state standards. Until the advent of the 2013 Next Generation Science Standards, little explicit attention has been given to particular standards of teaching evolution in secondary schools. This has led to the undermining of evolution instruction in terms of student understanding (Berkman & Plutzer, 2012). While natural section is clearly a basic aspect of evolution, it is also one of several evolutionary processes—thus, teaching only about natural selection provides a limited understanding of evolution.

Further, there has been a push, largely by politicians in some states, to allow the presentation of "alternatives" to evolution, as it was a concept that was debatable, rather the driving force behind our biological sciences. "A common technique—used to a greater or lesser extent by Colorado, Missouri, Montana, and West Virginia—is to direct students to study its "strengths and weaknesses" ("Undermining Evolution," 2012, p. 18). The Next Generation Science Standards serves to move us away from what has been termed the "cautious middle," advocating instead for evolution to be taught as it should be, as a unifying theme in Biology at the secondary school level (NGSS Lead States, 2013). So why evolution? Evolution is the driving dynamic force that unifies all of the life sciences and secondly, it provides educators with content that spurs scientific questions and hypothesis from their students.

With that being said, challenges such as antievolutionary ideals and lack of content knowledge arise. How do we prepare for the anticipated changes in science education?

Whether it is ambivalence or conflict avoidance, many science educators are not currently prepared to deliver evolution instruction to the degree necessary for student understanding. "Research suggests that many teachers do not feel that they have the expertise they need to confidently teach evolutionary biology in a rigorous and unapologetic manner" (Berkman & Plutzer, 2012, p. 29). With approximately one fourth of the core ideas of the framework focusing on evolution, we feel that educators should gain a deeper understanding of evolution to impart these essential building blocks to their students.

The EvoS Summer Institute at SUNY New Paltz offers educators an interdisciplinary approach to evolution, with SUNY faculty presenting evolution from numerous perspectives based on their area of expertise: molecular, psychological, health, and historical. As an educator, I found the experience invaluable. I was exposed to many perspectives on the role of evolution in not only our history but also in our present-day lives. It is that connection that I believe teachers need to make to deepen their understanding and, in turn, allow for their students to make their own connections as well. According to Martin Storksdiek, director of the Board on Science Education at the National Research Council, "independent of content knowledge, the teacher has to have the ability to make connections and excite students" (cited in Sawyers, 2011). The EvoS Institute allows for educators to identify their own misconceptions regarding evolution, address those issues through collaborative, interactive discussion with colleagues, and, perhaps most valuable, the development of lesson ideas to bring back to the classroom. The EvoS Institute is a very valuable forum for educators, addressing misconceptions, establishing connections, enhancing pedagogy, and, potentially, building a stronger, more competent educator.

Limitations and Future Directions

While the EvoS Institute at New Paltz provided a positive educational experience for science teachers, this study is clearly not without limitations. The fact that data were collected only from eight individuals, speaks to concerns around the generalizability of the findings. Further, several of these individuals paid for the experience, suggesting that they may have had a vested interest in providing positive feedback. Further, this particular study did not include measures of student learning outcomes. Future research into the positive outcomes associated with an intensive evolution-based education like the EvoS Institute would be wise to address these points. In a study of a 14-week-long intervention to educate science teachers about the nuances of evolutionary theory, Nehm and Schonfeld (2007) showed significant gains in evolution-based knowledge of the teachers. Future research using our one-week paradigm that examines learning outcomes could be conducted to see if our model leads to similar outcomes.

The EvoS Institute as a Model for Advancing Evolution

Evolution education in the United States is sorely in need of repair. One of the greatest problems, which is rarely discussed, is the fact that it is actually extremely difficult for people to become educated in evolution and its broad applications in most of our modern colleges and universities. The interdisciplinary EvoS model of evolution education has proven to be a powerful and effective method of evolution education for students from a variety of backgrounds (Chang et al., 2011).

The EvoS Summer Institute was designed to help bring the EvoS educational model that has been so successful for college students at places like Binghamton and New Paltz to middle- and high-school teachers. A goal here is to bring a powerful and broad evolution education, via highly skilled and motivated teachers, to students before they even walk onto a college campus.

Based on the feedback received from the participants of the 2015 EvoS Summer Institute at New Paltz, we believe that the condensed, multidisciplinary, intensive evolution education offered to teacher-participants was highly effective in terms of both effecting learning outcomes as well as in terms of developing in teachers a better appreciation for the broad-based applications of evolutionary principles.

In future offerings of the Institute, we hope to add a curriculum-development component to the sessions, to allow participants the opportunity to think of ways to integrate what they have learned into potential curricular materials.

Along with the expansion of the EvoS model of evolution education at the undergraduate level that is now offered at several colleges and universities, we believe that the week-long Summer Institute version of this curriculum designed for teachers provides yet another method for helping us as educators bring Darwin's extraordinary ideas to a broader audience.

References

Berkman, M., & Plutzer, E. (2012). An evolving controversy. *American Educator, 26*(2), 12–17, 20–23.

Chang, R., Geher, G., Waldo, J., & Wilson, D. S. (Eds.). (2011). EvoS Consortium [Special issue]. *Evolution: Education & Outreach, 4*(1).

Geher, G. (2014). *Evolutionary Psychology 101*. New York: Springer.

Glass, D. J., Wilson, D.S., & Geher, G. (2012). Evolutionary training in relation to human affairs is sorely lacking in higher education. *EvoS Journal, 4*(2), 16–22.

Hewson, P. W. (1992, June). *Conceptual change in science teaching and teacher education*. Paper presented at a meeting on "Research and Curriculum Development in Science Teaching," for the National Center for Educational Research, Documentation, and Assessment, Ministry for Education and Science, Madrid, Spain.

Johnson, R. L. (1985). *The acceptance of evolutionary theory by biology majors in colleges of the west north central states*. Unpublished doctoral dissertation. University of Northern Colorado, Greeley.

Nehm, R. H., & Reilly, L. (2007). Biology majors' knowledge and misconceptions of natural selection. *BioScience, 57*, 263–272.

Nehm, R. H., & Schonfeld, I. S. (2007). Does increasing biology teacher knowledge of evolution and the nature of science lead to greater preference for the teaching of evolution in schools? *Journal of Science Teacher Education, 18*, 699–723.

NGSS Lead States. (2013). *Next generation science standards: For states, by states*. Washington, DC: National Academies Press.

O'Brien, D. T., Wilson, D. S., & Hawley, P. H. (2009). "Evolution for Everyone": A course that expands evolutionary theory beyond the biological sciences. *Evolution: Education and Outreach, 2*, 445–457.

OECD. (2013). *PISA 2012 assessment and analytical framework: Mathematics, reading, science, problem solving and financial literacy*. Paris, France: Author.

Sawyers, Susan. (2011, January 25). What makes a good science teacher? *The Hechinger Report*. Retrieved from http://hechingerreport.org/what-makes-a-good-science-teacher/

Undermining evolution: Where state standards go wrong. *American Educator, 26*(2), 18–19.

Wilson, D. S. (2007). *Evolution for everyone: How Darwin's theory can change the way we think about our lives*. New York, NY: Delacorte.

Wilson, D. S., Geher, G., & Waldo, J. (2009). EvoS: Completing the evolutionary synthesis in higher education. *EvoS Journal, 1*(1), 3–10.

CHAPTER 20 | Teaching Evolution across the Curriculum

Beyond Campus-Wide Programs

DAVID SLOAN WILSON, GLENN GEHER,
ANDREW C. GALLUP, AND HADASSAH MATIVETSKY

THE CHAPTERS IN THIS book are based largely on a movement to teach evolution across the curriculum that began in 2003 when one of us (DSW) decided to initiate such a program at Binghamton University. The first sister program was started by another of us (GG) at SUNY New Paltz in 2007, followed by a National Science Foundation grant during 2008–2010 to develop our programs and create a multi-institution consortium. Groups at over 40 campuses expressed an interest in becoming involved. Of these, only a few ended up developing full-fledged campus-wide programs.

The EvoS Consortium movement can therefore be considered a glass that is part full but also part empty. As this volume shows, the need to teach evolution across the curriculum is greater than ever, and campus-wide programs work well when they can be established, but their establishment has become a limiting factor. In this final chapter, we will attempt to diagnose the limiting factors and suggest ways to teach evolution across the curriculum that include but go beyond campus-wide programs.

So You Want to Create an EvoS Program . . .

Imagine that you are a professor who truly understands the need to teach evolution across the curriculum. You know what's going on at the level of research and scholarship worldwide. You know how easy it is to excite the students in your own courses about the evolutionary perspective. You know that a number of like-minded faculty are sprinkled across departments at your university. You know that with a little organizing you can create a whole that greater than the sum of the parts. And you know that this has been done elsewhere, offering wonderful benefits to students and faculty alike. How likely are you to attempt starting a program? If you do, how likely are you to succeed?

It should be clear (especially if you really are a professor!) that severe obstacles would lie in your path. To begin, your effort would add to your current workload. It would be classified as service, which counts much less toward promotion and tenure than research and teaching. You would need to convince your administrators of the merits of the program and squeeze at least a little money out of them. These obstacles must be faced for the establishment of any program.

There are more obstacles when the program is transdisciplinary. No matter how hard a university tries to become transdisciplinary, a deep disciplinary structure gets in the way, especially when a program spans entire colleges (e.g., Liberal Arts, Education, Management, Engineering, Medicine, Law, etc.) in addition to departments within a college. Staff support doesn't come automatically the way it does for a department. Departments might not want to grant seats in their courses. Faculty might not be appropriately credited for their participation.

Finally, there are obstacles that come with the E-word per se. Administrators might not understand the relevance of teaching evolution across the curriculum, or they might worry about endorsing such a "controversial" topic, especially in geographical regions where religious creationism is common (Lynn, Glaze, Evans, & Reed, 2017). Faculty in social science and humanities departments might confuse a modern evolutionary perspective with social Darwinism, which is the use of evolution to justify inequality (see Wilson & Johnson, 2016, for detailed discussion).

Given all of these obstacles, it shouldn't be surprising that only a few of the many groups that expressed an interest in starting an EvoS program actually succeeded (Spaulding, Burch, & Lynn, 2014). Without wishing to discourage the formation and continuation of EvoS programs in any way, more is needed to teach evolution across the curriculum on a worldwide basis. How can that be accomplished?

Introducing the TVOL1000

Another possibility is to create a worldwide group that supports itself with a large number of small donations, similar social movements such as MoveOn.org and the Tea Party, which is also how innumerable church congregations support themselves. One thousand people donating an average of $10 per month provides an annual budget of $120,000 dollars, far more than any EvoS program can hope to achieve on a regular basis. Ten thousand people donating at the same rate would provide an annual budget of $1.2 million, enough to have a very substantial cultural impact.

One of us (DSW) has initiated such a group called the TVOL1000 (https://evolution-institute.org/article/tvol1000). TVOL stands for "This View of Life," from the final passage of Darwin's *Origin of Species* ("There is grandeur in this view of life. . . ."). One thousand is the initial recruitment goal. As soon as it is reached, the name will be changed to the TVOL10000, etc., on a logarithmic scale to emphasize that recruitment can be a geometric process, like bacteria multiplying in a petri dish.

TVOL1000 is administrated by the Evolution Institute (EI; https://evolution-institute.org), an independent nonprofit organization that DSW co-founded in 2007. Working through the EI eliminates the need to educate university administrations about the merits of an evolutionary perspective. The EI also publishes two online communication outlets: *This View of Life* (TVOL; https://evolution-institute.org/this-view-of-life/), which features articles accessible to an informed general public, and the *Social Evolution Forum* (SEF; https://evolution-institute.org/social-evolution-forum/), which features target articles and commentaries for a more professional audience. These are envisioned as part of a communication strategy called the Science to Narrative Chain, which notes that science is necessary but not sufficient to solve the problems of modern human existence. There must also be narratives capable of reaching mass audiences. Furthermore, the narratives must be connected to science with a chain of material providing intermediate detail, so that anyone can learn more no matter where they start on the chain. TVOL and the SEF are envisioned as links close to the science end of the chain, reaching far larger audiences than the academic literature and stimulating the creation of content that reaches still larger audiences.

Currently, TVOL and SEF receive approximately 30,000 pages views per month, a number that can be greatly increased with the investment of more time, effort, and money. With EI support, *Evonomics.com*, an online magazine started by Robert Kadar, the founding editor of TVOL, has achieved an astonishing circulation of 300,000 page views per month. The potential of reaching many more people than all EvoS programs combined should be obvious.

TVOL and SEF are run by a small paid staff and a number of topic editors who participate on a volunteer basis, as they would as editors of an academic journal. The authors of articles also receive no payment. Their incentive is the same as for publishing an article in an academic journal, enhanced by the knowledge that many, many more people will be reached. Finally, TVOL and SEF attempt to maintain the same standards of respectful and constructive discourse for science journalism as for academic science. Like a contact sport such as boxing, it is fine to hit hard as long as it is "above the belt," such as respecting the authority of empirical evidence, avoiding ad hominem attacks, and so on.

To summarize, the goals of TVOL and SEF are the same as the goals of an EvoS program—to teach "anything and everything from an evolutionary perspective," free of charge to the reader and without any of the baggage associated with an academic program. Of course, this means that the numerous *benefits* of an academic program are also lacking, such as a rigorous course of study, a group rather than a solitary experience, and credit for participation.

The TVOL1000 has the potential to provide benefits to members that go beyond the satisfaction of supporting a free service. TVOL members are also encouraged to get personally involved. Exactly what this means depends upon the qualifications and interests of the individual. It could include writing content, helping on the operations end, or initiating projects that promote "this view of life" in ways that go beyond TVOL and SEF (the EI has many such projects and is in a position facilitate others). One especially appealing prospect is the formation of local chapters—people who live close enough to actually get together to read, discuss, and act on the basis of an evolutionary world view.

Against this background, let's think about how the TVOL1000 could help our hypothetical professor contemplating whether to start an EvoS program. One possibility would be to get involved in the worldwide TVOL1000 community without trying to start anything at his or her university. Another possibility is to additionally create a local chapter without requiring the authorization of his or her administration.

The main "working parts" of an EvoS program are (a) a menu of courses and research opportunities made available to students, (b) a seminar series that features speakers from a diversity or biological and human-related disciplines from a common theoretical perspective, and (c) opportunities for training and collaborative research at the graduate student and faculty levels, including faculty who did not receive evolutionary training during their own higher education. The TVOL1000 offers rich opportunities for all three of these working parts. Consider a menu of course offerings that includes online courses and traveling to other campuses in addition to courses taught at the student's own university. Or consider a TVOL1000 seminar series where each speaker is hosted on a different campus and live-streamed to other campuses. The opportunities for faculty and graduate students can be much greater with a pool of collaborators that is not restricted to one's own location.

That said, there is no substitute for face-to-face interactions at a given location. EvoS programs are vibrant because faculty participants have reasons to get together socially and intellectually, undergraduate students become "hooked," taking multiple courses, becoming involved in research, and forming social relationships with each other, and so on. These benefits can be realized by forming a local chapter, without needing the permission of one's administration. A template could be developed and diplomas could be issued by the EI based on guidelines created by a governing board that could be just as good for career purposes as acknowledgment that one has completed a minor or certificate program on one's college diploma. Depending upon how large the self-supporting group becomes, it could even start providing financial support for speakers, pilot grants, and so on.

All of the directors of current EvoS programs have joined the TVOL1000, not as a way of replacing their programs, but as a way of strengthening and going beyond them. We encourage you, the readers of these words, to join us so that "this view of life" can become the new common sense much faster than it will otherwise.

References

Lynn, C. D., Glaze, A., Evans, W., & Reed, L. (2017; Eds.). *Evolution Education in the American South: Culture, Politics, and Resources in and around Alabama.* New York: Palgrave Macmillan.

Spaulding, K. N., Burch, R. L., & Lynn, C. D. (2014). Evolutionary studies' reproductive success and failures: Knowing the institutional ecology. *EvoS Journal: The Journal of the Evolutionary Studies Consortium,* 6(1), 18–38.

Wilson, D. S., & Johnson, E. M. (2016). Truth and reconciliation for Social Darwinism. *This View of Life.*

INDEX

abuse, familial, 47–48, 279–80
academia/academy, 21, *See also* higher education
   caste-like tenure system of, 145
   challenge of social coercion theory and, 149
   and change, 10
   and cooperation, 55–56
   and evolutionary psychology, 158
   interdisciplinary approaches in, 14
   oligarchs' attempts to control, 145, 145n17
academic employment, and beliefs about human behavioral sex differences, 162
academic freedom laws, 415
Aché foragers
   male mating strategies in, 281–82
   paternal investment and offspring survival in, 279
adaptation(s), 61–62 *See also* environment of evolutionary adaptiveness (EEA)
   vs. byproducts, 182, 183
   definition of, 67
   evolutionary psychology and, 163–64, 171*t*
   as proximate mechanisms, 67–69
   religion and, 180, 181–82, 182*t*

Adaptive Coloration in Animals (Cott), 100–1
adaptive fictions, meaning systems and, 187–88
adaptive flexibility, 257
   cultivation of, across diverse sociocultural contexts, 260
   and education for sustainable development, 250, 259, 264, 265*f*
   as foundation for sustainable development, 260
   and planetary sustainability, 259
   ultimate causation and, 256–57
adaptive hypotheses
   about humans, testing of, 85, 91–92
   alternative hypotheses in, 68*t*, 91–92
   controversy about, 85
   multiple hypotheses in, 91–92
   null hypothesis in, 92
   "beautiful," slain by ugly fact, 69, 70, 71
   multiple testable, formulation of, 67
   testing, 67–69, 68*t*
      direct, 68*t*
      indirect, 68*t*
Africa, climate instability in, and human evolution, 131n7
aggression
   familial, 47–48, 56–57
   interpersonal, time perspective and, 289
   SUNY EvoS seminar series on (2016), 47

aging, 194–95
agricultural revolution, 193–94
air quality
   contemporary concern about, 206
   physical culture movement's concern about, 206
Albright College
   EvoS program, 14, 29–31
      campus-wide events and, 34, 39
      and community colleges, 36, 39
      faculty participation in, 37–38
      local conferences, 36–37
      seminar series, 37
      student involvement in, 35, 38
      student recruitment for, 38
      website, 37, 39
   Experience Program, 35
   funding resources, 35
   implementation of EvoS curriculum, 31
alcohol, 200–1
   use, time perspective and, 289
alleles, and conflicts (or confluences) of interest, 120–21
ALLELE (Alabama Lectures in Life's Evolution) speaker series, 9, 14–15, 24–25
altruism, 126
   and global sustainability, 258–59
   social coercion theory and, 126
Alzheimer's disease, 286
American Association for the Advancement of Science (AAAS), Common Theme of Science, 369–70n2
American Humanist Association, manifesto of, 186
Amsterdam, Elana, 195–96
analog circuit design, genetic algorithms in, 228–33
Anaspidea. *See also Aplysia* spp.
   antipredator defenses of, 76*t*, 82–83
   chemical defense in, 71
   phylogeny of, 83–84, 84*f*
   purple ink release, 71–75, 74*f*, 84*f*
   red seaweed diet, and ink production, 74*f*, 76*t*

ancestral health movement, 199, *See also* Paleo diet; physical culture movement
   criticisms of, 197–98, 210
   disagreements in, 197
   global spread of, 210–11
   groups in, 196–97
   growing popularity of, 195–96
   historical perspective on, 213–14
   origins of, 198, 198n4
   and physical culture movement
      commonalities of, 199–200, 212
      differences between, 207
   scientific publications on, 209–10
   scientific support for, 209–10
   size of, 196–97
   stereotypes of, 197–98
Ancestral Health Society, 195–96, 210–11
Andrews, Alice, 19
animals. *See also* primates
   coloration, 97, 103
   patterns, analysis methods, 109
   research on, conceptual and methodological transitions in, 103
   weaponry used by, 47
anovulation, lactational, 345
antibiotic resistance, 194–95, 303–4
antidepressants, and breastfeeding mothers, 344
anxiety. *See also* epistemological and existential anxieties
   mismatch model and, 304
anxiety disorders, teaching and understanding, Tinbergean approach applied to, 309
Aplysia spp.
   antipredator defenses of, 75–83, 76*t*, 83*f*
   purple ink release, 71–75, 74*f*, 84*f*
   antipredator mechanisms of, 75–83, 76*t*
   hypotheses for, 75–82, 76*t*
   white ink secretion by, 75–82n4, 84*f*
aposematism, 76*t*, 97
Arizona State University, Center for Evolution and Medicine, 195–96
Armstrong, Neil, 332

artificial insemination, and risk of preeclampsia, 338
artificial intelligence. *See* genetic algorithm(s) (GAs)
Athletics for Physical Culture (Knauff), 200
Atlas, Charles, 199
Atran, Scott, 181
attachment, 15, 168
attentional bias, primate sexual signaling and, 105, 106–7
Australopithecus sediba, 131n6
authoritarian states
  coercive dismantling of, 147–49
  democratization of, 147–48
  economic productivity in, 148
  economic sanctions on, 147–48
  monopoly control of weaponry, 147
authority, in science vs. religion, 436, 437
autism, risk for, birth spacing and, 345
autoimmune disease, 194–95
  hygiene hypothesis for, 304
  mismatch model and, 304

Baker, Robin, *The Hitchhiker's Child,* 346
bald uakari, coloration, 100*f*
Beecher, Catharine, 199
behavior. *See also* evolutionary psychology; health-related behaviors and outcomes
  evolved, situationist perspective and, 166
  genetic variation and environmental conditions and, 289
  human (*see also* uniquely human panoply)
    adaptive value of, 90
    as biological vs. cultural, 90
    environmental pathogens and, 274–75
    objective study of, difficulty of, 119, 127–28
    ultimate-proximate distinction and, 126–27, 157
Behavioral and Brain Sciences (journal), evolution-themed articles, 4
  authors' evolutionary training and, 22–23
behavioral sciences, evolutionary approach to, and interdisciplinary studies, 15–16
beliefs, 434
  counterfactual, 184, 185
  and evolution controversy, 411–12
  intersection with knowledge and science education, 426–27
Bellah, Robert, 181
Betancourt, Kian, 51, 56
between-group selection, 181–82
Beyond Religion: Ethics for a Whole World (Dalai Lama), 188–89
Big Gods (Norenzayan), 181
Binghamton University
  Evolution for Everyone course, 372, 374*t*, 392, 397, 447–48
  EvoS courses, 371–72
    effects on students, 363
  EvoS program, 372, 447–48, 463
  EvoS revolution and, 3, 4, 13, 14, 21–22, 29–30, 37–38
  online library maintained by, 37
biological anthropology, modern evolutionary theory and, 64
biological evolution. *See also* evolution
  as "fact" with "proof," 63
  teaching and learning about, conceptual challenges of, 34
biological versus cultural, as false dichotomy, 90
biology
  proximate vs. ultimate explanations in, 67, 67*t*, 257
  and sociology, 65
biomedical research, eco-evolutionary principles and, 303, 316
biomedicine, modern evolutionary theory and, 64
birds, artificial lighting and, 337
birth outcomes
  adverse, long-term effects of, 281
  human life history and, 278
  male scarcity and, 280–81
birthweight, 277–78
  male scarcity and, 280–81
  paternal investment and, 339

Biston betularia, color morphs, adaptive hypotheses about, testing of, 68*t*
black-and-white snub-nosed monkeys, male, coloration, 105–6
body armor, and coercive power, 140–41
body hair, human, as deterrent to oral sex, 342
bottle-feeding
  consequences of, 342
  and effects on subsequent offspring, 345
Boyer, Pascal, 181
Boy Scouts, 199
brain, human
  communication-related areas, 132–33n9
  evolution of, 126–27, 129–31, 131n6, 133, 133n10
  lymphatic system of, 274–75
breastfeeding
  absence/cessation, health consequences of, 343–44
  in United States, 343
breast milk, pumping of, 344–45
breathing, physical culture movement's concern about, 206
Bruce effect, in rodents, 279–80
Buller, David, critique of evolutionary psychology, 166–67, 168
Burch, Rebecca, 9–10, 47–48, 56–57
Butler Act (1925), 413
byproduct mutualism, 120–21, 129, 132

Cacajao calvus. See bald uakari
Calvert, Alan, 206–7
Campbell, Joseph, 212
cancer, 194–95
  metastasis
    evolutionary framework for, 304
    evolution of, 312
    mechanism for, 312
    model for, 312
    therapeutic approaches for, 313
    Tinbergean approach to, 311
    vulnerability to, 312–13
    mortality rate for, 311–12
  treatment of, 311–12
  vulnerability to, evolution of, 304
Carmen, Rachel, 18
Carque, O., 207
celiac disease, 208
Cercocebus torquatus. See collared mangabey
Cercopithecus diana. See Diana monkey
Cercopithecus neglectus. See De Brazza's monkey
Cercopithecus pogonias. See crowned guenon
Chamberlin, T. C., on affection for the intellectual child, 66, 66n3
Champagne, Frances, 21–22
change, 409
  academia and, 10
character(s), phenotypic, 86–87
Checkley, Edwin, 206
chemotherapeutic resistance, 311–12
  evolutionary framework for, 304
childbirth, 277
childhood, human, social coercion theory of, 133
chimp-human divergence, 126–27n5
China, modern
  crony/criminal capitalist economy in, 144, 148
  state intimidation in, 147
Chomsky, Noam, 238
chord progression, 238–39
chords, musical, 236
  major, 236
  inversion, 237–38
Christian beliefs, and evolution controversy, 412, 417, 437
chromosome(s)
  genetic algorithms and, 220, 221
  human, 220
circle of fifths, in music, 238
circulation, physical culturalists' therapies for, 208
circumcision
  adaptive and unintended consequences of, 345
  and impregnation, 347

and loss of sensitivity of glans, 347
and protection from infection,
  330, 346–47
and semen displacement, 346–47
and sexual satisfaction of female
  partners, 347
classism, oligarchic interests
  and, 145–46
C major chord, 236
  inversion, 237–38
C major scale, 236
coercive threat from a distance, 122*b*
  conjoint projection, by humans
    evolution of, 117, 118, 121–22, 124–
      25, 124–25n4
    social implications of, 125–27
  increased scale of, and expansion of
    social scale, 137
  and nonkin cooperation, 126–27
cognition. *See also* higher-order cognitive
    skills (HOCS)
  sex differences in, 160
  smartphone use and, 350
  ultimate causation of, understanding of,
    and sustainable development, 259
cognitive change approaches, in
    education, 428
Cognitive Psychology (journal), 15
cognitive systems, human
  evolution of, 136
  social doubt and, 135, 135n11
cold exposure, physical culture movement
    and, 204, 204n8
collared mangabey, coloration, 100*f*
collective learning
  and adaptive flexibility, 257, 258
  as core human dynamic, in education
    for sustainable development,
    261, 262*t*
  and sustainable development, 249,
    250, 258
  ultimate causation of, understanding of,
    and sustainable development, 259
Colobus guereza. See mantled guereza
colonialism, Western, 201n6
The Colours of Animals (Poulton), 100–1

common squirrel monkey, coloration, 100*f*
communication
  electronic, shortcomings of, 351
  human, 131, 132–33n9
  nonhuman, 132
  oligarchs' attempts to control, 145
community colleges, and EvoS programs
  at liberal arts colleges, 36
competition, and conflicts (or
    confluences) of interest, 120
competitive parsimony
  of social coercion theory, 138
  of theories, 128, 149, 150
  social doubt and, 134
*The Complete Idiot's Guide to Eating
  Paleo* (Quinn & Glaspey), 195–96
complexity, 437
  and evolution, 438
  irreducible, 437, 438
Comte, August, 186
Conaway, Nolan, 19
conception, human
  factors affecting, 322–24
  risk of, rape and, 333
conditional strategy, 164
conflicts of interest. *See also* universal
    conflict of interest problem
  cost-effective coercive management of
    (*see* social coercion theory)
  nonkin animals and, 120–21
congestive heart failure (CHF)
  causes of, 307, 308
  evolutionary medicine for, change
    model for, 315*t*
  phylogenetic analysis of, 308–9
  teaching and understanding, Tinbergen
    approach applied to, 307
  treatment of, 307, 309
  vulnerability to, 308–9
    adaptive benefit of, 309
consciousness
  in eugenics vs. evolutionary
    psychology, 171*t*
  first-tier vs. second-tier, 189
contingent information, 132, 132n8
convergence, genetic algorithm and, 221

Index | 473

cooperation, 55–56, *See also* social
    cooperation
  and adaptive flexibility, 257, 258
  and conflicts (or confluences) of
      interest, 120
  education as form of, 263
  for global sustainability, 258–59
  multilevel
    as core human dynamic, in education
        for sustainable development,
        261, 262*t*
    ultimate causation of,
        understanding of, and sustainable
        development, 258
  non-kin, 126–27
  in animals, 120–21
  and sustainable development, 249, 250,
      255–56, 258
Cordain, Loren, 195–96n3, 196–97, 204
core human dynamics, and collaborative
    curriculum design in education
    for sustainable development,
    260, 262*t*
costumes (theatre), evolutionarily
    informed, 20
creationism, 21, 412–13, 414*t*
  teaching of, legal issues in, 413–15
  and Texas science standards, 439
creation science, 65, 414*t*, *See also*
    creationism
credit, easy/inflated, and adult pair
    bonding, 348
crested macaques, male, coloration, 105–6
critical thinking, 365
  assessment of, 376, 378*t*, 382*t*,
      387, 393
    with independent raters, 378*t*, 382,
        387, 387*t*, 389*f*, 390*f*, 393
    with lexical content analysis, 377–82,
        382*t*, 387, 392
  components/skills of, 365–66
  definition of, 365
  disposition toward, 366
    development of, 367
  and domain transfer of knowledge, 366
  evolution education's effects on,
      investigation of

    analysis of knowledge transfer in,
        385, 388*t*, 390, 391*f*, 394, 396
    course used for, 372, 374*t*
    critical thinking assessment in, 376,
        378*t*, 382*t*, 387
    Evolutionary Attitudes and Literacy
        Survey used in, 375, 382*t*,
        385, 392
    evolution literacy survey in, 398
    limitations of, 395
    measures used in, 375
    methods, 372
    participants, 374*t*, 375
    results, 385
  fostering
    education and, 364, 365
    through evolution education, 367
    improvement, instructional strategies
        and, 367
    science education and, 368
  critical thinking rubric, 382, 384*f*
crony/criminal capitalist economies, in
    modern states, 144, 148
CrossFit boxes, 204
CrossFit Games, 204–5, 204–5n9
crossover
  genetic algorithms and, 221, 223,
      225, 226*f*
  in nature, 225–26
crowned guenon, coloration, 100*f*
crypsis, 100–1
  primate coloration and, 108–9
cultural anthropology, modern
    evolutionary theory and, 64
cultural evolution
  and health, 195
  Huxley on, 250–51
culture, human
  social coercion theory of, 131
  "tight" vs. "loose," 185
curriculum (EvoS), 39
  course offerings, 32
  flexibility in, 33
  implementation of, at small liberal arts
      college, 31
  introductory classes, 33–34
  and major degree, 32–33

and minor degree, 31
prerequisites, 33
cycling, 203

Dalai Lama, 188–89
Dam1 complex, properties of, 19
Darwin, Charles, 3, 48–49, 180, 200–1, 369, 371, 410, 412
　as authority, 437
　bicentennial, college event celebrating, 34–35
　on falsifiability of his theory, 62–63
　sexual selection theory, 70–71, 72*t*, 97–98, 99
　study of animal coloration, 97
　study of primate coloration, 98*f*, 109
　and testing of his theory, 63
Darwinian medicine, 274
Darwinism, opposition to, 413
Darwin's Cathedral (Wilson), 181
Davenport, Demorest, 65
Dawkins, Richard, 166–67, 353, 425–26
De Brazza's monkey, coloration, 100*f*
defensible space, 291–92
degeneration, 200–1
democratization
　ascendancy of, 150
　of authoritarian states, 147–48
　continuing struggle for, 146
　and economic productivity, 148
　effect on wealth and power, 144–46
　global, 147–49
　social coercion theory on, 138
demonstration, teaching/learning by, human, 132–33n9
depression, mismatch model and, 304
The Descent of Man (Darwin), 97, 98–99, 200–1, 412
Deskbound: Standing Up to a Sitting World (Starrett), 206
diabetes, type 2, 89
diacylglycerol, 89
Diana monkey, coloration, 100*f*
Dicke, Willem Karel, 208
diet
　human
　　guidelines for, 282–83

　and health, 282
　in modern foraging societies, variability in, 282
　natural, 202
digital age, 193–94, 206n12
Digital Revolution, 199, 210–11, 212
Dinsmore, James, 18
disease, human
　chronic, 273–74
　evolutionary approach to, 193, 194–95
　lifestyle causes and, 273–74
　maternal–fetal environment and, 277–78
　in technologically advanced societies, 273–74
　Tinbergean framework applied to, 305–6, 316–17
disgust, as human disease-avoidance adaptation, 208–9, 293
Dmanisi
　fossil record, social coercion theory and, 130, 131n6
　social cooperation in, 130
Dobzhansky, Theodosius, 193, 369, 409–10, 457
Dr. Oz Show, The, 195–96, 195–96n3
drift, 182, 182*t*, 183
drills, male, coloration, 105–6
Drosophila melanogaster (fruit fly), sustained flight, diet and, 18
drugs
　patent, physical culture movement's opposition to, 205
　use, time perspective and, 289
Durkheim, Emile, on religion, 179–80, 181

Earth Charter, 252, 253–54, 255–56
Eaton, S. Boyd, 198, 198n4
ecology, evolution and, disciplinary divide between, 251
economics, modern evolutionary theory and, 63–64
economic theory, factual vs. practical realism in, 185
economy, modern, social coercion theory on, 139

ecosystems ecology, 251
education. *See also* evolution education; health education; higher education; medical education and curricula
  adaptive problems embedded in human history and, 352
  evolution and, 7
  as form of cooperation, 263
  intelligence and, 352
  modern evolutionary theory and, 64
  reform, science education and, 397–98
  repercussions of evolutionary conclusions for, 90–91
  in United States, local vs. national context and, 412
Education for Sustainable Development (ESD)
  and behavior change, 252
  challenges in, 250, 251
  core human dynamics of, and collaborative curriculum design, 260, 262*t*
  critiques of, 251
  curriculum design, core competencies in, 264–65, 265*f*
  discourse
    autonomy–cooperation dialectic in, 255
    dialectics of, 253
    diversity–universality dialectic in, 254
    reductionism–holism dialectic in, 254
    tradition–innovation dialectic in, 256
  evolutionary studies (EvoS) and, 249, 250, 260, 261–64, 265–66
  historical perspective on, 250
  global, 249, 255, 263, 265*f*
  learning outcomes, 264–65, 265*f*
  pedagogical approach
    transformation-oriented, emancipatory modes in, 255
    transmission-oriented, instrumental modes in, 255
  and transformative learning, 252
  understanding of ultimate causation and, 258
  values and attitudes in, 252
Edwards v. Aguillard, 414*t*
Eldredge, Niles, 41
electronic media, children's access to, negative effects of, 350
Eliot, Charles William, 211
elite throwing
  by humans, evolution of, 117–18, 121–22, 121–22n2, 128, 129–30
  fossil record and, 130–31
  limited maximal scale of, 135
Emlen, Doug, 41, 47
emotion, ultimate causation of, understanding of, and sustainable development, 259
emotional expressivity, human, 157, 168
emperor tamarin, coloration, 100*f*
  as research opportunity, 108
empirical evidence, in science vs. religion, 436
Empringham, James, 201–2
energy, intake and expenditure, and health, 282
Enlightenment, and science, 254–55
environment(s)
  ancient, diversity of, 195
  and food consumption, 289–90
  genetic variation and, interactions of, 289
  health challenges posed by, 276
  and health phenotype, 289
  human-built, and birth outcomes, 279
  modern, diversity of, 195
  naturalistic, health-promoting effects of, 290
  reshaping of, to promote health and longevity, 291
  and risk-taking behaviors, 287–88
  social effects of, 291–92
  unstable, health effects of, 287–88
Environmental Education (EE), and education for sustainable development, 250–51

environment of evolutionary adaptedness
(EEA), 69
   dietary, and human adaptation,
86–87, 90
epidemiology, 7
epigenetics, 21, 289
epistemological and existential
anxieties, 426–27
   evolution-related, 426–27, 429
epistemology, 185, 435
   pluralism in, education for sustainable
development and, 254–55
Epperson v. Arkansas, 414*t*
escapin, 75–82, 76*t*
essentialism, biological, 428
Establishment Clause, and teaching of
evolution, 414*t*
ethical concerns, social coercion theory
on, 140–41
ethology, 305
eugenics, 90–91, 275–76
   evolutionary psychology vs., 169,
171*t*, 174
   goals of, 169–70, 171*t*
   Huxley on, 250–51
   modern-day, 170–73
   sperm donation and, 172–73
evangelicals, Google Image results for,
187*f*, 187
evil
   definition of, 146
   social coercion theory on, 146
evolution
   acceptance of
      in academia, 410–11
      and belief preferences, 437
      factors affecting, 410–11, 426
      nature of science instruction
and, 371–72
      by public, 371, 410–11
   as common theme of science, 369–70,
369–70n2
   definition of, 409–10
   misconceptions about, 411, 428
   penchant for studying, intelligence
and, 353

   resistance to
      and education, 371–72, 458–59
      from the left vs. from the right,
163, 173–74
   theory of (*see* evolutionary theory)
   transdisciplinary applications of,
369–70, 409–10
*Evolution, Medicine & Public Health
Journal*, 209–10
Evolution: The Modern Synthesis
(Huxley), 69–70
evolutionary algorithms, 219–20
Evolutionary Attitudes and Literacy
Survey (EALS), 375, 382*t*,
385, 392
evolutionary biology, controversy in, 85
evolutionary developmental biology, 21
evolutionary health theories, 193, *See
also* ancestral health movement;
physical culture movement
   and diversity of environments, 195
   goals of, 212–13
evolutionary medicine. *See also*
Darwinian medicine
   change model for, 313, 315*t*
   culture of, cultivating, 313
   current state, 313, 314*t*
   eco-evolutionary approaches and,
303–4, 316
   future state vision for, 313, 314*t*
   Tinbergen framework applied to, 304–6
evolutionary mismatch, 182, 183
evolutionary principles, in
interdisciplinary studies, 15
evolutionary programming, 219
evolutionary psychology, 15, 21, 157, 397
   and academia, 158
   advances in (future directions for), 173
   applications of, 49–50
   and bad science controversy, 166, 174
   beneficial applications of, 174
   connections outside psychology, 174
   controversies about, 159
   and controversies about evolved
behavioral sex differences in
humans, 160, 174

evolutionary psychology (*Cont.*)
  criticisms of, responses to, 173
  and digestion of anomalies, 168–69
  eugenics controversy and, 169,
    171*t*, 174
  in EvoS curriculum, 32
  findings in, 6–7
  future of, 10
  and genetic determinism controversy,
    163, 173–74
  goals of, 170, 171*t*
  insights from, 157–58
  interdisciplinary approach and,
    10, 15–17
  and "just-so story," 167–68
  methods used in, 158–59
  naturalistic fallacy and, 166
  negative conceptions of, 53–54, 55, 56
  political pressure and, 10–11
  progressivity of, 168
  religion controversy and, 163, 173–74
  resistance to, 10, 163, 173–74
  as scientific framework, 158
  and situationist perspective, 165
Evolutionary Psychology (journal), 15
evolutionary religious studies (ERS), 180
  as discipline, current status of, 184
  integration with study of nonreligious
    meaning systems, 184
  origins of, 181
  theoretical framework for, 182–83
evolutionary strategies, 219
evolutionary studies (EvoS)
  basic programs in, 14–15
  controversies about, 8
  within disciplines, 5
  as educational initiative, 7
  and education for sustainable
    development, 249, 250, 260,
    261–64, 265–66
    historical perspective on, 250
  elements of, 5
  full-blown programs in, 14–15
  future of, 10
  and grey academic boundaries, 21
  interdisciplinary applications of, 13, 14
  as interdisciplinary program, 15, 22–23,
    30–31, 32–33, 49–50, 52, 55–56,
    447, 448–49
  media and, 11
  as model for education overall, 354
  outcomes with, 449
  popularity of, 11
  as principle-based approach, 15, 16–17
  program implementation, 31, 463
  student interest in, 11
  and student success, 17, 22–23
  transdisciplinary, obstacles to, 464
  TVOL1000 and, 466–67
evolutionary theory
  applications in nonbiological
    fields, 63–64
  controversy about, beliefs and, 411–12
  Darwin's, 48–49, 61–62
  as foundational, 23, 369, 371,
    409–10, 457
  and health practice, 293
  and health-related research, 274, 293
  hostility to, 426
  and instruction in nature of science, 370
  in introductory psychology texts, 33–34
  meaning system informed by, 186
  modern synthesis of, 63–64
  and need for "proof," 61–62
  as paradigm shift, 62–63, 64
  and physical culture movement, 200–1
  power of, 3
  relevance to everyday life, 4
  students' understanding of,
    improving, 365
  and SUNY New Paltz capstone
    course, 43
  support for, 63–64
  teaching of, 364–65
  and thinking outside the box,
    321–22, 353–54
  as true science, 65
evolution education
  across the curriculum, 463
  advancement of, EvoS Summer
    Institute as model for, 460
  challenges to, 458–59

cognitive change approaches, 428
conflict over, 410, 419
content-based approaches, 427
controversies about, 413, 415
current status of, 371, 415
disclaimers and, 414t
EvoS model of, 5
fostering critical thinking skills through, 367
future of, 416, 418
higher education and, 3–5, 13
historical perspective on, 411
human-related, benefits of, 365, 371, 397
legal issues in, 413–15, 414t
in modern academia, 22–23
Next Generation Science Standards and, 457–58
in public schools, interference with, 415
reform of, 417–18
state standards and, 416
Evolution for Everyone (book, Wilson), 3, 373, 447–48
Evolution for Everyone (course), 3, 5, 14–15, 24, 31, 370, 372, 374t, 392, 397, 447–48
Evolution Institute (EI), 465, 466, 467
evolution knowledge, undergraduates', 3, 4
evolution literacy survey, 398
evolution science, and education for sustainable development, 251
Evonomics.com, 465
EvoS. *See* evolutionary studies (EvoS)
EvoS Consortium, 23, 30, 293–94, 372, 448, 463
 affiliation with, benefits of, 36, 37, 38–39
 e-journal, 36, 39
 EvoS membership criteria, 14, 23–24
 national meetings and conferences associated with, 38–39
 seminars archived by, 35
 website, 448
"EvoS Effect," 363–64, 376, 392, 397

EvoS Journal, 448
exaptation, 71, 91
excreta
 ancestral health movement's view of, 208–9
 physical culturalists' concern about, 208–9
exercise
 functional, 204–5
 and health, 282
 natural forms of, 202–3
 physical culture movement and, 202, 203
 unnatural forms of, 203
exploration method, 219–20

facial expressions, human, 157
fact(s), 434, 440
 for atheists, 435
 for biblical literalists, 435, 435n1
factual realism, 184–85, 186
faith. *See also* religion
 and science, 442
 teachers', and teaching of evolution, 425–27, 429, 441–42
falsifiability/falsification
 and evolutionary psychology, 168
 of evolutionary theory, Darwin on, 62–63
 and large-scale paradigms, 168
 of scientific theory, 62–63, 65
 of theory, 118, 127–28, 149, 150
 social doubt and, 134
family resemblance, and familial aggression, 47–48, 56–57
family violence
 contextual factors and, 165
 genetic fitness perspective on, 165–66
 step-parenting and, 168–69
fasting, 202
 intermittent, 202
fear-conditioning, 311
feet, human, evolution of, social coercion theory and, 130–31, 131n6
female mate choice. *See* intersexual selection
feminism, critique of evolutionary psychology, 167

fertility, female human, effect on risky behavior, 334
fertilization by proxy, 328–30, 346
filicide, step-parents and, 6, 157–58, 165
fit individual, in human population, 221
fitness function, 221
   genetic algorithm and, 223
Fletcher, Horace, 208–9
Flint, Michigan, adverse birth outcomes in, 279
food. *See also* genetically modified food (GMOs)
   consumption, resource availability and, 289–90
   genetic engineering of, 331
   natural, 202
   in Paleo diet, 208, 208n13
foreskin. *See also* circumcision
   functions of, 329–30
fossil record, of human origins, social coercion theory and, 128
Fox News, 145
Frazer, J. G., 179
free riders
   coercion of, costs of, 122*b*, 122–23, 123*f*
   and economic systems, 124
   game theory of social coercion and, 124–25, 124–25n4
   human mating strategy and, 133–34
   ostracism of
      by humans, 118, 124–25, 124–25n4
      as individually adaptive, 126
   second-order, 122*b*, 125
Freiler v. Tangipahoa Parish Board of Education, 414*t*
funding, of EvoS program, 35, 37
future orientation
   effects on life history, 287
   and health promotion, 287
   in youth, socio-developmental environment and, 289

Gallup, Gordon, 9, 41, 42, 46, 55
gastropods, shells of, as antipredator defense, 82–83
Geher, Glenn, 8, 19–20, 30, 42, 44–45, 49, 51, 53, 54, 450, 463
gelada
   Bruce effect in, 279–80
   coloration, 100*f*
   in male, 105–6
gene(s)
   and conflicts (or confluences) of interest, 120
   confluent interests and, among close kin conspecifics, 120–21n1
   and environment, in development and selection of behavior, 85–86, 86*f*
   fit, 221
   genetic algorithms and, 221
   parasitic, 181–82, 183
   and personality traits, 289
   survival of, evolutionary perspective on, 286
generation(s), genetic algorithms and, 221, 223, 224*f*
genetic algorithm(s) (GAs), 219–20
   in analog circuit design, 228–33
      chromosome structure for, 228, 230*f*
      circuit generation, 230, 231*f*
      with differing low-pass design specifications, 233*f*, 233, 234*f*
      fitness evaluation, 231
      results, 232
   applications of, 227
   in circuit design, 220, 228
   crossover operator, 221, 223, 225, 226*f*
      arithmetic, 226, 226*f*
      two-point, 225–26, 226*f*
      uniform, 226*f*, 226
   fitness function, 221, 223
   flowchart for, 222*f*
   genetic operators, 223
   in music composition, 220, 234–42
      chromosome structure in Stage 1, 236, 237*f*
      chromosome structure in Stage 2, 241*f*, 241
      evaluation function for Stage 1, 237
      evaluation function for Stage 2, 241
      fitness evaluation, 235

and formal grammar rule, 237, 243–44
interval fitness and, 241
intervals evaluation function and, 240, 241
motif look-up table in Stage 1, 237*f*, 237
motif selection/generation, 242*f*, 242, 243*f*, 244*f*
ratios evaluation function in, 241
mutation operation, 226, 227*f*
mutation rates, 223, 226–27
origins of, 220
parameter setting and, 223
population size and, 223, 224, 225*f*
representation, 222, 223*f*
  binary-coded decimal (BCD), 222, 223*f*
  selection criterion, 223
  selection of next generation, 223
  standard, 220
  terminology for, 221
genetically modified food (GMOs), 331
  Greenpeace and, 331–32
genetic determinism, 21
  definition of, 163
  evolutionary psychology and, 163, 173–74
  naturalistic fallacy and, 166
genetic engineering
  of food, 331
  mate choice and, 331
genetic programming, 219
geology, 48–49
George V (King of England), 199
giraffe
  exaptation in, 71
  intersexual selection in, 71, 72*t*
  interspecific foraging competition hypothesis (IFCH) and, 69–71, 72*t*
  intraspecific foraging competition hypothesis (IntraFCH) and, 72*t*
  male–male encounters during mating season, 71, 74*t*
  male size, and mating success, 71, 74*t*
  neck of, 69–71, 72*t*

global optimization
  path-oriented methods, 219–20
    exploration method, 219–20
    prediction method, 219–20
  volume-oriented methods, 219–20
gluten, 208
God Is Watching You (Johnson), 181
golden snub-nosed monkey, coloration, as research opportunity, 108
The Gospel of Strength (Sandow), 206–7
Graham, Sylvester, 199
grammar(s)
  context-free, 238, 239–40
  formal, and music composition with genetic algorithm, 237, 243–44
Greagor, Stacey, 19
Great Firewall of China, 143
Green, William Scott, 181
Greenpeace, and GMOs, 331–32
green space, 290
group selection, 118, 126, 169–70
gymnasiums, 204
gymnastics, 199

Hackenschmidt, George, 206–7
Hadza people (Tanzania), 201
half step, in music, 236
handguns, and coercive power, 146, 146n19
hands, human, evolution of, social coercion theory and, 130–31, 131n6
hand washing, health messaging about, 293
Hartwig, Dallas and Melissa, 196–97
Harvard
  Hemenway Gymnasium, 211
  and physical culture movement, 211
Haselton, Aaron, 18
health, human, 193, *See also* evolutionary health theories
  across life course, 277
  evolution and, 85, 213
  evolutionary approach to, 193, 194–95, 273, 293 (*see also* ancestral health movement; physical culture movement)

Index | 481

health, human (*Cont.*)
  goals of, 212–13
  historical perspective on, 213–14
  as "unifying field theory" for human well-being, 194–95, 213
  food and, 202 (*see also* Paleo diet)
  holistic approach to, 195
  improvements, in technologically advanced societies, 273
  men's, issues specific to, 283
  risk-taking behaviors and, 281
  women's, evolutionary insights on, 284
health education
  evolutionarily informed, 292
  evolutionary approach to, 195
health interventions, life history framework for, 286, 293
health messaging, evolutionarily informed, 292
health promotion
  environmental changes and, 291
  future orientation and, 287
  limits to, 286
health-related behaviors and outcomes
  and disease, 273–74
  evolutionary approach to, 273, 293
  life history framework for, 286, 293
health research
  evolutionary approaches in, 275–76, 293
  evolutionary theory and, 274
health science curricula, Tinbergean framework applied to, 304–5, 306
Hemenway Gymnasium (Harvard), 211
heritability
  and natural selection, 85–87, 85–86n5
  of obesity, in humans, 88
  of primate coloration, 107
hierarchy
  and control of coercive power, 146
  in elite male-dominated states, 142
higher education, evolution education and, 3–5, 13, 448
higher-order cognitive skills (HOCS). *See also* critical thinking; knowledge transfer

fostering, education and, 364
science education and, 364
history
  factual vs. practical realism in, 185
  predictive theory of, from social coercion theory, 138
  theories of, vs. social coercion theory, 137–38
The Hitchhiker's Child (Baker), 346
Holland, John, 220
Holler, Richard H., 48
homicide, steep future discounting and, 287–88
Homo naledi, fossil record, social coercion theory and, 130–31, 131n6
homophobia, oligarchic interests and, 145–46
hostile manipulation problem, 132
human adaptation, testing
  controversy about, 85
  difficulty of, 86
humanism, 186–87
  evolutionary, Huxley on, 250–51
  and factual realism, 186–87
  Google Image results for, 187*f*, 187
  as meaning system, 187–88
hunter-gatherers, 193–94
  physical culture movement and, 201
Huxley, Julian, 250–51
Huxley, Thomas H., 69
hydropathy, 208
hygiene hypothesis, 304
hymenoptera
  social, communication by, 132
  social cooperation in, 123n3
hypothesis, 436
hypothesis testing, 65, *See also* adaptive hypotheses; strong inference
  alternative hypotheses in, 68*t*, 91–92
  null hypothesis in, 92

identity, and evolution-related epistemological and existential anxieties, 429
illiteracy, 143
indigenous culture

traditions of, and innovation in globalized world, 256
and Western culture, 256
Industrial Revolution, 193–94, 199
infancy, human, social coercion theory of, 133
infant, bottle-feeding of, consequences of, 342
infantile crying, 168
infant mortality
paternal investment and, 339
preeclampsia and, 339
infections, sexually transmitted, circumcision and, 330, 346–47
information exchange
elite control of, 148–49
oligarchic interests and, 150–51
power and, 143
In Gods We Trust (Atran), 181
intelligence. *See also* IQ
and childlessness, 353
domain-specific types of, 352
and education, 352
and evolutionarily novel pursuits, 352, 353
and individual characteristics, 352
and penchant for studying evolution, 353
intelligent algorithms. *See* genetic algorithm(s) (GAs)
intelligent design, 414*t*
teaching of, legal issues in, 413–15
intelligent design movement, 429–31
rhetoric of, 431, 436, 437, 438
deconstruction of, 433–34
interdisciplinary studies
evolutionary studies as, 15, 22–23, 30–31, 32–33, 49–50, 52, 55–56, 447, 448–49
EvoS programs and, 5, 13, 14
shared content in, 15
intersexual selection, 71, 72*t*
definition of, 98–99
evolutionary consequences, 99
and primate coloration, 104, 109–10
in primates, experimental research on, 104

interspecific foraging competition hypothesis (IFCH), 69–71, 72*t*
interspirituality, 189
interval, musical, 236
intrasexual selection, 70–71, 72*t*
definition of, 98–99
evolutionary consequences, 99
and primate coloration, 104, 109–10
in primates, experimental research on, 104
intraspecific foraging competition hypothesis (IntraFCH), and giraffe, 72*t*
intuition, in science vs. religion, 436, 437
IQ
and problem-solving, 352
and scheduling in artificial light, 336–37

Japanese macaques, female
coloration, 107–8
color ornamentation, 107
Johnsen, Laura, 20
Johnson, Dominic, 181
Jones, E. J., 208n13
Journal of Evolution and Health, 195–96, 209–10
Journal of Evolutionary Economics, 63–64
Journal of Evolutionary Medicine, 209–10
The Journal of the Evolutionary Studies Consortium, students' publications in, 45
journals
evolution-based, and authors' disciplines, 16–17
literature cited, in evolution-based vs. non-evolution-based, 16
Judson, Olivia, 85–86n5

Kadar, Robert, 465
Kellogg, John Harvey, 199
Kellogg, W. K., 275–76
Keynes, John Maynard, 275–76
keys, musical, 235, 238
kin-selection theory, 120, 128
kissing, as mate choice mechanism, 325

Kitavans (Paua New Guinea), 201
Kitzmiller v. Dover Area School District, 414*t*, 415
Knauff, Theodore, 200
knowledge building, as core human dynamic, in education for sustainable development, 261, 262*t*
knowledge transfer
  across disciplines and domains, critical thinking and, 366
  in evolution education, analysis of, 385, 388*t*, 390, 391*f*, 394, 396
  failure of, 366–67
  fostering, education and, 364
Konner, Melvin, 198
Kresser, Chris, 195–96n3, 196–97
Kuhn, Thomas, 62–63, 127
  on ordinary science, 64
Kurtz, Abigail, 19–20

lactase, persistence, 194, 282
  convergent selection on, 87
lactose intolerance, 87, 276
Lamarck, 69–70
law(s), in science, 436
law enforcement
  as Darwinian adaptation, 123
  projectile weapons and, 117–18, 124, 136
learning vs. unlearning, 310–11
Lebanon Valley College, Evolution for Everyone course, 14–15
Lemur catta. *See* ring-tailed lemur
leptin, 89, 90
liberal arts colleges, smaller, EvoS programs at, 29, *See also* Albright College
  advertising, 38, 39
  barriers to, 37
  campus-wide events and, 34, 39
  and community colleges, 36, 39
  faculty participation in, 37–38
  goals of, 39
  and outreach beyond the institution, 38

student recruitment for, 38
  website, 37, 39
Liberty University, 9
life expectancy, 273, *See also* longevity extension, limits to, 286
life history/life history theory, 273, 276–77
  as framework for health behaviors, 286, 293
  as framework for health interventions, 286, 293
  human, 277
  slower
    benefits of, 286–87
    effects on health behaviors, 288–89
    promotion of, 286–87
life history strategy
  evolutionary perspective on, 286–87
  human, sex differences in, 160
light, artificial, effects on natural processes, 336
Linguistic Inquiry and Word Count (LIWC) analysis, 377–82, 387, 387*t*, 392
lipodystrophy, 88–89
Living Paleo for Dummies (Joulwan & Petrucci), 195–96
logic, in science vs. religion, 436
The Logic of Scientific Discovery (Popper), 62–63
longevity. *See also* life expectancy
  environmental changes and, 291
Longo, Valter, 202
Louisiana Creationism Act, 414*t*
low-pass filter, 229*f*
  circuit design, genetic algorithm in, 228
Lynn, Chris, 9, 14–15

Macaca fuscata. *See* Japanese macaques
Macaca mulatta. *See* rhesus macaque
Macaca nigra. *See* crested macaques
Macfadden, Bernarr, 199, 202, 203, 203n7, 204n8, 204–5n10, 206–7
male–male (intrasexual) competition. *See also* giraffe; intrasexual selection

and human mortality rates, 6
Managing the Body: Beauty, Health, and Fitness in Britain 1880–1939 (Zweiniger-Bargielowska), 198n5
Manders, Leah, 19
mandrills, coloration, 100*f*
  in female, 107–8
  in male, 98*f*, 105–6
Mandrillus leucophaeus. See drills
Mandrillus sphinx. See mandrills
mantled guereza, coloration, 100*f*
Mao Zedong, 142
Mark's Daily Apple, 196–97
Mass, Spencer, 8
mastication, physical culturalists' concern about, 208–9
mate choice
  in human population, 221
    easy/inflated credit and, 348
    and genetic engineering, 331
    provisioning hypothesis for, 348
  mechanisms, 325
  nonrandom, 173
materialism, methodological vs. philosophical, 438
maternal investment, human, environment and, 287
mathematics, modern evolutionary theory and, 63–64
mating strategy, human
  sex differences in, 160
  social coercion theory of, 133–34
MCAT, questions on evolution, 313–15
McLean v. Arkansas Board of Education, 414*t*
meaning system(s)
  and adaptive fictions, 187–88
  definition of, 184
  evolutionary approaches to health and, 212
  evolutionary science and, 186
  factual and practical realism in, 185, 186
  functional role of, 184
  informed by evolutionary theory, 186
  informed by science, 188

nonreligious, evolutionary religious studies and, 184
medical education and curricula, 7
  change model/template for, 313, 315*t*
  eco-evolutionary principles and, 303, 316
  ecologists and evolutionary biologists in, 316–18
  evolutionary approach to, 195
  morning reports and rounds in, 318
medicine. *See also* Darwinian medicine; evolutionary medicine
  advances in, 273–74
  conventional, physical culture movement's criticism of, 205
  evolutionary approaches to, 274–75
  preventive, physical culture movement's support of, 205
  transdisciplinary approach and, 274–75
meme, 181–82
men. *See also* sex differences
  and ancestral health movement, 205n11
  mortality among, across lifespan, factors affecting, 284
  and physical culture movement, 205
  reproductive success, factors affecting, 284
  social status and economic power of, and reproductive success, 281
menopause, 323
  evolutionary perspectives on, 285–86
men's health, 283
metabolic syndrome (MetS), 87
  treatment of, Paleolithic nutrition and, 209–10
microbiome, 194–95, 201–2, 274–75
might and right, 146, 147
mismatch, 258–59
  and disease, 277–78, 304
mismatch hypothesis, 193, 194, 200, 210, 282
modernization, response to, 199
molecular biology, 65
money, and power, 144
monkeys. *See also* primates; *specific monkey*
  coloration

Index | 485

monkeys (*Cont.*)
    Darwin on, 97–98
    sexual selection and, 98
monogamy, human, social coercion theory of, 133–34
moon, and life on Earth, 332
morning sickness, as evolutionary adaptation, 284–85
motivation, and health behavior, 292
Murdoch, Rupert, 145
music, modern evolutionary theory and, 63–64
music composition
    automated, 235
    autonomous system for, 235
    fundamentals of, 235
    generative, 234–35
    genetic algorithms in, 234–42
    transformative, 234–35
mutation, genetic algorithms and, 221, 223, 226, 227f

National Research Council (NRC)
    Crosscutting Concepts, 369–70n2
    Unifying Concepts, 369–70n2
National Science Foundation (NSF)
    and EvoS Consortium, 30, 463
    faculty–student research funded by, 17
    grant to expand EvoS in higher education, 13–14, 448, 463
Natterson-Horowitz, Barbara, 195–96
Natural Foods: The Safe Way to Health (Carque), 206–7
naturalistic fallacy, 166, 188, 200
natural selection, 48–49, 61–62
    in evolutionary psychology, 170, 171t
    and health, 195
    heritable variation and, 85–87, 85–86n5
    and primate coloration, 97–98
natural theology, 62–63
Natural Theology (Paley), 61–62
natural way of life, evolutionary approaches to health and, 200–1, 212
nature/nurture debate, 21–23
    and human behavioral sex differences, 161–62
nature of science (NOS)
    instruction in
        and acceptance of evolution, 371–72
        benefits of, 365, 368–69, 371–72
        evolutionary theory in, 370
        misconceptions about, 368–69
    skills in, improvement of,
        understanding of evolutionary theory and, 365
    understanding of
        and acceptance of evolution, 410–11
        problems with, 417–18
naturism, 207–8
Nazi state, 143
Neolithic revolutions, 136–37, 137n14
neonatal mortality, 277–78
Nesse, Randolph, 195–96, 197
New Atheist Movement, 186–87
New Imperialism, 201n6
New York State public school health education criteria, evaluation of, 19–20
Next Generation Science Standards, 369–70n2, 416, 418–19, 457–58
Noctuid moths, hearing/ears, adaptive hypotheses about, testing of, 68t
Nolen, Tom, 8
nonconditional strategy, 164
Norenzayan, Ara, 181
North Korea, 150
nuclear receptors, evolution of, 17–19
nudism, 207–8
nutrition, scientific approach to, 202
Nye, Bill, 9, 14–15
Nystrom, Ken, 8

obesity, in humans, 283, 289–90, 291
    artificial light and, 337
    evolutionary hypotheses for, 90
    genetics and, 88
    heritability of, 88
    mismatch model and, 304
    and public health policy, 88–89, 90
    sleep duration and, 337

types (proximate causes) of, 88–89
obesity-associated metabolic disorders (OAMD), 87, 90
objectivism, 186–87
octave, 236
offspring, 221
oligarchs
   authoritarian external, and domestic aspiring, collusion between, 148–49
   hierarchical rule of, 150
   relations to wealth and power, 144–46
olive baboons, sexual swellings, 107
Olmstead, Fredric Law, 290
On the Origin of Species (Darwin), 3, 61, 62–64, 98–99, 410
ontogeny, 69–70
   Tinbergen's four questions and, 67*t*
opaline, 75–82, 83*f*, 83–84
oral sex, as artifact of personal hygiene, 341
ovulation, coitus-induced, 334–35
ovulation induction factors (OIFs), in seminal fluid, 334–35
ovulatory cycle, and humor preferences in females, 18

Paleo Cooking from Elana's Pantry (Amsterdam), 195–96
Paleo diet, 282
   clinical trials of, 209–10
   foods allowed in, 208, 208n13
   growing popularity of, 195–96
   origins of, 198, 198n4
   scientific publications on, 209–10
   vegetarian versions, 208n13
The Paleo Diet (Cordain), 195–96n3
Paleoista (Stephenson), 195–96n3
Paleolithic diet. *See* Paleo diet
Paleo Physicians Network, 195–96
Paleo Solution (Wolf & Cordain), 204
The Paleo Solution (Wolf), 195–96
Paleo Vegan: Plant-Based Primal Recipes (Jones & Roettinger), 208n13
Paley, William, 61–62
Panda, Satchidananda, 202

Pandora's Box or Our Invisible Foes and How to Conquer Them and What to Eat and Why (Empringham), 201–2, 203–4
Papio anubis. *See* olive baboons
parental investment, human
   and behavioral sex differences, 160
   environment and, 287
parental status, and beliefs about human behavioral sex differences, 162
parsimony, of theory, 118, 128, *See also* competitive parsimony
paternal investment
   effects on offspring survival and reproduction, 279–81, 338–39
   environment and, 287
path-oriented methods, 219–20
pedagogy
   social coercion theory of, 132
   social doubt and, 135, 135n11
   transformation-oriented, emancipatory modes in, 255
   transmission-oriented, instrumental modes in, 255
pedophiles, female pubic hair removal and, 327–28
Peloza v. Capsitrano School District, 414*t*
penis, human. *See also* circumcision; foreskin; prepuce
   and fertilization by proxy, 328–30
   as semen displacement device, 328
peppered moths. See *Biston betularia*
personal hygiene, oral sex as artifact of, 341
phycobilisomes, 74*f*, 76*t*
phylogeny, Tinbergen's four questions and, 67*t*
Physical Culture (magazine), 202, 204–5n10, 206–7
Physical Culture Exhibition, 204–5, 204–5n10
physical culture movement, 198–99, *See also* ancestral health movement
   and ancestral health movement
   commonalities of, 199–200, 212
   differences between, 207

physical culture movement (*Cont.*)
  historical perspective on, 213–14
  literature on, 198n5
  locations important to, 210–11
  principles of, 199, 201–2
  proponents of, 199
  publications, 206–7
  use of media, 206
physical education programs, for primary and secondary schools, 211
piggybacking hypothesis, of fertilization by proxy, 328–30, 346
Pigliucci, Massimo, 21–22
pitch
  musical, 235–36
  of sound, 235
Plant-Based Paleo: Protein-rich Vegan Recipes for Well-Being and Vitality (Zoe), 208n13
Poe's law, 351
Polar Bear Club, 204n8
politics
  and evolutionary psychology, 10–11
  social coercion theory on, 140, 142
Popper, Karl, 62, 127, 168
  critical rationalism of, 63, 85
  on empirical method, 63, 66
  on falsifiability of scientific theory, 62–63, 63n1
  on modifications of theory, 63n2
  on "no conclusive disproof" of theory, 63n1, 70
population(s)
  genetic algorithms and, 220, 221
  human, 220–21
Post, Charles, 199
postpartum depression, in bottle-feeding mothers, 343–44
posttraumatic stress disorder (PTSD)
  cause of, 310
  evolution of, 310–11
  teaching and understanding, Tinbergean approach applied to, 309
  treatment of, 309–10
  vulnerability to, 310
    adaptive function of, 311

posture, physical culture movement's concern about, 206
power
  actual locus of, in short versus long term, 143
  Mao on, 142
  social coercion theory on, 139
  wealth and, 144
power-scavenging, elite throwing by humans and, 118, 121–22, 129, 131n7
Practical Paleo (Sanfilippo), 195–96
practical realism, 184–85
predator avoidance, primate coloration and, 108–9
prediction method, 219–20
preeclampsia
  risk factors for, 339–40
  risk of
    artificial insemination and, 338
    semen change and, 325
    semen familiarity and, 339–40
pregnancy, 277, 324
  maternal investment in
    factors affecting, 278
    male scarcity and, 280–81
pregnancy sickness, as evolutionary adaptation, 284–85
premenstrual syndrome (PMS), evolutionarily informed model of, 285
prepuce, male, functions of, 329–30
present orientation, in youth, and health-related behaviors and outcomes, 289
press, oligarchs' attempts to control, 145
preterm birth, 277–78
  male scarcity and, 280–81
  paternal investment and, 339
prevention, primary, 275
Priessnitz, Vincent, 208
Primal Blueprint (Sisson), 204
primal diet. *See* Paleo diet
primary schools
  evolution education in, 7, 8
  physical education programs for, 211
primates

coloration, 100f
  across species, as research
    opportunity, 108–9
  comparative analyses of, 108–9
  and crypsis, 108–9
  Darwinian thinking on, 109
  experimental research on,
    104, 105–6
  future research directions, 108
  natural selection and, 97–98
  observational studies of, 105–6
  and predator avoidance, 108–9
  quantification of, 104, 105
  recent studies of, 104
  research on, conceptual and
    methodological transitions in, 103
  research on, history of, 99
  sexual selection and, 97–98, 100–1
  as signal, 107–8
  and social status, 105–6
  structures and patterns associated
    with, as research opportunity, 109
dietary variability in, 282
diversity, selective forces for, as
  research opportunity, 108–9
female
  color ornamentation, 107
  facial coloration, 107
  fertility, coloration and, 107
  mate choice, coloration and, 105–6
intrasexual selection in, 99
male, facial coloration, 105–6
male–male interactions, coloration
  and, 105–7
mate choice, coloration and, 105–6,
  107, 108
natal coat colors, 108–9
sexual dichromatism, 108–9
sexual selection in, 100f (see also
  intersexual selection; intrasexual
  selection)
  future research directions, 108
  research on, 109
sexual skin, coloration of,
  101–3, 107–8
sexual swellings, 101, 102
  experimental research on, 104

visual phenotypes, as research
  opportunity, 109
Primitive Man and His Food
  (Vries), 198n4
probiotics, 201–2
progressivity, of evolutionary
  psychology, 168
promiscuity, human
  female, contextual factors and, 165
  social coercion theory of, 133–34
*Promotion and Conservation of Health,
  Strength and Mental Energy*
  (Strongfort), 206–7
proof. *See also* strong inference
  extraordinary, for extraordinary claims,
    61–62, 166–67
  in science, 62
prostitution, social coercion theory
  of, 133–34
Protestant church, and evolution
  controversy, 412
provisioning hypothesis, for mate
  selection, 348
proximal killers, game theory of social
  coercion and, 122–23, 123f, 125
proximate causation
  in medicine, 306
  of traits, 256–57
proximate explanations, in biology, 67,
  67t, 257
psychology. *See also* evolutionary
  psychology
  human sexual, social coercion theory
    of, 134
  modern evolutionary theory and, 63–64
pubic hair
  as deterrent to oral sex, 342
  functions of, 326–27
  removal, 326, 327–28
publications. *See also specific publication*
  physical culture movement and, 206–7
  scientific
    on Paleo diet/lifestyle, 209–10
    social doubt and, 134
    by students at SUNY New Paltz, 45
public health
  evolutionary approaches to, 275, 291

public health (*Cont.*)
  repercussions of evolutionary conclusions for, 91
  and strong inference, 87

racism, oligarchic interests and, 145–46, 150–51
Rand, Ayn, 186–87
rape
  reproductive conditions of, simulation by artificial insemination, 338
  and risk of conception, 333
  social coercion theory of, 133–34
rape victims, and tonic immobility, 55
reactance, 428
realism
  factual, 184–85, 186
  practical, 184–85
Reed, Aaron, 18–19
Reinking, Jeffrey, 8, 17–19
religion. *See also* faith
  as adaptation versus byproduct, 180, 181, 182*t*
  and cooperation among believers, 183
  and counterfactual beliefs, 184, 185
  Darwin's theory of evolution and, 61–62
  enigmas of, 179
  and evolution, 425
  evolutionary psychology and, 163, 173–74
  evolutionary studies of (*see* evolutionary religious studies (ERS))
  and evolution-related epistemological and existential anxieties, 429
  functional basis of (Durkheim), 179–80, 181
  as naive scientific theory, 179
  and science, 411–12, 417, 425–26, 436, 442
  secular utility of, 179–80
  six major evolutionary hypotheses about, 181–83, 182*t*
  and teaching of evolution, 414*t*, 425
  unifying theoretical framework for, 181
  as wasteful byproduct, 180

Religion Explained (Boyer), 181
Religion in Human Evolution (Bellah), 181
religiosity, and attitudes toward evolution, 410–11, 429
religious intolerance, oligarchic interests and, 145–46
remote killers, game theory of
  social coercion and, 122–23, 123*f*, 124–26
reproduction, sex and, connection between, human knowledge of, 322
research
  evolutionary studies and, 18–20, 57
  undergraduate, 17–20, 35–36, 39, 41, 42
revelation, in science vs. religion, 436
rhesus macaque
  female
    color ornamentation, 107
    facial coloration, 107
  male, coloration, 106–7
Rhinopithecus bieti. *See* black-and-white snub-nosed monkeys
Rhinopithecus roxellana. *See* golden snub-nosed monkey
right
  democratized might as, 146
  social coercion theory on, 140
ring-tailed lemur, coloration, 100*f*
risk-taking behaviors
  and health, 281
  steep future discounting and, 287–88
  uncertain environment and, 287–88
Roettinger, A., 208n13
Rolon, Vania, 53
romantic relationships, electronic media and, 351
roots, of musical scale, 236
Roulette wheel, selection method, for genetic algorithms, 223, 224*f*, 224
r-phycoerythrin, 75–82, 76*t*
r-phycoerythrobilin, 75–82, 76*t*
Russia, modern, crony/criminal capitalist economy in, 144, 148

Sagan, Carl, 61–62
Saguinus imperator. *See* emperor tamarin
Saimiri sciureus. *See* common squirrel monkey
sanctions
  economic, coercive power of, 147–48
  trade, coercive power of, 148
Sandow, Eugene, 199, 206–7
Sanfilippo, Diane, 195–96
Sanger, Margaret, 275–76
Sargent, Dudley Allen, 199, 211
Savanah principle of human intelligence, 352
scale, musical, 236
  degrees, 238
  major, 236
Scheepers, Lue, 70–71
schema, genetic algorithm and, 222–23
Schwarzenegger, Arnold, perspective on, in eugenics vs. evolutionary psychology, 171*t*
science(s). *See also* nature of science (NOS)
  creative destruction of older theory in, 150
  education for sustainable development and, 254–55
  hard vs. soft, 65
  language of, 436
  mature, creative activity in, 150
  meaning system informed by, 188
  misconceptions about, 417–18
  public distrust of, 416–17
  as reductionist, 254–55
  and religion, 411–12, 417, 425–26, 436, 442
  social coercion theory of, 134
  and sociocultural narratives, 264
  successful, fundamentals of, 127
science denialism, 416–17
science education, 397–98
  and critical thinking, 368
  future of, 418
  and higher-order cognitive skills, 364
  legal issues in, 413–15, 414*t*
  reform of, 417–18
  state standards and, 416

  in United States, in mid-20th century, 413
science of humans, 119
science-to-narrative chain, 264, 266
Science to Narrative Chain, 465
scientific framework
  evolutionary psychology as, 158
  good, characteristics of, 157–58
scientific literacy, 426
  factors affecting, 416–18
  promotion of, 419
scientific method, 134, 368–69
Scientific Revolution, social doubt and, 135
scientific theory, 436
  falsifiability of, 62–63, 65
  modification of, 63n2
  testing of, 63
scientists, public distrust of, 416–17
Scopes "monkey" trial, 413
Scott, Eugenie, 436, 437
Scott, Jamie, 203
sea hares. *See Aplysia* spp.
secondary schools
  evolution education in, 7, 8
    Next Generation Science Standards and, 457–58
  physical education programs for, 211
sedentary lifestyle, 208
  contemporary concern about, 206
  Eliot on, 211
  physical culture movement's concern about, 206, 211
selection, 68–69, *See also* intersexual selection; intrasexual selection; natural selection; sexual selection
  between-group, 181–82
  genetic algorithms and, 221, 223
  group, 118, 126, 169–70
  in human population, 221
  level of, in eugenics vs. evolutionary psychology, 169–12, 171*t*
  Roulette wheel method, for genetic algorithms, 223, 224*f*, 224
  species-level, 169–70, 171*t*
  within-group, 181–82

selector, in eugenics vs. evolutionary psychology, 170, 171*t*
self-domestication, 126–27
Selman v. Cobb County School District, 414*t*
semantics, 436
  and debate about evolution, 434
semen familiarity, 339–41
  and preeclampsia, 339–40
semen sampling, as mate choice mechanism, 325–26
semen signature, 325
seminal fluid
  chemistry, contextual factors and, 335
  effects on female reproductive tract, 334–35
  immunologic properties of, and implantation and fetal maturation, 340–41
seminal plasma hypersensitivity, 326
senescence, 286
Senko, Corwin, 18
sex, and reproduction, connection between, human knowledge of, 322
sex differences
  in behavior
    biological, 10, 161–62
    socialization and, 161–62
  evolved
    vs. cultural differences, 54–55, 56
    in humans, 10
  evolved behavioral, in humans, controversies about, 160, 174
  as focus of evolutionary theory, 54–55, 56
  in human desire for sexual variety, 6–7, 157–58
  in human mortality rates, 6, 157–58, 281–82, 284
  in math abilities, 90–91
  in reproductive life span, 323
  in risk-taking behaviors, 281
sexism, oligarchic interests and, 145–46, 150–51
sex ratio(s), in human populations, effects on sexual interactions, 280–81

sexual dichromatism, in primates, 108–9
sexual dimorphism, 99
sexual hair, human, 327, *See also* pubic hair
sexual hyperactivity, human, artificial light and, 337
sexual jealousy, male
  eugenics view of, 169
  evolutionary psychology's view of, 169
sexual psychology, human, social coercion theory of, 134
sexual/reproductive behavior, human, social coercion theory of, 133
sexual selection
  and coloration in primates, 97–98, 100–1
  and health, 195
  in primates, 100*f* (*see also* intersexual selection; intrasexual selection)
  research on, 109
  and reproductive success, 98–99
  and sex differences in mortality rates, 286
sexual selection theory, 70–71, 72*t*, 97–98, 99
sexy son hypothesis, 331
sickle cell anemia, 276
signing, human, 132–33n9
Simmons, Robert, 70–71
Sisson, Mark, 196–97, 204
situationism
  evolutionary, 165–66
  and evolutionary psychology, 165–66
  in social sciences, 165
sleep
  artificial light and, 336–37, 338
  duration, and obesity in humans, 337
smartphones, use of
  cognitive deficiencies associated with, 350
  social consequences of, 351
  thinking style and, 350
smoking, health messaging about, 293
social coercion
  game theory of, 122*b*, 122–23, 123*f*, 124–25, 124–25n4

in nonhuman animals, 123n3
social coercion theory, 117, 118
   and altruism, 126
   on contemporary power and massive human coalitions, 139
   disagreements with, 120
   and explanation of human panoply, 119, 128
   first draft of, 124
   fundamentals of, 120, 122*b*
   on human adaptive revolutions, 135
   on human biological properties, 131
   on human origins and fossil record, 128
   on humans as democratic animals, 138
   on humans as scientific animals, 134
   on humans as xenophobic animals, 138
   implications for academy, 149
   implications for human future, 150
   as "individual selection" theory, 126, 128–29, 135
   instruction in, 150n20
   on political enfranchisement of women, 142n16
   on power, 139
   power to unify, simplify, and empower, 118
   predictive theory of history from, 138
   scientific revolution enabled by, 118–19
   on social psychology, 138
   theory of evil, 146
   vs. traditional theories of history, 137–38
social consensus, in modern states, 143
social constructivism, 251
social cooperation
   human
      and adaptive revolutions, 137
      democratized, 138
      evolution of, 124, 128
      as individually adaptive, 123*f*, 126, 135
      limited maximal scale of, 135
      rapid evolution of, 126–27, 126–27n5
      scale of, and social adaptation, 136
      social coercion and, 117–18
      in hymenoptera, 123n3
      kinship-independent ("public"), 120–21, 122*b*, 129, 141, 142
      evolution of, 118
      in nonhuman animals, 123n3
      nonkin, 126–27
   in animals, 120–21
social Darwinism, 464
social doubt
   and consensual beliefs about the "right," 140
   social coercion theory of, 134
social-emotional reasoning, as core human dynamic, in education for sustainable development, 261, 262*t*
Social Evolution Forum (SEF), 465–66
social isolation, 291–92
socialization, environmental changes and, 291
social psychology, social coercion theory on, 138
social revolution(s), human, 136–37
social sciences, 21
   modern evolutionary theory and, 63–64
   repercussions of evolutionary conclusions for, 91
social unit
   ancestral human, size of, 135–36
   elite male warriors and, 141–42
   human, limited maximal scale of, 135
   size of, and social coercion, 135–36, 135–36n13
sociobiology, 276
socioeconomic status, and perceived extrinsic mortality risk, effects on health behaviors, 288–89
sociology, academic employment in, and beliefs about human behavioral sex differences, 162
soft sciences, vs. hard sciences, 65
Southeastern Evolutionary Perspectives Society (SEEPS), 9, 14–15
spandrel, 182
Spaulding, Kristina, 9
special creation, 62–63

species, new, evolution of, 61
species-level selection, 169–70, 171*t*
speech, human, 132–33n9
Spencer, Herbert, 180, 186
sperm donation, 172–73
Stapell, Hamilton, 8
Starrett, Kelly, 206
states. *See also* authoritarian states
   archaic, 141–42, 141n15
   coercive pan-global coalitions of, 147
   and distribution of coercive power, 147
   exploitative behavior of, 147
   hierarchical control of coercive power within, 147
State University of New York (SUNY) New Paltz, 463
   and evolutionary psychology, 159–60
   EvoS program, 21–22, 37–38, 448
      benefits of, 56
      capstone course, 43
      education provided by, 44
      hikes related to, 45
      inspiring aspects of, 41–42
      outcomes, 45
      personal perspectives on, 42
      students' publications from, 45
      and teaching assistants, 41, 42, 48
   EvoS Speaker (Seminar) Series, 41, 43, 45, 52
   EvoS Summer Institute at, 7–8, 449
      advances in (future directions for), 459
      analysis of, 451, 452*t*, 459
      costs of, 450
      coursework in, 449
      field trip in, 450
      interdisciplinary approach in, 459
      as model for advancing evolution education, 460
      participants, 450, 451
      secondary teacher's perspective on, 457
      survey (evaluation), by participants, 450, 451, 452*t*
      topics covered in, 449
   EvoS Summit (2012), 17, 21
   implementation of EvoS curriculum, 31–32
State University of New York (SUNY) Oswego, 9–10
STEAM (science, technology, engineering, arts, and mathematics) approach, 21–22
stepchildren
   abuse of, 47–48, 279–80
   genetic fitness perspective on, 165–66
Stephenson, Nell, 195–96n3
step-parents
   and family violence, 168–69
   and filicide, 6, 157–58, 165
   genetic fitness perspective on, 165–66
The Stone Age Diet: Based on In-depth Studies of Human Ecology and the Diet of Man (Voegtlin), 198n4
Storksdiek, Martin, 459
strategic pluralism, 164, 165
Strongfort, Lionel, 206–7
strong inference, 65, 85
   elements of, 66
   public health and, 87
   steps in, systematic application of, 66
Structure of Scientific Revolutions (Kuhn), 62–63
substance abuse, time perspective and, 289
sun exposure
   health messaging about, 292–93
   physical culture movement and, 203–4
sunfish, male, nonconditional (mating) strategy of, 164
SUNY. *See* State University of New York (SUNY)
Super Strength (Calvert), 206–7
sustainability, 7, *See also* Education for Sustainable Development (ESD)
   contextual diversity and, 253
   need for generalized theory base for, 253
sustainable development. *See also* Education for Sustainable Development (ESD)

adaptive flexibility as foundation
for, 260
challenges of, 249
human dynamics in, ultimate causation
of, 258
local context and, 258–59
tabula rasa approach, 21
teachers
epistemological anxieties of, 425–26
EvoS Summer Institute for (SUNY
New Paltz), 449
faith/religious beliefs of, and teaching
of evolution, 425–27, 429, 441–42
science, professional development
workshop for, 426, 430
additional resources supplied in, 439
biologist's contribution to, 439
conduct (method) of, 432
evolution-related definitions for, 433
handbook/workbook for, 433,
441, 441*t*
module 1: what you know, how you
know it, and what it means, 434
module 2: science and faith as
epistemologies, 436
module 3: does accepting evolution
mean you are an atheist?, 437
module 4: setting the record
straight, 438
open discussion of fears in, 432
outcomes of, 440
participant response to, 440
pilot group for, 432
setting the tone for, 432
Texas science teacher's contribution
to, 439

teacher training
in education for sustainable
development, 264
in evolution
cognitive change approaches, 428
content-based approaches, 427
teaching. *See* evolutionary studies (EvoS);
evolution education; teachers
technology. *See* genetic
algorithm(s) (GAs)

terrestrial organisms, evolution of, 332–12
Texas science standards, 439
Thayer, A., 100–1
theory. *See also* scientific theory;
sexual selection theory; social
coercion theory
definition of, 369n1, 436
of evolution (*see* evolutionary theory)
role in scientific method, instruction
about, 369
theory of mind assessment, scores on,
mating-relevant scenario and, 19
theory testing, 149, 150
process of, 127–28
social doubt and, 134
Theropithecus gelada. See gelada
This View of Life (TVOL), 465–66
threat from a distance. *See* coercive threat
from a distance
throwing. *See also* elite throwing
human capacity for, 121–22, 121–22n2
implications (threat) of, human
recognition of, 121–22n2
as skill, 135–36, 135–36n12
targets of, risk/cost to, 121–22n2
Tiger, Lionel, 159–60
Tilden, John Henry, 205
time perspective(s)
effects on health, 287
effects on life history, 287
in youth, and health-related behaviors
and outcomes, 289
Tinbergen, Niko, 67–68, 67*t*, 257, 305
four questions posed by, 67*t*, 305–6
application to medicine, 306
tobacco, 200–1
use, time perspective and, 289
tonic immobility, 46, 55
traits
ecological mechanism of, 256–57
evolutionary function of, 256–57
proximate causation of, 256–57
ultimate causation of, 256–57
tree frogs, male
conditional (mating) strategy
of, 164–65
strategic pluralism of, 164–65

truth(s), 434, 435n1, 440
  for atheists, 435
truthiness, 69–70
truth violations, Christian view of
    evolution and, 435
Truzzi, Marcello, 61–62
Turner Clubs, 199
TVOL1000, 464
Tylor, E. B., 179

ultimate causation
  and adaptive flexibility, 256–57
  and education for sustainable
    development, 250, 264
  of human dynamics in sustainable
    development, 258
  of traits, 256–57
  understanding of, and education for
    sustainable development, 258
ultimate explanations, in biology, 67,
    67t, 257
ultimate-proximate distinction, and
    human behavior, 126–27, 157
uncertainty
  health effects of, 288
  and risk-taking behaviors, 287–88
UNESCO: Its Purpose and Its Philosophy
    (Huxley), 250–51
uniquely human panoply, 119
  evolution of, 121
  explanation of, social coercion theory
    and, 119, 128
United Kingdom, wealth and power
    in, 144
United Nations, 189
United Nations Educational, Scientific,
    and Cultural Organization
    (UNESCO), and education
    for sustainable development,
    249, 250–51
United States
  republican governance vs. attempted
    hierarchical power grab in, 150–51
  wealth and power in, 144
universal conflict of interest
    problem, 120–21

University at Albany, 9
University of Alabama. *See also* ALLELE
    (Alabama Lectures in Life's
    Evolution) speaker series
  evolutionary studies in, 9
  EvoS Club, 14–15
  EvoS minor curriculum, 14–15, 24–27
  EvoS program, 14–15
University of California, Los
    Angeles (UCLA)
  Evolutionary Medicine Program,
    195–96, 317
  evolutionary medicine programs, 317
University of Missouri, EvoS program, 14
University of Wisconsin–Eau Claire, 4

vaccines
  as evolutionarily consonant
    invention, 205
  physical culture movement's opposition
    to, 205
Valid Assessment of Learning in
    Undergraduate Education
    (VALUE) critical thinking
    rubric, 382–83
Varga, Andrea, 8
vegetarianism, 208, 208n13
vervet monkey, male, coloration, 105–7
virulence, 194–95
Vitality Supreme (Macfadden), 206–7
Voegtlin, W. L., 198n4
volume-oriented methods, 219–20
Vries, A. de, 198n4

Waldo, Jennifer, 8, 19
walking
  benefits of, 283
  as most natural form of exercise,
    202–3, 203n7
warfare, evolutionary perspective on, 166
Warinner, Christina, TED talk on Paleo
    diet, 210, 210n14
The Way to Live (Hackenschmidt), 206–7
wealth, and power, 144
weaponry. *See also* handguns
  animal and human, 47

coercive, and human historical transitions, 137
intrasexually selected, 99
  in giraffes, 70–71, 72t
monopoly control of, exploitation by authoritarian regimes, 147
projectile
  body armor and, 140–41
  and law enforcement, 117–18, 124, 136
  and social coercion, 122b
  shock, and coercive power, 140–41, 141n15
website, on evolutionary topics, 448
Webster v. New Lenox School District, 414t
Wedberg, Nicole, 42
Wells, H. G., 275–76
Wells, Minnie, 205
Western culture, and indigenous culture, 256
Whole 30 program, 196–97
whole step, in music, 236
wicked problems, of 21st century, 249
Williams, Stephen, 20
Wilson, David Sloan, 10, 29–30, 181, 447–48, 463, 465
  and EvoS revolution in higher education, 3, 4
Wilson, E. O., 9
Windship, George Barker, 211, 211n15

within-group selection, 181–82
Wolf, Robb, 195–97, 204
women. *See also* sex differences
  and ancestral health movement, 205n11
  and physical culture movement, 205
  political enfranchisement of, social coercion theory on, 142n16
women's health, evolutionary insights on, 284
women's health movement, 283–84
women's studies, academic employment in, and beliefs about human behavioral sex differences, 162
Wright, Patricia, 41

xenophobia
  oligarchs' exploitation of, 145–46, 145–46n18, 150–51
  social coercion theory on, 138

YMCA, 199

Zakirova, Zuchra, 17–19
Zoe, J., 208n13
Zoobiquity: The Astonishing Connection between Human and Animal Health (Natterson-Horowitz), 195–96
Zuk, Marlene, 210
Zweiniger-Bargielowska, Ina, 198n5